More Than Just a Textbook

Internet Resources

Step 1 Connect to **Math Online** glencoe.com

Step 2 Connect to online resources by using **QuickPass** codes. You can connect directly to the chapter you want.

IM7055c1

Enter this code with the appropriate chapter number.

For Students

Connect to **StudentWorks Plus Online** which contains all of the following online assets. You don't need to take your textbook home every night.

- Personal Tutor
- Chapter Readiness Quizzes
- Multilingual eGlossary
- Concepts in Motion
- Chapter Test Practice
- Test Practice

For Teachers

Connect to professional development content at **glencoe.com** and **eBook Advance Tracker** at **AdvanceTracker.com**

For Parents

Connect to **glencoe.com** for access to **StudentWorks Plus Online** and all the resources for students and teachers that are listed above.

Glencoe McGraw-Hill

IMPACT
Mathematics

Mc Graw Hill **Glencoe**

New York, New York   Columbus, Ohio   Chicago, Illinois   Woodland Hills, California

COURSE
3

About the Cover

A drum is an instrument that creates sound by striking an object against a stretched membrane. Drums date back to 6000 B.C. Cylindrical drums, like the Tom Quad pictured on the cover, originated in ancient Mesopotamia.

In 1995, a machine was developed that calculated the number of strokes in a 90-second time span. This invention paved the way for speed drumming competitions. In January 2005, Mike Mangini set the current world record with 1,203 single stroke rolls in one minute, or 20.05 strokes per second.

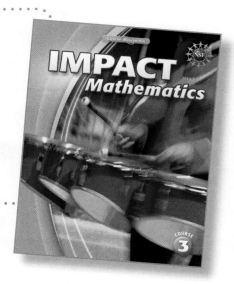

These materials include work supported in part by the National Science Foundation under Grant No. ESI-9726403 to MARS (Mathematics Assessment Resource Service). Any opinions, findings, and conclusions or recommendations expressed in this material are those of the authors and do not necessarily reflect the views of the funding agencies. For more information on MARS, visit http://www.nottingham.ac.uk/education/MARS.

The McGraw·Hill Companies

 Macmillan/McGraw-Hill
Glencoe

The algebra content for *IMPACT Mathematics* was adapted from the series *Access to Algebra*, by Neville Grace, Jayne Johnston, Barry Kissane, Ian Lowe, and Sue Willis. Permission to adapt this material was obtained from the publisher, Curriculum Corporation of Level 5, 2 Lonsdale Street, Melbourne, Australia.

Send all inquiries to:
Glencoe/McGraw-Hill
8787 Orion Place
Columbus, OH 43240-4027

ISBN: 978-0-07-888705-5
MHID: 0-07-888705-4

Printed in the United States of America.

1 2 3 4 5 6 7 8 9 10 055/079 17 16 15 14 13 12 11 10 09 08

COURSE 3

Contents in Brief

Focal Points and Connections
See pages vi and vii for key.

Principal Investigator

Faye Nisonoff Ruopp
Brandeis University
Waltham, Massachusetts

Consultants and Developers

Consultants

Frances Basich Whitney
Project Director, Mathematics K–12
Santa Cruz County Office of Education
Santa Cruz, California

Robyn Silbey
Mathematics Content Coach
Montgomery County Public Schools
Gaithersburg, Maryland

Dr. Selina Vásquez Mireles
Associate Professor of Mathematics
Texas State University—San Marcos
San Marcos, Texas

Teri Willard
Assistant Professor
Central Washington University
Ellensburg, Washington

Special thanks to:

Peter Braunfeld
Professor of Mathematics Emeritus
University of Illinois

Sherry L. Meier
Assistant Professor of Mathematics
Illinois State University

Judith Roitman
Professor of Mathematics
University of Kansas

Developers

Senior Project Director
Cynthia J. Orrell

Senior Curriculum Developers
Michele Manes, Sydney Foster, Daniel Lynn Watt, Ricky Carter, Joan Lukas, Kristen Herbert

Curriculum Developers
Haim Eshach, Phil Lewis, Melanie Palma, Peter Braunfeld, Amy Gluckman, Paula Pace

Special Contributors
Elizabeth D. Bjork, E. Paul Goldenberg

Project Reviewers

Glencoe and Education Development Center would like to thank the curriculum specialists, teachers, and schools who participated in the review and testing of the first edition of *IMPACT Mathematics*. The results of their efforts were the foundation for this second edition. In addition, we appreciate all of the feedback from the curriculum specialists and teachers who participated in review and testing of this edition.

Debra Allred
Math Teacher
Wiley Middle School
Leander, Texas

Tricia S. Biesmann
Retired Teacher
Sisters Middle School
Sisters, Oregon

Kathryn Blizzard Ballin
Secondary Math Supervisor
Newark Public Schools
Newark, New Jersey

Linda A. Bohny
District Supervisor of Mathematics
Mahwah Township School District
Mahwah, New Jersey

Julia A. Butler
Teacher of Mathematics
Richfield Public School Academy
Flint, Michigan

April Chauvette
Secondary Mathematics Facilitator
Leander ISD
Leander, Texas

Amy L. Chazaretta
Math Teacher/Math Department Chair
Wayside Middle School, EM-S ISD
Fort Worth, Texas

Franco A. DiPasqua
Director of K–12 Mathematics
West Seneca Central
West Seneca, New York

Mark J. Forzley
Junior High School Math Teacher
Westmont Junior High School
Westmont, Illinois

Virginia G. Harrell
Education Consultant
Brandon, Florida

Lynn Hurt
Director
Wayne County Schools
Wayne, West Virginia

Andrea D. Kent
7th Grade Math & Pre-Algebra
Dodge Middle School, TUSD
Tucson, Arizona

Russ Lush
6th Grade Teacher & Math Dept. Chair
New Augusta–North
Indianapolis, Indiana

Katherine V. Martinez De Marchena
Director of Education 7–12
Bloomfield Public Schools
Bloomfield, New Jersey

Marcy Myers
Math Facilitator
Southwest Middle School
Charlotte, North Carolina

Joyce B. McClain
Middle School Mathematics Consultant
Hillsborough County Schools
Tampa, Florida

Suzanne D. Obuchowski
Math Teacher
Proctor School
Topsfield, Massachusetts

Michele K. Older
Mathematics Instructor
Edward A. Fulton Jr. High
O'Fallon, Illinois

Jill Plattner
Math Program Developer (Retired)
Bend La Pine School District
Bend, Oregon

E. Elaine Rafferty
Retired Math Coordinator
Summerville, South Carolina

Karen L. Reed
Math Teacher—Pre-AP
Chisholm Trail Intermediate
Fort Worth, Texas

Robyn L. Rice
Math Department Chair
Maricopa Wells Middle School
Maricopa, Arizona

Brian Stiles
Math Teacher
Glen Crest Middle School
Glen Ellyn, Illinois

Nimisha Tejani, M.Ed.
Mathematics Teacher
Kino Jr. High
Mesa, Arizona

Stefanie Turnage
Middle School Mathematics
Grand Blanc Academy
Grand Blanc, Michigan

Kimberly Walters
Math Teacher
Collinsville Middle School
Collinsville, Illinois

Susan Wesson
Math Teacher/Consultant
Pilot Butte Middle School
Bend, Oregon

Tonya Lynnae Williams
Teacher
Edison Preparatory School
Tulsa, Oklahoma

Kim C. Wrightenberry
Math Teacher
Cane Creek Middle School
Asheville, North Carolina

The Curriculum Focal Points identify key mathematical ideas for this grade. They are not discrete topics or a checklist to be mastered; rather, they provide a framework for the majority of instruction at a particular grade level and the foundation for future mathematics study. The complete document may be viewed at www.nctm.org/focalpoints.

G8-FP1 Algebra: Analyzing and representing linear functions and solving linear equations and systems of linear equations

Students use linear functions, linear equations, and systems of linear equations to represent, analyze, and solve a variety of problems. They recognize a proportion ($y/x = k$, or $y = kx$) as a special case of a linear equation of the form $y = mx + b$, understanding that the constant of proportionality (k) is the slope and the resulting graph is a line through the origin. Students understand that the slope (m) of a line is a constant rate of change, so if the input, or x-coordinate, changes by a specific amount, a, the output, or y-coordinate, changes by the amount ma. Students translate among verbal, tabular, graphical, and algebraic representations of functions (recognizing that tabular and graphical representations are usually only partial representations), and they describe how such aspects of a function as slope and y-intercept appear in different representations. Students solve systems of two linear equations in two variables and relate the systems to pairs of lines that intersect, are parallel, or are the same line, in the plane. Students use linear equations, systems of linear equations, linear functions, and their understanding of the slope of a line to analyze situations and solve problems.

G8-FP2 Geometry and Measurement: Analyzing two- and three-dimensional space and figures by using distance and angle

Students use fundamental facts about distance and angles to describe and analyze figures and situations in two- and three-dimensional space and to solve problems, including those with multiple steps. They prove that particular configurations of lines give rise to similar triangles because of the congruent angles created when a transversal cuts parallel lines. Students apply this reasoning about similar triangles to solve a variety of problems, including those that ask them to find heights and distances. They use facts about the angles that are created when a transversal cuts parallel lines to explain why the sum of the measures of the angles in a triangle is 180 degrees, and they apply this fact about triangles to find unknown measures of angles. Students explain why the Pythagorean theorem is valid by using a variety of methods—for example, by decomposing a square in two different ways. They apply the Pythagorean theorem to find distances between points in the Cartesian coordinate plane to measure lengths and analyze polygons and polyhedra.

G8-FP3 Data Analysis and Number and Operations and Algebra: Analyzing and summarizing data sets

Students use descriptive statistics, including mean, median, and range, to summarize and compare data sets, and they organize and display data to pose and answer questions. They compare the information provided by the mean and the median and investigate the different effects that changes in data values have on these measures of center. They understand that a measure of center alone does not thoroughly describe a data set because very different data sets can share the same measure of center. Students select the mean or the median as the appropriate measure of center for a given purpose.

G8-FP4C   **Algebra:** Students encounter some nonlinear functions (such as the inverse proportions that they studied in grade 7 as well as basic quadratic and exponential functions) whose rates of change contrast with the constant rate of change of linear functions. They view arithmetic sequences, including those arising from patterns or problems, as linear functions whose inputs are counting numbers. They apply ideas about linear functions to solve problems involving rates such as motion at a constant speed.

G8-FP5C   **Geometry:** Given a line in a coordinate plane, students understand that all "slope triangles"—triangles created by a vertical "rise" line segment (showing the change in y), a horizontal "run" line segment (showing the change in x), and a segment of the line itself— are similar. They also understand the relationship of these similar triangles to the constant slope of a line.

G8-FP6C   **Data Analysis:** Building on their work in previous grades to organize and display data to pose and answer questions, students now see numerical data as an aggregate, which they can often summarize with one or several numbers. In addition to the median, students determine the 25th and 75th percentiles (1st and 3rd quartiles) to obtain information about the spread of data. They may use box-and-whisker plots to convey this information. Students make scatterplots to display bivariate data, and they informally estimate lines of best fit to make and test conjectures.

G8-FP7C   **Number and Operations:** Students use exponents and scientific notation to describe very large and very small numbers. They use square roots when they apply the Pythagorean theorem.

Table of Contents

Focal Points
and Connections
See pages vi and vii
for key.

G8-FP2

Focal Points and Connections
See pages vi and vii for key.

G8-FP4C

Exponents and Exponential Variation **144**

Focal Points and Connections
See pages vi and vii for key.

G8-FP4C

G8-FP7C

Focal Points
and Connections
See pages vi and vii
for key.

G8-FP1

G8-FP2

Focal Points and Connections
See pages vi and vii for key.

G8-FP2

⑧ Quadratic and Inverse Relationships

Focal Points and Connections
See pages vi and vii for key.

G8-FP4C

Focal Points
and Connections
See pages vi and vii
for key.

G8-FP4C

10 Functions and Their Graphs 522

Focal Points and Connections
See pages vi and vii for key.

G8-FP1

Focal Points and Connections
See pages vi and vii for key.

G8-FP3

G8-FP6C

Focal Points
and Connections
See pages vi and vii
for key.

G8-FP1

G8-FP4C

Linear Relationships

Real-Life Math

It's Only Natural! Linear relationships can be found in a variety of situations in nature. One of the foremost Renaissance artists, Leonardo da Vinci, believed that in the perfect body, the parts should be related by certain ratios. For instance, the length of the arm should be three times the length of the hand. This relationship can be expressed by the linear equation $a = 3h$, where a is arm length and h is hand length.

Think About It Use a tape measure to find the lengths of your arm and hand. What is the ratio r of these two lengths? Write an equation that expresses the relationship between your hand length h and arm length a.

Contents in Brief

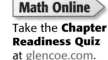

Math Online

Take the **Chapter Readiness Quiz** at glencoe.com.

Dear Family,

The class is about to begin an exciting year of mathematics. The first chapter is about *linear relationships*. These are relationships in which one amount, or variable, changes at a constant rate as another variable changes. Linear relationships can be represented using words, equations, graphs, and tables, as shown below.

Key Concept—Linear Relationships

Randall earns $8 per hour. This is a linear relationship between the variable *hours worked H* and the variable *dollars earned D*. For every hour that Randall works, his earnings increase by $8.

Randall's earnings are represented by the following equation, graph, and table.

$$\text{Equation: } D = 8 \cdot H$$

Graph

Table

Hours	Dollars Earned
0	0
1	8
2	16
3	24
4	32

Many commonplace situations are linear relationships, such as the number of miles traveled at a constant rate of speed. Sometimes a relationship may not be exactly linear, but it may be close enough to use a linear model to make predictions and estimates.

Chapter Vocabulary

coefficient

slope-intercept form

direct variation

standard form

Home Activities

Look for real-world situations that can be modeled using linear relationships.

- time to type a book report if you type 40 words per minute
- number of babysitting hours to earn $100 if you earn $5 per hour
- number of hours to travel 330 miles if you travel 60 miles per hour

LESSON 1.1

Direct Variation

Randall earns $8 per hour in his after-school job. He drew this graph to show the relationship between the number of hours he works and the number of dollars he earns.

Earnings

Algebra is a useful tool for investigating relationships among *variables*, quantities that may vary or represent a number. Randall's graph shows that there is a relationship between the variable *hours worked* and the variable *dollars earned*.

The graph of this relationship is a line. A straight-line graph indicates a *constant* rate of change. In the graph, each time the number of hours increases by 1, the number of dollars earned increases by 8. Relationships with straight-line graphs are called are *linear relationships*.

In this chapter, you will explore graphs, tables, and equations for linear relationships. You will start by making a "human graph."

> **Explore** ....
>
> Select a team of nine students to make the first graph. The team should follow these rules.
>
> • Line up along the *x*-axis. One student should stand on −4, another on −3, and so on, up to 4.
>
> • Multiply the number on which you are standing by 2.

- When your teacher says "Go!" walk forward or backward to the *y* value equal to the result you found in the previous step.

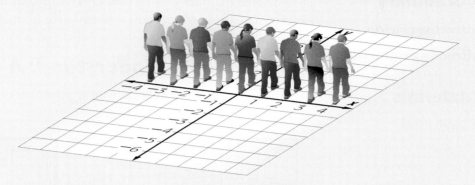

Describe the resulting "graph."

With the students on the first team staying where they are, select another team of nine students. The second team should follow these rules.

- Line up along the *x*-axis. One student should stand on −4, another on −3, and so on, up to 4.

- Multiply the value you are standing on by 2 and then add 3.

- When your teacher says "Go!" walk forward or backward to the *y* value equal to the result you found in the previous step. You may have to go around someone from the first team.

Are both graphs linear?

Does either graph pass through the origin? If so, which one?

Write an equation for each graph.

Explain why the two graphs will never intersect.

Investigation Direct Linear Variation

Vocabulary

direct variation

directly proportional

Materials

• graph paper

The two human graphs you created illustrate two types of linear relationships, which you will investigate in the following exercise sets.

✅ Develop & Understand: A

One weekend, Mikayla delivered pamphlets explaining her town's new recycling program.

1. Copy and complete the table to show the number of pamphlets that Mikayla delivered on Sunday.

Hours Worked on Sunday, h	0	1	2	3	4	5	6
Sunday Deliveries, s	0	150	300	450	600	750	900
Total Deliveries, t	350	500	650	800	950	1,100	1,250

2. Look at your completed table.

a. As the number of hours Mikayla worked doubled from 1 to 2, did the number of Sunday deliveries also double? As the number of hours worked doubled from 2 to 4, did the number of Sunday deliveries also double?

b. As the number of hours worked doubled from 1 to 2, did the total number of deliveries also double? As the number of hours worked doubled from 2 to 4, did the total number of deliveries also double?

c. As the number of hours worked tripled from 1 to 3, did the number of Sunday deliveries also triple? As the number of hours worked tripled from 2 to 6, did the number of Sunday deliveries also triple?

d. As the number of hours worked tripled from 1 to 3, did the total number of deliveries also triple? As the number of hours worked tripled from 2 to 6, did the total number of deliveries also triple?

3. Look at the first two rows of your table. Write an equation describing the relationship between the number of Sunday deliveries *s* and the number of hours worked on Sunday *h*.

4. Look at the first and third rows of your table. Write an equation describing the relationship between the total number of deliveries *t* and the number of hours worked on Sunday *h*.

Think about how the number of hours Mikayla worked and the number of Sunday deliveries are related. When you multiply the value of one variable by a quantity such as 2, 30, or 150, the value of the other variable is multiplied by the same quantity. That means the number of pamphlets delivered is **directly proportional** to the number of hours worked.

Here is another way to show that the ratio of Sunday deliveries to hours worked is constant.

$$\frac{\text{Sunday deliveries}}{\text{hours}} = \frac{s}{h} = \frac{150}{1} = \frac{300}{2} = \frac{450}{3} = 150$$

Real-World Link

In the United States, about 1 billion trees' worth of paper are thrown away each year. About 250 million trees would be saved each year if Americans recycled one-tenth of their newspapers.

A linear relationship in which two variables are directly proportional is a **direct variation**. The equation for any direct variation can be written in the form $y = mx$, where x and y are variables and m is a constant.

Not all linear relationships are direct variations. For example, the relationship between the number of hours Mikayla worked and the total number of deliveries is linear, but it is *not* a direct variation.

You will now examine graphs of the relationships related to Mikayla's recycling pamphlet deliveries.

✅ *Develop & Understand: B*

5. On a grid like the one below, graph the equation that you wrote in Exercise 3 showing the relationship between the number of hours Mikayla worked and the number of Sunday deliveries. Label the graph with its equation.

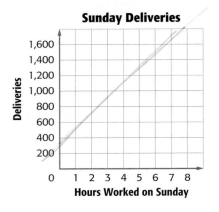

Sunday Deliveries

Math Link

The origin is the point (0, 0).

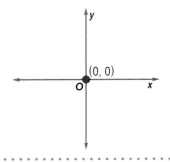

6. On the same grid, graph the equation you wrote in Exercise 4 showing the relationship between the number of hours Mikayla worked and the total number of deliveries.

7. How are the graphs similar? How are they different?

8. Think about the situation each graph represents.

    a. Explain why the graph for Sunday deliveries passes through the origin.

    b. Explain why the graph for the total number of deliveries does not pass through the origin.

9. Explain why the graphs will never intersect.

10. The equation for the number of pamphlets delivered on Sunday is $s = 150h$.

    a. What does 150 represent in the situation?

    b. How would changing 150 to 100 affect the graph?

    c. How would changing 150 to 200 affect the graph?

11. Must the graph of a direct variation pass through the origin? Explain.

✅ Develop & Understand: C

These graphs show the relationship between the number of pamphlets delivered and the number of hours worked by five students.

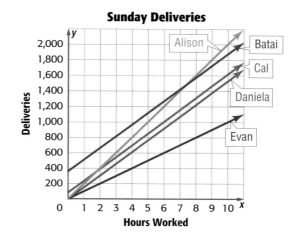

Sunday Deliveries

12. Whose graphs show the same delivery rate? Explain how you know.

13. Whose graphs show a direct variation? Explain how you know.

14. Whose graphs show a relationship that is not a direct variation? Explain how you know.

Share & Summarize

1. Describe, in words, two more linear situations, one that is a direct variation and one that is not.

2. How will the graphs for the two relationships differ?

3. How will the equations for the two relationships differ?

Investigation ② Decreasing Linear Relationships

Materials

• graph paper

In Investigation 1, you looked at how the number of pamphlets Mikayla delivered increased with each hour she worked. You will now examine this situation by thinking about it in another way.

Mikayla started with a stack of pamphlets to deliver. For every hour she worked, the number of pamphlets in her stack decreased.

✅ Develop & Understand: A

Suppose Mikayla started with 1,000 pamphlets. On Saturday, she delivered 350 of them. On Sunday, she delivered the remaining pamphlets, at a constant rate of 150 an hour.

1. How many pamphlets did Mikayla have left to deliver when she began work on Sunday?

2. Copy and complete the table to show the number of pamphlets that Mikayla has left to deliver after each hour of work on Sunday.

Hours Worked on Sunday, h	0	1	2	3	4
Pamphlets Remaining, r					

3. Write an equation to describe the relationship between the number of pamphlets remaining r and the number of hours worked on Sunday h.

4. After how many hours did Mikayla run out of pamphlets? Explain how you found your answer.

5. Draw a graph of your equation. Is the relationship linear?

6. How is your graph different from the graphs you made in Investigation 1? What about the situation causes the difference?

7. Is the number of pamphlets remaining directly proportional to the number of hours Mikayla worked on Sunday? That is, is the relationship a direct variation? Explain how you know.

8. Consider the equation you wrote in Exercise 3.

   a. In your equation, you should have added $-150h$ or subtracted $150h$. What does the negative symbol or the minus sign before $150h$ indicate about the situation? How does it affect the graph?

   b. Your equation should also have the number 650 in it. What does 650 indicate about the situation? How does it affect the graph?

9. Luisa delivered pamphlets more slowly than Mikayla. She started with 1,000 pamphlets, delivered 200 on Saturday, and then delivered 100 pamphlets per hour on Sunday. Write an equation for the relationship between the number of hours Luisa worked on Sunday and the number of pamphlets she had left.

You will use what you have learned about decreasing linear relationships as you work on the next exercise set.

✅ Develop & Understand: B

10. Invent a situation involving a decreasing linear relationship. Your situation should *not* be a direct variation.

    a. Describe your situation in words.

    b. Describe your situation with a table.

    c. Describe your situation with an equation.

    d. Describe your situation with a graph.

11. **Challenge** Invent a situation involving a decreasing linear relationship that *is* a direct variation.

Consider these six graphs.

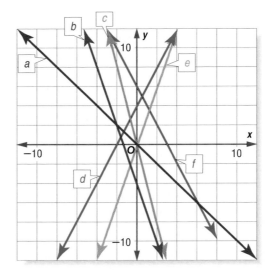

1. Sort the graphs into groups in at least two ways. Explain the criteria for each of your groups.

2. Draw a Venn diagram that illustrates the groups you used to sort the graphs.

Investigation 3 Recognize Direct Variation

In this investigation, you will practice identifying direct variations by examining written descriptions, graphs, tables, and equations.

Think & Discuss

Describe how you can tell that a relationship is linear.
- From its equation.
- From a table of values.
- From a description in words.

✅ *Develop & Understand: A*

Exercises 1–4 describe how the amount of money in a bank account changes over time. For each exercise, do Parts a–c.

    a. Determine which of these descriptions fits the relationship.

      • a direct variation

      • linear but not a direct variation

      • nonlinear (in other words, not linear)

    b. Explain how you decided which kind of relationship is described.

    c. If the relationship is linear, write an equation for it.

1. At the beginning of school vacation, Eli had nothing in the bank. He then started a part-time job and deposited $25 a week.

2. At the beginning of her vacation, Tanya had $150 in the bank. Each week, she deposited another $25.

3. At the beginning of school, Bret had $150 in the bank that he had earned over the summer. During the first week of school, he withdrew one-fifth of his savings, or $30. During the second week of school, he withdrew one-fifth of the remaining $120. He continued to withdraw one-fifth of what was left in the account each week.

4. At the beginning of school, Ginger had $150 in the bank. Each week, she withdrew $25.

Real-World Link

From 1996 to 2005, the number of ATM terminals in the United States increased from 139,134 to 396,000. Yet, during that same time period, the per-ATM transaction volume has decreased from 6,399 to 2,214.

✅ *Develop & Understand: B*

For each graph, determine which of these descriptions fits the relationship. Explain how you decided.

- a direct variation
- linear but not a direct variation
- nonlinear

5.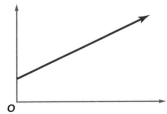

6.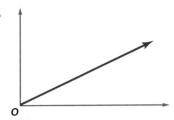

7.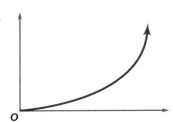

8.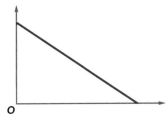

9.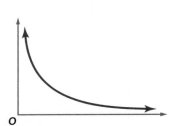

10.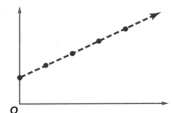

✅ *Develop & Understand: C*

For each equation, determine which of these descriptions fits the relationship. Explain how you decided.

- a direct variation
- linear but not a direct variation
- nonlinear

11. $y = 6p$

12. $y = 6p + 1$

13. $s = 5.3r$

14. $s = 5.3r - 2$

15. $t = 4s^2$

16. $y = 12 - 5p$

17. $y = \dfrac{5p}{2}$

18. $y = \dfrac{5}{2p}$

✅ *Develop & Understand: D*

For Exercises 19–22, do Parts a and b.

a. Determine whether the table could describe a linear relationship. Explain how you decided.

b. If the relationship could be linear, determine whether it is a direct variation. Explain how you decided.

19.

x	1	2	3	4	5
y	3	7	11	15	19

20.

u	1	2	3	4	5
w	1	3	9	27	81

21.

p	2	4	6	8	10
q	26	46	66	86	106

22.

t	2	6	9	10	25
r	6	18	27	30	75

Share & Summarize

Copy and complete the table by telling how you can identify each type of relationship for each type of representation.

	Words	Graph	Equation	Table
Nonlinear	Does not have a constant rate of change	not a straight line		
Direct variation		A line that passes through the origin	$y = mx$ or kx	
Linear but not a direct variation			Can be represented in the form $y = mx + b$ with $b \neq 0$	

Draw a Venn diagram that illustrates the groups you used to sort the graphs.

Practice & Apply

1. Carlos and Shondra were designing posters for the school play. During the first two days, they created 40 posters. By the third day, they had established a routine. They calculated that together they would produce 20 posters an hour.

   a. Make a table like the following that shows how many posters Carlos and Shondra make as they work through the third day.

Hours Worked, h							
Posters Made, p							

Math Link

When graphing an ordered pair, put the first number on the horizontal axis and the second number on the vertical axis.

   b. Draw a graph to represent the number of posters Carlos and Shondra will make as they work through the third day.

   c. Write an equation to represent the number of posters that they will make as they work through the third day.

   d. Make a table to show the *total number* of posters that they will have as they work through the third day.

Hours Worked, h							
Total Posters, t							

   e. Draw a graph to show the total number of posters Carlos and Shondra will have as they work through the third day.

   f. Write an equation that will allow you to calculate the total number of posters that they will have based on the number of hours worked.

   g. Explain how describing just the number of posters created the third day is different from describing the total number of posters created. Is direct variation involved? How are these differences represented in the tables, the graphs, and the equations?

2. **Economics** The Glitz mail order company charges $1.75 per pound for shipping and handling on customer orders.

    The Lusterless mail order company charges $1.50 per pound for shipping and handling plus a flat fee of $1.25 for all orders.

    a. For each company, make a table showing the costs of shipping items of different whole-number weights from 1 to 10 pounds.

    b. Write an equation for each company to help calculate how much you would pay for shipping C on an order of any weight W.

    c. Draw graphs of your equations. Label each with the name of the corresponding company.

    d. Which company offers the better deal on shipping?

    e. Describe how the graphs you drew could help you answer Part d.

    f. How would the Lusterless company have to change its rates so that they vary directly with the weight of a customer's order?

3. **Business** Trevor handed out advertising flyers last weekend. He distributed 400 flyers on Saturday and 200 per hour on Sunday.

    a. Write an equation for the relationship between the number of hours Trevor worked on Sunday h and number of flyers that he distributed on Sunday s.

    b. Write an equation for the relationship between the number of hours Trevor worked on Sunday h and total number of deliveries t.

4. Which of these graphs represent decreasing relationships? Explain how you know.

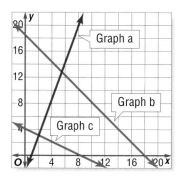

5. Which of these tables represent decreasing relationships? Explain how you know.

Table A

x	1	2	3	4	5	6	7	8	9
y	19	18	17	16	15	14	13	12	11

Table B

x	1	2	3	4	5	6	7	8	9
y	−2	1	4	7	10	13	16	19	22

Table C

x	1	2	3	4	5	6	7	8	9
y	1.5	1	0.5	0	−0.5	−1	−1.5	−2	−2.5

6. Which of these equations represent decreasing relationships? Explain how you know.

a. $y = -x + 20$　　　　**b.** $y = 3x - 5$　　　　**c.** $y = -\frac{1}{2}x + 2$

7. Linear relationships for five businesses are described here in words, equations, tables, and graphs. Determine which equation, table, and graph match each description. Then tell whether or not the relationship is a direct variation. Record your answers in a table like the one below.

Business	Equation Number	Table Number	Graph Number	Type of Relationship
Rent You Wrecks				
Get You There				
Internet Cafe				
Talk-a-Lot				
Walk 'em All				

Descriptions of the Businesses
- The Rent You Wrecks car rental agency charges $0.25 per mile plus $3.00 for a one-day rental.
- The Get You There taxi company charges a rider $2.00 plus $0.10 per mile.
- The Internet Cafe charges $5.00 plus $0.30 per minute for Internet access.
- The Talk-a-Lot phone company charges $0.75 plus $0.10 per minute for one call.
- The Walk 'em All pet-walking service charges $0.10 per minute to care for your dog.

Equations

i. $y = 0.3x + 5$

ii. $y = 0.25x + 3$

iii. $y = 0.1x + 2$

iv. $y = 0.1x$

v. $y = 0.1x + 0.75$

Tables

i.

x	1	2	3	4	5	6
y	0.85	0.95	1.05	1.15	1.25	1.35

ii.

x	1	2	3	4	5	6
y	2.10	2.20	2.30	2.40	2.50	2.60

iii.

x	1	2	3	4	5	6
y	3.25	3.50	3.75	4.00	4.25	4.50

iv.

x	1	2	3	4	5	6
y	5.30	5.60	5.90	6.20	6.50	6.80

v.

x	1	2	3	4	5	6
y	0.1	0.2	0.3	0.4	0.5	0.6

Graphs

i.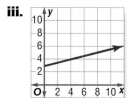
ii.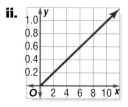

iii.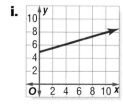
iv.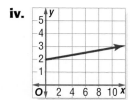
v.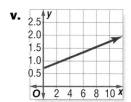

Connect & Extend

8. Reed is hiking in Haleakala Crater on the island of Maui. For the first hour, he walks at a steady pace of 4 kph (kilometers per hour). He then reaches a steeper part of the trail and slows to 2 kph for the next 2 hours. Finally, he reaches the top of the long climb and hikes downhill at a rate of 6 kph for the next 2 hours.

   a. Make a graph of Reed's trip showing distance traveled d and hours walked h. Put distance traveled on the vertical axis.

   b. Does the graph represent a linear relationship?

9. A blue plane flies across the country at a constant rate of 400 miles per hour.

   a. Is the relationship between hours of flight and distance traveled linear?

   b. Write an equation and sketch a graph to show the relationship between distance and hours traveled for the blue plane.

   c. A smaller red plane starts off flying as fast as it can, at 400 miles per hour. As it travels, it burns fuel and gets lighter. The more fuel it burns, the faster it flies. Will the relationship between hours of flight and distance traveled for the red plane be linear? Why or why not?

   d. On the axes from Part b, sketch a graph of what you think the relationship between distance and hours traveled for the red plane might look like.

10. Three navy divers are trapped in an experimental submarine at a remote location. They radio their position to their base commander, calling for assistance and more oxygen. They can use the radio to broadcast a signal to help others find them, but their battery is running low. The base commander dispatches three vehicles to help.

- a helicopter that can travel 45 miles per hour and is 300 miles from the sub

- an all-terrain vehicle that can travel 15 miles per hour and is 130 miles from the sub

- a boat that can travel 8 miles per hour and is 100 miles from the sub

Each vehicle is approaching from a different direction. The commander needs to determine which vehicle will reach the submarine next so he can tell the sub to turn its radio antenna toward that vehicle.

a. To assist the base commander, create three graphs on one set of axes that show the distance each vehicle is from the sub over time. Put time on the horizontal axis. Label each graph with the vehicle's name.

b. Use your graphs to determine when the commander should direct the submarine to turn its antenna towards the helicopter, the all-terrain vehicle, and the boat.

c. Write an equation for each graph that the commander could use to determine the exact distance d each vehicle is from the submarine at time h.

11. One day, Molly walked from Allentown to Brassville at a constant rate of four kilometers per hour. The towns are 30 kilometers apart.

a. Write an equation for the relationship between the distance Molly traveled d and the hours she walked h.

b. Graph your equation to show the relationship between hours walked and distance traveled. Put distance traveled on the vertical axis.

c. How many hours did it take Molly to reach Brassville?

d. Now write an equation for the relationship between the hours walked h and the distance remaining to complete the trip r.

e. Graph the equation you wrote for Part d on the same set of axes you used for Part b. Label the vertical axis for both d and r.

f. How can you use your graph from Part e to determine how many hours it took Molly to reach Brassville?

12. Three cellular telephone companies have different fee plans for local calls.

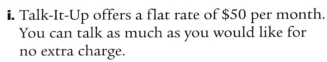

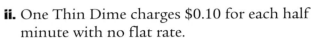

   i. Talk-It-Up offers a flat rate of $50 per month. You can talk as much as you would like for no extra charge.

   ii. One Thin Dime charges $0.10 for each half minute with no flat rate.

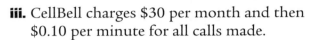

   iii. CellBell charges $30 per month and then $0.10 per minute for all calls made.

   a. For each company, write an equation that relates the cost of the phone service c to the number of minutes a customer talks during a month t.

   b. Any linear equation can be written in the form $y = mx + b$. Give the value of m and b for each equation that you wrote in Part a.

   c. For each company, make a graph that relates the cost of the phone service to the number of minutes a customer talks during a month.

   d. Where do the values of m and b appear in the graph for each phone company?

Math Link

The area of a circle is the number pi, represented by π, multiplied by the radius of the circle squared.

$$A = \pi r^2$$

---

13. Geometry You have studied formulas to calculate the area, perimeter, and circumference of various shapes. Some of these formulas are linear and some are not. Tell whether the formula for each measurement below is linear or not. Explain your answer.

   a. area of a circle

   b. circumference of a circle

   c. area of a square

   d. perimeter of a square

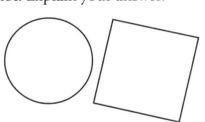

14. In Your Own Words Explain how you can tell whether a linear relationship is a direct variation from a written description, from an equation, and from a graph.

Mixed Review

15. Candace's grandfather gave her four marbles. One was blue, one was yellow, and two were green. She wants to give half of the marbles to her friend Marina for her birthday. Candace puts all the marbles in a leather bag and lets Marina choose two of them without looking. What is the probability that Marina will choose both green marbles?

16. Physical Science The table lists the elements that make up Earth's crust.

Earth's Crust

Element	Mass (%)
Oxygen	49.13
Silicon	26.00
Aluminum	7.45
Iron	4.20
Calcium	3.25
Sodium	2.40
Potassium	2.35
Magnesium	2.35
Hydrogen	1.00
Others	1.87

Source: *Ultimate Visual Dictionary of Science*

a. If you had a 1,000-kg sample of Earth's crust, how many kilograms of it would you expect to be iron?

b. What percentage of Earth's crust make up the three components that are in the greatest abundance?

c. Compare the amount of calcium in Earth's crust to the amount of hydrogen in two ways.

Find each percent.

17. 32% of $100

18. 80% of $200

19. 2.5% of $10,000

LESSON 1.2

Slope

People in many professions work with the concept of steepness. Highway engineers may need to measure the steepness of hills for a proposed highway. Architects may need to describe the steepness of a roof or a set of stairs. Ladder manufacturers may need to test the stability of a ladder as it relates to its steepness as it leans against a wall.

Think & Discuss

Think about one of the situations described above, the steepness of the roof of a house, which is called the roof's *pitch*. Suppose you want to precisely describe the steepness of each of these three roofs.

- Would it help to measure only the length of the roof from the peak down to one edge of the roof? Explain.

- Can you think of another way you might measure steepness?

Investigation ① Describe Slope

Vocabulary

slope

Materials

- graph paper
- metric ruler

In this investigation, you will explore slope, a common way to describe steepness.

✓ Develop & Understand: A

1. A ladder leans against a wall. The top of the ladder is 10 feet up the wall, and the base is 4 feet from the wall. In this scale drawing, 10 mm represents 1 foot of actual distance.

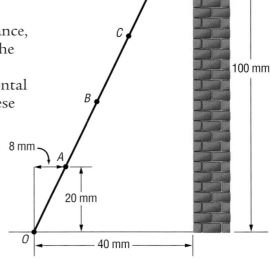

Notice that the vertical distance, or *rise,* between point *O* on the ground and point *A* on the ladder is 20 mm. The horizontal distance, or *run,* between these two points is 8 mm.

a. What is the vertical distance, or rise, in the drawing from point *O* to point *E*? What is the horizontal distance, or run, from point *O* to point *E*?

b. Copy and complete the table by measuring the rise and run between the given points on the scale drawing.

Points	A to B	A to C	B to C	A to D	B to D	D to E	O to E
Rise	20						
Run	8						

The steepness of the ladder or of any line between two points can be described by the ratio $\frac{\text{rise}}{\text{run}}$.

c. Add another row to your table. Label it $\frac{\text{rise}}{\text{run}}$. Compute the ratio for each pair of points in the table.

d. Choose any two unlabeled points on the ladder. Find the ratio $\frac{\text{rise}}{\text{run}}$ for your points. How does this ratio compare to the ratios for the points in the table?

2. Here is a scale drawing of a second ladder positioned with the top of the ladder 8 feet up the wall and the base 4 feet from the wall.

Select at least three pairs of points on this ladder. Calculate the ratio $\frac{\text{rise}}{\text{run}}$ for each pair. What do you find?

3. How does the $\frac{\text{rise}}{\text{run}}$ ratio for the first ladder compare to the $\frac{\text{rise}}{\text{run}}$ ratio for the second ladder? Which ladder appears to be steeper?

4. Imagine a third ladder positioned higher, 11 feet up the wall and 4 feet from the wall at its base. How would its $\frac{\text{rise}}{\text{run}}$ ratio compare to the ratios for the first two ladders?

5. Is $\frac{\text{rise}}{\text{run}}$ a good way to describe steepness? Explain.

✅ Develop & Understand: B

What happens if you try to find the ratio $\frac{\text{rise}}{\text{run}}$ for a curved object? The drawing to the right shows a cable attached to a wall.

6. Calculate the ratio $\frac{\text{rise}}{\text{run}}$ for each pair of points. What do you find?

- Points P and Q
- Points Q and R
- Points P and R

7. Describe the difference between the steepness of a ladder and the steepness of a curved cable. Be sure to discuss the ratio $\frac{\text{rise}}{\text{run}}$ for the two situations.

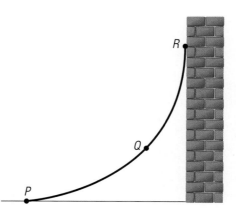

Using the ratio $\frac{\text{rise}}{\text{run}}$ is a good way to describe steepness for a ladder but not for a curved cable. Since a ladder is straight, you can calculate $\frac{\text{rise}}{\text{run}}$ between any two points. The ratio will be the same regardless of which points you choose.

The ratio $\frac{\text{rise}}{\text{run}}$ is also used to describe the steepness of a line. The ratio $\frac{\text{rise}}{\text{run}}$ for a line is called the line's **slope**.

✅ Develop & Understand: C

Consider the line in this graph.

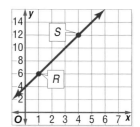

8. What are the coordinates of points R and S?

9. Find the slope of the line through points R and S.

You might have found the slope of the line by subtracting coordinates. The rise from point R to point S is the difference in the y-coordinates for those points. The run is the difference in the x-coordinates. However, you might not have thought about whether the *order* in which you subtract the coordinates affects the value of the slope.

10. In Parts a–c, find the slope of the line again by subtracting the coordinates for point R from the coordinates for point S.

    a. Find the rise by subtracting the y-coordinate of point R from the y-coordinate of point S.

    b. Find the run by subtracting the x-coordinate of point R from the x-coordinate of point S.

    c. Use your answers to calculate the slope of the line through points R and S.

11. Would you find the same slope if you subtracted the coordinates of point S from the coordinates of point R? Try it and see.

12. Would you find a different value for $\frac{\text{rise}}{\text{run}}$ if you used a different pair of points on the line? Explain.

13. Adam calculated the slope of the line as -2. His calculation follows.

$$\frac{\text{rise}}{\text{run}} = \frac{12 - 6}{1 - 4} = \frac{6}{-3} = -2$$

What was Adam's mistake?

14. Review your answers to Exercises 11–13. Is the order in which you subtract the coordinates important? Explain.

Graph the line through each pair of points. Find the slope of each line.

15. (−3, 4) and (−7, 2)   **16.** (2, 4) and (3, 3)

17. (3, 5) and (4, 5)   **18.** (−3, 4) and (−4, 6)

19. Look back at your work in Exercises 15–18. Two of the lines have a negative slope. What do you notice about these lines?

20. One of the lines in Exercises 15–18 has a slope of 0. What do you notice about that line?

21. A line has slope $-\frac{2}{3}$. One point on the line is (4, 5). Find two more points on the line. Explain how you found them.

22. Consider the line passing through the points (2, 4) and (2, 7).

     a. Graph the line. What does it look like?

     b. Try to find the slope of the line. What happens?

     c. What is the *x*-coordinate of every point on the line?

     d. What is an equation of the line?

Real-World Link

Surveyors can use instruments to determine the slope of a certain section of land.

· ·

Share  Summarize

1. Two lines with positive slope are graphed on one set of axes. Explain why a greater slope for one line means that line will be steeper than the other.

2. What does a negative slope tell you about a line?

3. What does a slope of 0 tell you about a line?

4. Give the coordinates of two points so that the line connecting them has a positive slope.

Investigation ② Slope and Scale

Materials

- graph paper
- graphing calculator

Slope is a good measure of the steepness of objects such as ladders. However, when you are using a graph to find or to show the slopes of lines, you need to be careful.

Explore

Copy the grids below onto graph paper. Graph the equation $y = 2x + 1$ on each grid.

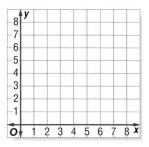 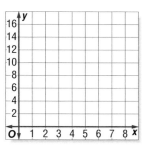

Describe the difference between the graphs. What do you think causes this difference?

In addition to slope, what other factor affects how steep the graph of a line looks?

✅ Develop & Understand: A

1. Ava's mother thinks Ava is spending too much money on DVDs. Ava says that since DVDs cost only $10 at Deep Discount Sounds, the amount she spends does not increase very quickly.

   a. Ava decides to make a graph showing how the total amount spent changes as she buys more DVDs. She thinks that if she chooses her scales carefully, she will convince her mother that the amount increases at a slow rate. Draw a graph that Ava might use. (Hint: Use your observations from Explore to create a graph that does not look very steep.)

   b. Ava's mother knows about graphing, too. She wants to make a graph to convince Ava that the total cost increases quickly as she adds to her DVD collection. Draw a graph that Ava's mother might use.

2. Imagine that you are using a graph to keep track of the amount of money remaining in a bank account. You start with $200 and withdraw $5 a week. Suppose you graph the time in weeks on the horizontal axis and the bank balance in dollars on the vertical axis.

   a. What will be the slope of the line?

   b. Draw the graph so it appears that the balance is decreasing very rapidly.

   c. Draw another graph of the same relationship so it appears that the balance is decreasing very slowly.

You can graph an equation on a graphing calculator and then adjust the window settings to change the appearance of the graph.

✅ Develop & Understand: B

Set the window of your graphing calculator to the standard window settings, x and y values from -10 to 10.

3. Graph the equation $y = x$. Sketch the graph.

4. Now change the window using the square window setting on your calculator. This adjusts the scales so the screen shows 1 unit as the same length on both axes. Graph $y = x$ using the new setting. Sketch the graph.

5. Compare the graphs that you made in Exercises 3 and 4.

6. Adjust the window settings to make the line appear steeper than both graphs. Record the settings that you used.

7. Adjust the window settings to make the line appear less steep than the other graphs. Record the window settings that you used.

Share & Summarize

Work with a partner. One of you should use your calculator to graph the equation $y = 3x + 2$ so that it looks very steep. The other should graph the same equation so that it does not look very steep.

1. Together, write a description of what each of you did to make your graphs look as they do.

2. Try to explain why your method works.

Practice & Apply

1. Consider these tables of data for two linear relationships.

Relationship 1	
x	**y**
1	4.5
2	6
3	7.5
4	9

Relationship 2	
x	**y**
−3	1
−1	3
1	5
3	7

a. Use the (x, y) pairs in the tables to draw each line on graph paper.

b. Find the slope of each line by finding the ratio $\frac{\text{rise}}{\text{run}}$ between two points.

c. Did you use the tables or the graphs in Part b? Does it matter which you use? Explain.

d. Check your results by finding the slope of each line again, using points different from those you previously used.

2. Look at the roofs on these three barns. All measurements are in feet.

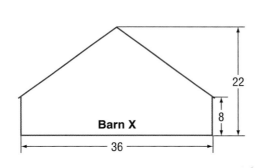
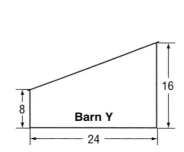
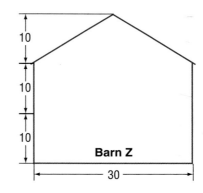

Barn X — 22, 8, 36

Barn Y — 16, 8, 24

Barn Z — 10, 10, 10, 30

a. Which roof looks the steepest?

b. Find the *pitch*, or slope, of the roof of each barn. Was your prediction correct?

3. Create an equation for a linear relationship.

a. Construct a table of (x, y) values for your equation. Include at least five pairs of coordinates.

b. Draw a graph using the values in your table. If your graph is not a line, check that the values in your table are correct.

c. Choose two points from the table, and use them to determine the slope of the line. Check the slope using two other (x, y) pairs.

In Exercises 4–7, you are given a slope and a point on a line. Find another point on the same line. Then draw the line on graph paper.

4. slope: $\frac{1}{2}$; point: (3, 4)   **5.** slope: -1; point: (2, 5)

6. slope: 3; point: (2, 8)   **7.** slope: $\frac{1}{4}$; point (4, 5)

In Exercises 8–11, find the slope of the line by identifying two points on the line and using them to determine the slope.

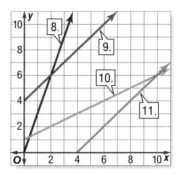

12. Here are two linear relationships, $y = 3x + 2$ and $y = 2x + 3$. Graph both relationships on copies of the grids shown below. Each grid will have two graphs.

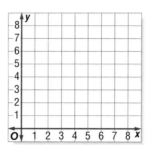   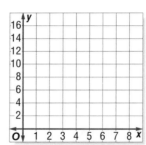

a. What aspects of the graphs depend on the grid on which you drew them?

b. What parts of the graphs are not affected by the scale of the grid? For example, do the points of intersections with the *x*- and *y*-axes change from one grid to the other?

13. Study these graphs.

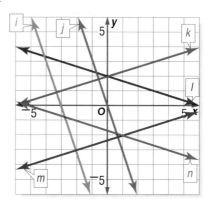

a. Which pairs of lines have the same slope?

b. Find the slope of each line.

Connect & Extend

For Exercises 14–23, answer Parts a and b.

a. What is the constant difference between the *y* values as the *x* values increase by 1?

b. What is the constant difference between the *y* values as the *x* values decrease by 2?

Math Link

Lines have a *constant difference* in their *y* values. When the *x* values change by a certain amount, the *y* values also change by a certain amount.

14. $y = x$

15. $y = x + 2$

16. $y = 3x - 3$

17. $y = -2x + 12$

18. $y = 5x$

19. $y = \frac{1}{2}x$

20. $y = 23x - 18$

21. $y = -x$

22. $y = -2x + 6$

23. $y = -\frac{1}{2}x + 4$

24. In Your Own Words How can you determine the slope of a line from a graph? If you are given the slope of a line, what else do you need to know before you can graph the line?

Math Link

A right angle has a measure of 90°.

25. Isi looked at the equations $y = \frac{3}{2}x - 1$ and $y = -\frac{2}{3}x + 2$ and said, "These lines form a right angle."

a. Graph both lines on *two different* grids with the axes labeled as shown here.

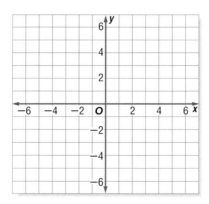 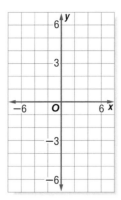

b. Compare the lines on each grid. Do both pairs of lines form a right angle?

c. What kind of assumption must Isi have made when she said the lines form a right angle?

d. What is the relationship of the slopes of these two lines?

Mixed Review

Tell whether each set of ordered pairs could describe a linear relationship.

26. (4, 15); (7, 24); (3, 12)   **27.** (0, −5); (2, 1); (1, −2); (3, 4)

28. Irene made a bead necklace by alternating groups of three blue beads and two red beads. She used a total of 75 beads. How many beads of each color are in her necklace? Explain how you found your answer.

State whether the data in each table could be linear. Explain.

29.

a	−4	−3	−2	−1	0	1
b	−32	−13.5	−4	−0.5	0	0.5

30.

a	−4	−3	−2	−1	0	1
b	−12.1	−9.6	−7.1	−4.6	−2.1	0.4

Write Equations

In Lessons 1.1 and 1.2, you used verbal descriptions, equations, graphs, and tables to represent linear relationships.

Think & Discuss

Suppose you want to determine if a relationship is linear. Would you rather use a description, equation, graph, or table to make your determination? Why?

Investigation 1   The Method of Constant Differences

Vocabulary

coefficient

constant term

Materials

• graph paper

Ciera wanted to determine whether a set of values could be linear.

Table 1

Input	1	2	3	4
Output	4	7	10	13

Table 2

Input	0	1	2	3
Output	1	6	11	16

Table 3

Input	0	1	2	3
Output	29	27	25	23

Table 4

Input	2	3	4	5
Output	10	15	21	28

Think & Discuss

Ciera concluded that tables 1, 2, and 3 could be linear.

• What pattern might have led Ciera to her conclusion?

• Write a symbolic rule that fits all input/output pairs for each table that could be linear. Use a letter for each input and output.

• Draw a graph of each table. Extend your graphs and estimate the output for each input of −3. Which estimates do you think are the most accurate? Why?

Ciera made her table using consecutive integers as inputs. Then she may have found the difference between neighboring pairs of outputs. When the differences between all neighboring pairs are the same, they are *constant differences*.

✅ Develop & Understand: A

Below are several tables that you can use to try the method of constant differences. For each, if you think the relationship could be linear, find a rule and check it with the pairs in the table. If you do not think the relationship is linear, draw a graph to check.

1.

Input, x	−1	0	1	2	3	4	5	6
Output, y	−7	3	13	23	33	43	53	63

2.

Input, x	0	2	1	5	4	3
Output, y	−2	16	7	43	34	25

3.

Input, d	3	2	1	5	4
Output, s	15	25	35	45	55

4.

Input, t	−2	−1	0	1	2	3	4	5
Output, u	1	0.5	0	−0.5	−1	−1.5	−2	−2.5

5. Suppose you already know that the table of values is linear. This means you can write the rule as $y = ax + b$ with numbers in place of a and b and with x as the input and y as the output. What input would you use so you could find the value of b? Why would you use that input?

✅ Develop & Understand: B

You will now think some more about *why* the method of constant differences works.

6. The differences in the output values in this table are all 3.

x	1	2	3	4	5
y	4	7	10	13	16

Think of a situation, like those you have studied in this chapter, that might have this table. Describe what the difference of 3 means for your situation. Does it make sense that the relationship between these variables could be linear? Explain.

Real-World Link

In downhill skiing, ski slopes are rated by their difficulty. Beginner's slopes are indicated by green dots, intermediate slopes by blue squares, and advanced slopes by black diamonds.

7. If you made a graph showing how y changes as x changes, what would be its slope? What does this slope mean for your situation in Exercise 6?

8. Think about the definition of *slope*. How does using consecutive integers for your inputs help you find the slope?

✅ Develop & Understand: C

Use the method of constant differences to analyze the pattern in each table. Predict the missing outputs.

9.

Input, x	1	2	3	4	5	6	7	8
Output, y	1	2	4	8	16	32		

10.

Input, s	1	3	4	8	11	12	13	14
Output, t	8	18	23	43	58	63		

11.

Input, a	1	2	3	4	5	6	7	8
Output, b	1	2	4	7	11	16		

12.

Input, g	1	2	3	4	5	6	7	8
Output, h	7.5	6	4.5	3	1.5	0		

13.

Input, m	−3	−1	1	3	5	7	9	11
Output, n	2	−2	−6	−10	−14	−18		

14. In which tables do the differences tell you the relationships could be linear? Write rules for those relationships.

15. Plot the data in each table. For the graphs that seem to be linear, draw a line through the points. Extend the linear graphs to show the value of the output when the input is 0, −1, and −2. Label those points with their coordinates.

16. Find the slope of each line drawn in Exercise 15. Compare the slopes to the rules written for Exercise 14.

17. In Exercise 14, did you choose all the relationships that could be linear? Did you write a correct rule for each of them? If not, figure out what went wrong, and write new rules for those relationships.

Nearly all of the rules that you have encountered in this chapter can be written in the form $y = ax + b$.

The outputs y and the inputs x are called *variables* because their values can vary. The other two numbers also have special names.

A number that is multiplied by the input, like a, is called a **coefficient**. In the equation $y = ax + b$, we say that "*a* is the coefficient of *x*."

A number that stands by itself, like b, is often referred to as the **constant term**.

Example

$y = 2x + 6$ The 2 is a coefficient, and the 6 is the constant term.

$y = 4x - 3$ The 4 is a coefficient, but what is the constant term? It helps to rewrite the rule in the form $y = ax + b$. Since subtracting a number is the same as adding its opposite, this rule can be rewritten as $y = 4x + (-3)$. So, the constant term is -3.

$y = x + 12$ You can see that the constant term is 12, but there is not a number being multiplied by x. Is there a coefficient? You can also rewrite this rule as $y = 1x + 12$. Here, a is 1, so 1 is the coefficient.

Share & Summarize

1. Explain how to determine whether the input/output pairs in a table could be related in a linear way.

2. Using the new vocabulary terms on the previous page, explain how to write a symbolic rule for a table that you have determined could describe a linear relationship.

Investigation 2 Understand the Symbols

Materials

• graph paper

In this chapter, you have looked at relationships that describe real situations. Most of these relationships were *linear*. They could be graphed as a line and expressed in a rule that looks like $y = ax + b$.

Now you will extend what you know about the coefficient and the constant term, a and b, by investigating how they affect the quadrants through which the graph passes.

✓ Develop & Understand: A

1. What does the coefficient a tell you about the graph of $y = ax + b$?

2. What does the constant term b describe about a graph of $y = ax + b$?

Math Link

The four quadrants of a graph are numbered.

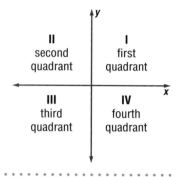

II second quadrant	**I** first quadrant
III third quadrant	**IV** fourth quadrant

3. For each equation in the table below, $a = 0$ and $b \neq 0$.

   a. Complete the table. Choose two more equations for which $a = 0$ and $b \neq 0$. Add them to your table.

x	−2	−1	0	1	2	3
$y = 0x - 2$						
$y = 3$						
?						
?						

   b. On one set of axes, graph all four equations.

4. Through how many quadrants does each graph pass? Do you think this will always be true for $y = ax + b$ when $a = 0$ and $b \neq 0$? Test several more rules of your own to check.

5. For each equation in the table below, $a \neq 0$ and $b = 0$.

   a. Complete the table. Choose two more equations for which $a \neq 0$ and $b = 0$. Add them to your table.

x	−2	−1	0	1	2	3
$y = x$						
$y = -2x$						
?						
?						

   b. On one set of axes, graph all four equations.

6. Through how many quadrants does each graph pass? Do you think this will always be true for $y = ax + b$ when $b = 0$ and $a \neq 0$? Test some more rules to check.

7. Through how many quadrants do you think $y = ax + b$ will pass if $a = 0$ and $b = 0$? Graph the equation $y = 0$. Through how many quadrants does the graph pass?

8. Look back at the graphs in this chapter. Through how many quadrants can $y = ax + b$ pass *at the most*? Explain your reasoning.

9. Through how many quadrants can $y = ax + b$ pass *at the least*? Explain your reasoning.

10. Look at your graphs for this exercise set for graphs that do not cross the horizontal axis. What do their rules have in common? Do you think all such rules will have graphs that do not cross the horizontal axis?

11. Are there any graphs in your examples or elsewhere in this chapter that do not cross the vertical axis? Can you think of a rule in the form $y = ax + b$ with such a graph?

The outputs in the two tables below start out the same way. Both begin with 1, 2, 4, and 8, but they continue differently.

Input	1	2	3	4	5	6	7	8
Output	1	2	4	8	16			

Input	1	2	3	4	5	6	7	8
Output	1	2	4	8	15			

Even though these two tables match in four places, they do not match in the fifth place and maybe will never match again. This shows that if someone gives you a sequence of numbers like 1, 2, 4, 8, ..., you cannot reliably say what number will come next without knowing more.

✅ Develop & Understand: B

In sequences like the one just mentioned, it helps to have information about what kind of relationship is being described. For example, if you know that a relationship is linear, you can find the rule from just a few points.

12. This table describes a linear relationship.

Input	1	2	3	4	5
Output	3	1	−1	−3	−5

a. Think about how many points you need to plot in order to draw the line for this relationship. What is the *fewest* number of points needed? Explain.

b. Graph the relationship using no more points than you need.

c. Use your graph to find a rule for this relationship.

Not all linear rules are in the form $y = ax + b$. If you can rearrange a rule into this form, you know that its graph will be a line.

13. Consider the equation $2y = 4x + 5$.

a. Graph the equation to show that it is a linear rule.

b. Write a new rule, using the form $y = ax + b$, that has the same graph as $2y = 4x + 5$.

1. Explain how the graph of a linear relationship can help you to guess the rule.

2. Explain how the method of constant differences can help you to guess the rule for a relationship.

Investigation 3 Slope-Intercept Form

Vocabulary

slope-intercept form

Materials

• graph paper

In the previous two investigations, you wrote equations or rules in the form $y = ax + b$ for linear relationships. The form $y = mx + b$, or the form $y = ax + b$, is called the **slope-intercept form** of a linear equation.

The chart below summarizes the relationships that you have discovered among the numbers in an equation in the form $y = mx + b$, entries in an input/output table, and the features of a graph.

	Equation $y = mx + b$	Input/Output Table	Graph
Slope	The value of m, the coefficient of x, represents the slope.	As the input value x changes by 1 unit, the output value y changes by m units.	The number m is the slope, $\frac{\text{rise}}{\text{run}}$, of the line. It shows the rate at which y changes as x changes.
Constant Term	The value of b represents the constant term.	When the input value of $x = 0$, the output value of $y = b$.	The constant term b shows where the line crosses the y-axis. Thus, b is known as the y-intercept. It is located at $(0, b)$.

You know that you can use an equation to complete a table. You can also plot points using the x and y coordinates from your table to see if the equation might be linear.

In this investigation, you will examine how to use the slope-intercept form of a linear equation, $y = mx + b$, to plot points without creating a table, going directly from the equation to the graph.

Math Link

You have been using the letter a to represent the rate of change, or slope, in linear relationships. Commonly, you will see the letter m used in the same way. The two forms $y = ax + b$ and $y = mx + b$ are interchangeable.

Think & Discuss

Darnell and Maya were trying to make a graph of the equation $y = 3x - 1$.

I can tell from the equation that the point $(0,1)$ will be on our graph.

Then we can use the slope to figure out the other points on the graph.

$y = 3x - 1$

- Is Maya correct? How can she know the point $(0, -1)$ will be on the graph?
- How might you use the slope to determine other points on the graph as Darnell suggests?
- How many points would you need to plot to be able to draw the line?

✅ Develop & Understand: A

1. Earlier, you learned to think of the slope as $\frac{\text{rise}}{\text{run}}$. How would you describe the rise and run for a slope of -2?

2. How would you describe the rise and run for a slope of $\frac{1}{2}$?

3. For each graph below, determine the coordinates $(0, b)$ for the y-intercept.

a.

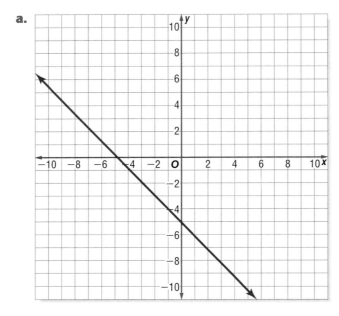

b.

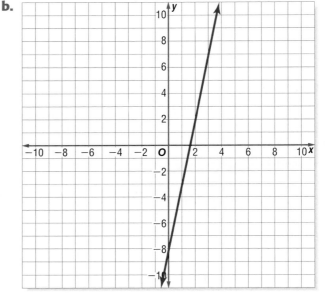

c.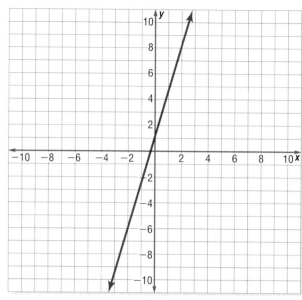

d.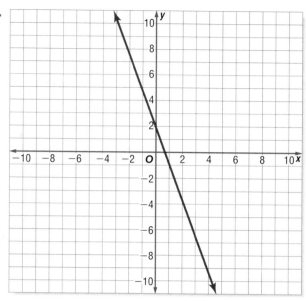

4. For the graphs in Exercise 3, determine the slope, $m = \frac{rise}{run}$, for each line.

5. Use the y-intercepts from Exercise 3 and the slopes from Exercise 4 to write the equation of each line in slope-intercept form.

6. For each equation, use a grid to plot the y-intercept $(0, b)$. Then use the slope to plot two more points. Connect the points to form a graph of the equation.

a. $y = 3x - 1$ **b.** $y = \frac{1}{2}x + 4$

c. $y = -2x - 1$ **d.** $y = -4x + 5$

7. Describe the graph when the slope m is positive.

8. Describe the graph when the slope m is negative.

9. Verify the graphs you created in Exercise 6 by completing a table like the one below for each equation. Check to see that the coordinate pairs in the table are on the line you graphed.

x	−2	−1	0	1	2
y					

Real-World Link

Cell phone plans can be analyzed using equations and graphs. Use the monthly fee as the y-intercept and the rate per minute as the slope.

Share & Summarize

1. Explain how to use a linear equation in slope-intercept form to graph the equation without first creating a table.

2. Explain how to write a linear equation in slope-intercept form by examining a coordinate graph of the line.

3. In this investigation, you used a known point, the y-intercept at $(0, b)$, and the slope to graph a line. Do you think you can graph a line given a different point, not the y-intercept, and the slope?

Use Points and Slopes to Write Equations

Materials

- graph paper

You will now learn how to find equations for lines when you know two points on the line or when you know the slope of the line and one point.

Explore

The table describes a linear relationship.

x	−2	−1	0	1	2	3
y	−3	−1	1	3	5	7

What is an equation of the line described by the table? Explain how you found the equation.

Use two data pairs (x, y) to find the slope of this line. How is the slope used in the equation you wrote?

Graph the equation. What is the y value of the point at which the graph crosses the y-axis? How is this value used in the equation?

In Explore, you were probably able to find the values for m and b fairly easily. But now look at this table, which also shows data pairs for a linear relationship.

x	−6	−4	−1	$1\frac{1}{2}$	3	7
y	$-3\frac{3}{4}$	$-2\frac{1}{4}$	0	$1\frac{7}{8}$	3	6

Finding an equation of the line for these data is a slightly more complex task. You could calculate the slope, but the y-intercept is not given. You cannot be certain what it is by graphing the data pairs and looking at the graph.

However, you *can* determine an equation of a line if you know the *slope* and *one point* on the line. The fact that linear equations take the form $y = mx + b$ helps you do this.

Example

What is an equation of the line that has slope 3 and passes through the point (2, 5)?

Start with the fact that the equation of a line can be written in the form $y = mx + b$. The slope is 3, so $m = 3$. This gives the equation $y = 3x + b$.

Because the point (2, 5) is on this line, substituting 2 for x and 5 for y will make the equation a true statement. We say that the point (2, 5) *satisfies* the equation $y = 3x + b$.

$$y = 3x + b$$
$$5 = 3(2) + b$$
$$5 = 6 + b$$
$$-1 = b$$

Now you know that the value of the y-intercept, b, is -1. You can write the final equation.

$$y = 3x - 1$$

✅ Develop & Understand: A

1. What is an equation of the line that has slope 4 and passes through the point (1, 5)?

2. What is an equation of the line that has slope 3 and passes through the point (2, 4)?

3. What is an equation of the line that has slope -2 and passes through the point (8, -12)?

4. What is an equation of the line that has slope 0 and passes through the point (3, 5)?

✅ Develop & Understand: B

Suppose you know only two points on a line but not the slope. How can you find an equation for the line? For example, suppose you want to write an equation of the line that contains the points (1, 3) and (3, 11).

5. What is the slope of the line connecting these points? Show how you find it.

6. If the equation of this line is in the form $y = mx + b$, what is the value of m?

7. Now find the value of *b*, the *y*-intercept, without drawing a graph. Show your work. (Hint: Look back at your work in Exercises 1–4 if needed.)

8. Write an equation of the line.

9. Check that the point (3, 11) satisfies your equation by substituting 3 for *x* and 11 for *y* and then evaluating. Also, check that the point (1, 3) satisfies your equation. If either does not, was your error in determining the value of *m* or *b*? Write down what you find. Adjust the equation if necessary.

✅ Develop & Understand: C

Find an equation of the line through each pair of points. Plot the points and draw the line to check that the equation is correct. Review the process you followed in Exercises 5–9 if necessary.

10. (3, 7) and (8, 12)

11. (6, 11) and (18, 17)

12. (0, 0) and (100, 100)

13. (3, 5) and (−1, 5)

✅ Develop & Understand: D

Researchers have discovered that people shake salt over their food for about the same amount of time regardless of how many or how large the holes are in the saltshaker.

When you use a saltshaker with large holes, you will probably use more salt than when you use one with small holes. In fact, there appears to be a linear relationship between the average amount of salt people sprinkle on their food and the total area of the holes in their saltshakers. The following data were collected.

Total Area of Holes (mm^2), *a*	4.5	8
Average Amount of Salt Applied (g), *s*	0.45	0.73

Assume the researchers are correct and that the amount of salt is linearly related to the total area of the holes. You can use the small amount of data in the table to estimate how much salt would be shaken onto food from shakers with different-sized holes.

14. Plot the two points from the table with total area on the horizontal axis. Use them to graph a linear relationship. Think carefully about your axes. Size the scales so they can be read easily.

15. Use your graph to estimate the amount of salt delivered from a shaker with a total hole area of 6 mm^2.

16. If each hole has a radius of 1.1 mm, what is the area of one hole?

17. Use your graph and your answer to Exercise 16 to estimate the amount of salt delivered by a shaker having ten such holes.

18. If you want to limit your salt intake at each meal to 0.5 g, what should be the total area of the holes in your saltshaker?

19. Find an equation of the line that passes through the points (4.5, 0.45) and (8, 0.73).

20. Use the equation to find the amount of salt delivered by the given total hole areas.

  a. 2.7 mm^2, approximately the smallest total hole area used in commercial saltshakers

  b. 44.7 mm^2, approximately the largest total hole area used

Share & Summarize

1. Without using numbers, describe a general method for writing an equation for a line if all you know is the slope of the line and one point on the line.

2. Without using numbers, describe a general method for writing an equation for a line if all you know is two points on the line.

. . . .
Real-World Link
The average adult human body contains approximately 250 grams of salt.
. . . .

Investigation 5 — What Does Standard Form Tell You?

Vocabulary

standard form

Materials

• graph paper

In this investigation, you will examine another equation form to determine if an equation is linear.

Think & Discuss

Consider the following situation.

Jeremy has cows and chickens on his farm. The animals have a total of 46 legs. How many cows and chickens might Jeremy have?

There is not enough information given to figure out exactly how many cows and how many chickens Jeremy has. But there is a set of possible answers. To analyze the relationship between cows, chickens, and legs, discuss the following.

If x stands for the number of cows, how can you express the number of cow legs?

If y stands for the number of chickens, how can you express the number of chicken legs?

✓ Develop & Understand: A

1. From the information in Think & Discuss, write an equation to show the total number of legs.

2. Assume that Jeremy has three cows. Describe or show how you would determine the number of chickens.

3. Complete the following table.

Cows (x)	1	2	3	4	5	6	7	8	9	10	11
Chickens (y)											

4. Plot the points from the table to create a graph. According to the graph, does this appear to be a linear relationship? Explain why the table seems to support the conclusion you drew from the graph.

5. How are the two variables related?

6. The situation states that Jeremy has cows and chickens. Therefore, the table did not include an entry for $x = 0$, which would mean Jeremy has zero cows. However, if Jeremy did have zero cows, how many chickens would he have?

7. Using methods you learned in this chapter, write the slope-intercept form of the equation for this graph and table.

Both of the following equations represent the same linear relationship.

$$4x + 2y = 46 \qquad \text{and} \qquad y = -2x + 23$$

The form $y = -2x + 23$ is written in *slope-intercept form*. The form $4x + 2y = 46$ is known as **standard form** for a linear equation.

> ## Think & Discuss
>
> You will find that one form of a linear equation may be more convenient to use in certain situations.
>
> The equation $4x + 2y = 46$ shows that four legs for each cow plus 2 legs for each chicken will equal a total of 46 legs. In the equation $y = -2x + 23$, the slope of -2 tells that the two chickens are removed for every additional cow. The y-intercept of 23 tells the number of chickens you would have if there were no cows.
>
> Both equations are correct. However, which form is more helpful in making sense of the situation?
>
> In another situation, Joey earns \$10 for set-up plus \$8 per hour x working as a disc jockey. The slope-intercept equation for his total pay y is $y = 8x + 10$. Joey's total pay is calculated by multiplying his number of hours of work by 8 and adding the set-up fee.
>
> The equivalent standard form equation would is $8x - y = -10$. This equation indicates that multiplying each hour of work by 8 and subtracting Joey's total pay leaves a deficit of \$10.
>
> Both of these descriptions are correct. Which form is helpful in making sense of this situation?

Math Link

Follow these steps to find the standard form equation representing Joey's pay.

- Start with the slope-intercept equation $y = 8x + 10$.
- Subtract $8x$ from both sides to get $-8x + y = 10$.
- Divide both sides of $-8x + y = 10$ by -1.

✅ Develop & Understand: B

8. Create a table for each equation in standard form.

   a. $8x - 2y = -2$ **b.** $3x + y = 2$

   c. $2x - y = 4$ **d.** $4x + 8y = -8$

x	−2	−1	0	1	2	3	4
y							

9. Plot points from the tables to create a graph for each of the equations in Exercise 8. What do all of the graphs have in common?

10. Which equations represent a line that is increasing from left to right? What does this tell you about the slope?

11. Which equations represent a line that is decreasing from left to right? What does this tell you about the slope?

Math Link

Remember, to solve equations, you must keep things balanced.

Each linear equation written in standard form can be written in slope-intercept form by solving the equation for y.

Example

$$8x - 2y = -2$$
$$-2y = -8x - 2 \qquad \text{Add } -8x \text{ to both sides.}$$
$$y = 4x + 1 \qquad \text{Divide both sides by } -2.$$

✅ Develop & Understand: C

12. In the example, you see that the equation in standard form, $8x - 2y = -2$, is equivalent to the equation in slope-intercept form, $y = 4x + 1$. Show that $y = -4x + 1$ and the table you created in Part a of Exercise 8 represent equivalent relationships.

13. Rewrite each of the other equations from Exercise 8 in slope-intercept form.

$$3x + y = 2 \qquad 2x - y = 4 \qquad 4x + 8y = -8$$

14. What does the table indicate about the slope of the line given in slope-intercept form?

Standard form allows you to quickly find an equation's intercepts. Consider the equation $4x + 8y = -8$. Substituting 0 for x results in a value of -1 for y. Substituting 0 for y results in a value of -2 for x. If you graphed the line $4x + 8y = -8$, you should notice the intercepts at $(0, -1)$ and $(-2, 0)$.

Real-World Link

Taxi fares often vary based on the number of people in the cab and the number of miles driven. Equations can be used to determine the final fare.

✅ Develop & Understand: D

Find the x- and y-intercepts for each line.

15. $2x + 10y = 20$

16. $3x + 12y = -24$

17. $x + 2y = 1$

18. $3x - 7y = 10$

Share & Summarize

1. Chairs & Things manufactures three-legged and four-legged stools. Suppose the company wants to use up the 60 legs it currently has in-stock. Make a table to show the number of each type of stool the company can manufacture. Write an equation that represents this situation. Is your equation in standard or slope-intercept form?

2. Explain how the standard form of a linear equation and the slope-intercept form of a linear equation are related.

3. The two forms of an equation are beneficial in different situations. Give examples of when you would use each form.

Inquiry

Investigation 6 Linear Designs

Materials

• graphing calculator

Math Link

The four quadrants of a graph are numbered.

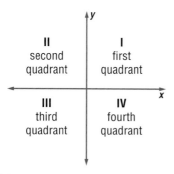

You can make some very interesting designs with linear equations on your graphing calculator.

Try It Out

See if you can produce a starburst design, like the one below, by graphing four linear equations in one window. Your design does not have to look exactly like this one. But it should include two lines that pass through quadrants I and III and two lines that pass through quadrants II and IV. Make a sketch of your design.

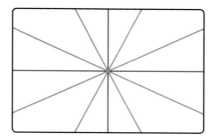

1. Record the equations you used.

2. What do you notice about the *y*-intercepts in your equations?

3. What do you notice about the slopes in your equations?

Make It Rain

On weather maps, a set of parallel lines is often used to symbolize sheets of rain. On your calculator, try to create four parallel lines that are evenly spaced, similar to the design below. Sketch your design.

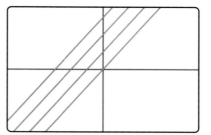

4. Record the equations that you used.

Go on ▶

5. What do you notice about the slopes in your equations? How is this reflected in your design?

6. What do you notice about the *y*-intercepts in your equations? How is this reflected in your design?

Making Diamonds

Try to make your own diamond shape, like the one below, on your calculator. Make a sketch of your design.

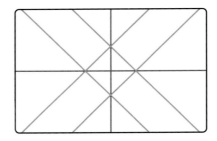

7. Record the equations that you used.

8. What do you notice about the slopes in your equations?

9. What do you notice about the *y*-intercepts in your equations?

What Have You Learned?

10. Think about the equations that you wrote to make parallel lines, lines radiating from a central point as in a starburst, and intersecting lines that form squares or diamonds. Make a design of your own with at least four lines.

    Share what you have learned by preparing a written report about making linear designs on a graphing calculator. Include sketches of your designs and the equations that can reproduce them.

Practice  Apply

Decide whether each relationship could be linear. If you think it could be, find a rule and write it in symbols. Then check your rule with each input/output pair.

1.

x	4	3	2	5	1
y	−7	−5	−3	−9	−1

2.

z	−1	1	3	5	7	9
w	−12	−6	0	6	12	18

3.

x	3	−1	0	2	4	5	1
y	10	2	1	5	17	26	2

4.

m	4	7	5	3	6
n	31	19	26	35	22

5.

u	10	12	8	13	7
z	64	76	52	82	46

6. The telephone service PhoneHome charges a fee of *r* dollars per minute. Customers also pay a service charge of *s* dollars every month, regardless of how long they use the phone. If *m* is the number of minutes you use the phone, then your monthly bill *b* in dollars can be computed with the rule $b = rm + s$.

a. Think about the four variables in this situation. What kinds of values can each variable have? For example, can *r* be 5? Can it be −3? Can *s* be 3.24? Can it be 0? Can *m* be 5?

b. PhoneHome offers a special calling plan, the Talker's Delight plan. This plan has a monthly rate of *s* dollars and no additional charge per minute, no matter how long you talk. In this arrangement, what is the value of *r*? Through how many quadrants does the graph of $b = rm + s$ pass for this charge plan?

c. PhoneHome also offers the TaciTurn plan. There is no monthly fee, but you *are* charged by the minute. What is the value of *s* in this pricing plan? Through how many quadrants does the graph of $b = rm + s$ pass?

d. Assuming any values of *r* and *s*, through how many quadrants can the graph of $b = rm + s$ pass *at the most*, if *m* is time in minutes and *b* is your monthly phone bill? Explain your reasoning.

e. Through how many quadrants must $b = rm + s$ pass *at the least*? Explain your reasoning.

7. Four graphs are drawn on the axes.

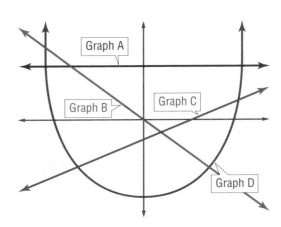

a. Point *a* has a negative first coordinate and a positive second coordinate. On which graphs could point *a* be located?

b. Point *b* has two negative coordinates. On which graphs could point *b* be located?

c. Which graphs contain points in both quadrants II and IV?

d. Which graphs contain points on both the *x*-axis and the *y*-axis?

e. Which graph could contain points $(2, -6)$ and $(-2, -6)$?

In Exercises 8 and 9, you are given a slope and a *y*-intercept for a line. Write the slope-intercept form of the equation for this line. Identify another point that would be on the line.

8. slope: -2; *y*-intercept $(0, 3)$ **9.** slope: $\frac{3}{2}$; *y*-intercept $(0, -1)$

Math Link

Lines have a constant difference in their *y* values. When the *x* values changes by a certain amount, the *y* values also change by a certain amount.

In Exercises 10 and 11, you are given a graph of a line. Write the slope-intercept form of the equation for each line.

10. 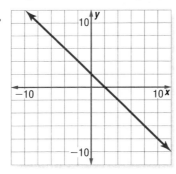   **11.**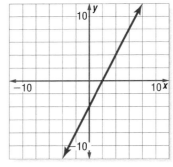

Find an equation of each line.

12. a line with slope -1 and passing through the point $(1, 4)$

13. a line with slope $\frac{1}{3}$ and passing through the point $(3, 3)$

14. a line with slope -2 and passing through the point $(3, 6)$

15. a line with a slope 0 and passing through the point $(3, 4)$

Find an equation of the line passing through the given points.

16. $(3, 4)$ and $(7, 8)$ **17.** $(2, 7)$ and $(6, 6)$ **18.** $(3, 5)$ and $(9, 9)$

In Exercises 19–21, you are given an equation for a line in standard form. For each equation, find the x and y intercepts.

19. $5x - 2y = 10$ **20.** $2x + y = -3$ **21.** $4x + 2y = 6$

In Exercises 22–24, match each equation in standard form with the equivalent equation written in slope-intercept form.

a. $y = -x + 2$ **b.** $y = 2x - 4$ **c.** $y = -\frac{1}{2}x + 3$

22. $2x - y = 4$ **23.** $x + 2y = 6$ **24.** $x + y = 2$

Connect & Extend   **25.** Ivan created this pattern of triangular shapes made from squares.

a. Complete the table. The *first* differences d are the differences between adjacent values of A. The differences between adjacent values of d are the *second* differences.

Base Length, *b*	1	2	3	4	5	6	7
Total Area, *A*	1	3	6				
First Differences, *d*		2	3				
Second Differences, *s*			1				

b. Is the relationship between b and A linear? How do you know?

c. Are the second differences constant?

d. Here is another geometric pattern. Make a table of base length, area, and first and second differences. Do you find any constant differences?

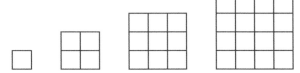

e. Write an equation for the area of the squares in Part d.

f. The equation for the triangle numbers in Part a is $A = \frac{b^2 + b}{2}$. What do you think constant second differences might tell you about a relationship's equation?

26. The last few lessons have been mostly about linear relationships. There are several very important and common sequences of output numbers that are not linear. Below are tables of some of these sequences. You will continue to see these patterns as you further your studies in mathematics.

    a. Find the pattern and complete each table.

i.

Input	1	2	3	4	5	6	7	8	9	10	11
Output	1	4	9	16							

ii.

Input	1	2	3	4	5	6	7	8	9	10	11
Output	1	3	6			21					

iii.

Input	1	2	3	4	5	6	7	8	9	10	11
Output	1	2	3	5	8		21				

iv.

Input	0	1	2	3	4	5	6	7	8	9	10
Output	1	2	4		16						

    b. These sequences are so important that each has its own special name. One sequence is related to triangular patterns, so it is called the *triangular numbers*. Another is related to square patterns and is called the *square numbers*. The other two are called *powers of 2* and *Fibonacci numbers*. Try to match each name to its sequence.

    c. The first differences of each sequence also have names. Try to name each set of first differences.

27. Consider the slope-intercept form of the equation of a line, $y = 3x - 2$.

    a. Identify two points that would lie on this line. Explain how you determined each point.

    b. Plot the two points and draw a sketch of the line on a copy of the grid shown to the right.

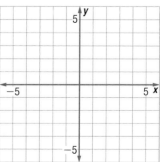

c. Which situations below could be represented with the equation $y = 3x - 2$? Support your answers.

    i. Juan paid a three-dollar registration fee and two dollars per t-shirt that he sold for the fundraiser. The equation shows how much money he raised.

    ii. Lakesha spent two dollars on supplies and then made three dollars for each quart of lemonade she sold. The equation shows her profit.

    iii.

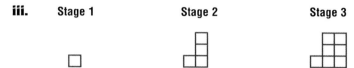

In Exercises 28–30, match each graph with the situation in Parts a–c that would produce the graph.

a. On Saturday, Stacy started with 24 completed pages in her scrapbook. She worked on new pages all day, completing two pages each hour.

b. The temperature at sundown was 36 degrees. The temperature overnight dropped three degrees per hour.

c. Beng built squares with toothpicks. For the first square, he used four toothpicks. To add another square, he used three toothpicks.

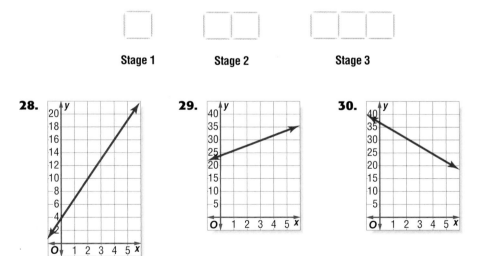

For each equation, identify the slope and the *y*-intercept.

31. $y = 2x + 0.25$

32. $y = -x + 5$

33. $y + 3 = x$

34. $-0.5y = x$

35. $\frac{1}{2}x = y - \frac{3}{4}$

36. $3x = y$

37. The table shows *x* and *y* values for a particular relationship.

x	6	3	1	2.5
y	7	1	−3	0

   a. Graph the ordered pairs (x, y). Make each axis scale from −10 to 10.

   b. Could the points represent a linear relationship? If so, write an equation for the line.

   c. From your graph, predict the *y* value for an *x* value of −2. Check your answer by substituting it into the equation.

   d. From your graph, find the *x* value for a *y* value of −2. Check your answer by substituting it into the equation.

   e. Use your equation to find the y value for each of the *x* values 0, −1, −1.5, −2.5. Check that the coordinates all lie on the line.

   f. Write the equation of the line in standard form.

38. Consider these four equations.

   i. $y = 2x - 3$

   ii. $y = -\frac{1}{2}x - 6$

   iii. $y = \frac{2}{5}x + 4$

   iv. $y = -\frac{5}{2}x$

   a. Graph the four equations on one set of axes. Use the same scale for each axis. Label the lines with the appropriate roman numerals.

   b. What is the slope of each line?

   c. What do you notice about the angle of intersection between lines i and ii? Between lines iii and iv?

   d. What is the relationship between the slopes of lines i and ii? Between the slopes of lines iii and iv?

   e. Make a conjecture about the slopes of perpendicular lines.

   f. Create two more lines with slopes that fit your conjecture. Are they perpendicular?

   g. Write an equation for the line that passes through the point $(-1, 4)$ and is perpendicular to $y = \frac{1}{3}x + 4$. Check your answer by graphing both lines on one set of axes. Use the same scale for each axis.

Math Link

A *conjecture* is a statement someone believes to be true but has not yet been proven true.

In Exercises 39–41, match a–c with the correct linear equation.

   a. Adults and children attended the children's matinee on Saturday. Tickets for adults were $3 each, and tickets for children were $2 each. The theatre sold $80 worth of tickets.

   b. Jay bought 80 new holiday ornaments. They came in boxes of either three or four ornaments.

   c. Lisbeth held a grand opening at her new coffee shop. She charged a special price of $2 for each person for any drink on the menu. Some people won door prizes that cost her $3 each. At the end of the day, she had made $80.

39. $3x + 2y = 80$

40. $2x - 3y = 80$

41. $3x + 4y = 80$

42. In Your Own Words How can you determine the slope of a line from a graph? If you are given the slope of a line, what else do you need to know before you can graph the line?

Mixed Review

Write an equation for each situation.

43. the number of houses T Geoff sells in a year if he sells h houses per month

44. the distance d traveled in h hours by a plane averaging 200 miles per hour

45. the width of a room if the ratio of width w to length l of the room is 3:2

List the following numbers from least to greatest.

46. $\sqrt{25}, 525\%, 4.75, \dfrac{9}{2}$

47. $-1, \left(\dfrac{1}{2}\right)^4, -\sqrt{4}, -0.5, (-2)^4$

Review & Self-Assessment

Chapter Summary

In this chapter, you looked at *linear relationships,* relationships with straight-line graphs. You investigated examples of increasing and decreasing linear relationships. In some cases, these relationships were *direct variations,* or *directly proportional* relationships.

You learned that the steepness of a line is called the *slope.* You used the ratio $\frac{\text{rise}}{\text{run}}$ to represent the slope of a line. You learned how the *scale* of a graph can affect the graph's appearance. In addition to graphs, you used verbal descriptions, equations, and tables to represent linear relationships.

Lastly, you found *constant differences* to determine if a relationship could be linear. You used slopes and points to write linear equations. You also learned to express linear equations in *slope-intercept* and *standard* forms.

Vocabulary

coefficient

constant

direct variation

directly proportional

slope

slope-intercept form

standard form

Strategies and Applications

The questions in this section will help you review and apply the important ideas and strategies developed in this chapter.

Recognizing linear relationships and writing linear equations

In Questions 1–6, tell which of the following descriptions fit the relationship. Explain how you decided.

- a direct variation
- linear but not a direct variation
- nonlinear

1. Aisha paid $2.50 admission to a carnival and $1.25 for each ride. Consider the relationship between the total spent for admission and rides and the number of rides taken.

2. The Scrooge Loan Company charges a penalty of $10 the first time a borrower is late making a payment. The penalty is $20 for the second late payment, $40 for the third late payment, and so on, doubling with each late payment. Consider the relationship between the total penalty and the number of late payments.

3. $d = 65t$

4. $y = 13 = 12x$

5.

x	0	10	20	30	40	50	60
y	5	55	105	155	205	255	305

6.

x	40	30	20	15	10	5	0
y	184	138	92	69	46	23	0

Understanding the connection between a linear equation in the form $y = mx + b$ and its graph

7. Consider the equation $y = 300 - 25x$.

a. How would changing -25 to -30 affect the graph of this equation?

b. How would changing -25 to -20 affect the graph of this equation?

c. How would changing -25 to 25 affect the graph of this equation?

d. How would changing 300 to -100 affect the graph of this equation?

Using a linear graph to gather information or to make predictions

8. There is a linear relationship between temperature in degrees Fahrenheit and temperature in degrees Celsius. Two temperature equivalents are $0°C = 32°F$ and $30°C = 86°F$.

a. Make a four-quadrant graph of this relationship. Plot the points $(0, 32)$ and $(30, 86)$. Connect the points with a line.

b. Use the line joining the two points to convert these temperatures.

   i. $5°F$

   ii. $20°C$

   iii. $-30°C$

Demonstrating Skills

In Questions 9–11, estimate the slope of the line.

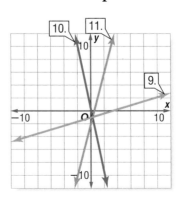

Find the slope of the line that passes through each given pair of points.

12. $(5, -3)$ and $(-1, 9)$ **13.** $(3, 10)$ and $(5, 20)$

14. $(3, 4)$ and $(-1, -2)$ **15.** $(-6, -4)$ and $(-2, 5)$

16. Find an equation of the line that has slope -2 and passes through the point $(-1, -1)$.

17. Find an equation of the line that passes through the points $(4, 4)$ and $(8, -2)$.

18. If the difference between two neighboring numbers in a set of output values is the same no matter which pair you choose, they are _____ .

Use the method of constant differences to predict the missing outputs in each table.

19.

Input, x	1	2	3	4	5	6
Output, y	4	6	8	10		

20.

Input, x	0	1	2	3	4	5	6
Output, y	0	1	8	27	64		

21.

Input, x	-2	0	2	4	6	8	10
Output, y	-7	-5	-3	-1	1		

22. In which tables do the differences tell you the relationships could be linear? Write rules for these relationships.

23. Write the rules above in the form $y = ax + b$.

Identify the coefficient and the constant of each equation.

24. $y = 3x + 5$ **25.** $y = 2x - 12$

26. $y = -x + 1$ **27.** $y = 3x$

28. Through how many quadrants will the equation $y = 5$ pass?

29. Through how many quadrants will the equation $y = 5x$ pass?

30. Graph the linear equation $3y = 6x + 3$. Then rewrite the equation using the form $y = ax + b$.

31. $y = mx + b$ is known as the _____ form of a linear equation.

Graph the following equations.

32. $y = \frac{1}{2}x + 3$ **33.** $y = 2x - 1$

34. $y = -x$ **35.** $y = x + 2$

36. What is the slope and y-intercept of the equation $x - 3y = -6$?

37. Find the slope and y-intercept for the line shown in the graph. Write an equation for the line in slope-intercept form.

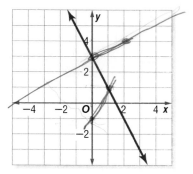

Write each equation from slope-intercept form into standard form.

38. $3y = -2x + 7$ **39.** $4y = 21x$

Write each equation from standard form into slope-intercept form.

40. $6x + 2y = 14$ **41.** $2x - 4y = 16$

Test-Taking Practice

SHORT RESPONSE

1 Find a formula to approximate converting from centimeters to inches given that 2 cm ≈ 0.8 in. and 10 cm ≈ 4 in.

Show your work.

Answer _____

MULTIPLE CHOICE

2 David began graphing a linear relationship by plotting the points (2, 1) and (4, 2). What point will he plot when $x = 6$?

A (6, 0)

B (6, 3)

C (6, 4)

D (6, −3)

3 In the equation $y = 3x + 5$, what is the change in y-values as x-values change by 1 unit?

F 3

G 4

H 5

J 6

4 Which of the following points lies on the line $y = 5x + 2$?

A (0, 5)

B (5, 0)

C (0, 2)

D (2, 0)

5 Write the equation $x + 3y = 12$ in slope-intercept form.

F $3y = -x + 12$

G $x = -3y + 12$

H $y = -\frac{1}{3}x + 4$

J $y = -3x + 4$

Lines and Angles

Real-Life Math

Temperature Tony says that you can tell the temperature outside by listening to the chirps of the crickets. Do you agree with Tony's statement? The frequency of a cricket's chirps varies according to temperature. To get a rough estimate of the temperature in degrees Fahrenheit, count the number of chirps in 15 seconds and then add 37. The number you get will be an approximation of the outside temperature. If x is the number of chirps a cricket makes every 15 seconds, the equation $y = x + 37$ can help you estimate the outside temperature in degrees Fahrenheit.

Think About It Tony counts the chirps of a cricket one evening. He notices that the crickets chirp 41 times in 15 seconds. What is the value of x in this scenario? Write the equation Tony will use to estimate the temperature. What will Tony estimate the outside temperature to be?

Math Online
Take the **Chapter Readiness Quiz** at glencoe.com.

Dear Family,

In Chapter 2, the class will learn more about the characteristics of functions and the relationships of angles. Students will learn about families of lines, which are groups of lines that share a common characteristic, such as slope or *y*-intercept. Finally, students will practice using a compass and a straight edge to construct geometrical figures.

Key Concept—Functions

Knowing the effect of the rate of change can help interpret different situations. For example, if you need to save money for a vacation next summer, how much could you save? If you start with $0 and save a certain amount per week, the function is linear. The line will go through (0,0) and will have a certain slope. If you save $20 each week, it will have one slope. If you save $30 per week, the slope will increase.

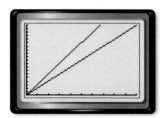

Chapter Vocabulary

alternate exterior angles	line of best fit
alternate interior angles	outlier
arc	perpendicular
bisect	perpendicular bisector
collinear	straightedge
complementary angles	supplementary angles
exterior angles	transversal
interior angles	vertical angles

Home Activities

- Discuss how changes in a situation change the related graph. In the scenario above, how much money would you have if you started with $50 instead of $0? $25?
- Look for intersections of lines and discuss the angles formed.
- In buildings or at home, discuss the angles formed by walls, in furniture, or in landscaping.
- Discuss the relationship among the angles and how this knowledge can be helpful.

Lines

You have worked with equations in the form $y = mx + b$, where m is the slope of the line and b is the y-intercept. Both m and b are constants, while x and y are variables.

The values of m and b affect what the graph will look like. In this lesson, you will study these effects to help you analyze patterns in graphs and equations.

Materials

- graphing calculator

Explore

Each group contains equations in the form $y = mx + b$.

Group I	**Group II**
$y = x + 3$	$y = -3x - 2$
$y = 2x + 3$	$y = -3x$
$y = -2x + 3$	$y = -3x + 1$
$y = \frac{1}{2}x + 3$	$y = -3x + 3$

- Graph the four equations in Group I in a single viewing window of your calculator. Make a sketch of the graphs. Label the minimum and maximum values on each axis.

  What do the four equations in Group I have in common? Give another equation that belongs in this group.

- Now graph the four equations in Group II in a single viewing window. Sketch and label the graphs.

  What do the four equations in Group II have in common? Give another equation that belongs in this group.

- In one group, the equations have different values for m but the same value for b. If you start with a specific equation and change the value of m, how will the graph of the new equation be different?

- In the other group, the equations have different values for b but the same value for m. If you start with a specific equation and change the value of b, how will the graph of the new equation be different?

The lines in Group I are a *family of lines* that all pass through the point (0, 3). The lines in Group II are a family of lines with slope -3.

Parallel Lines and Collinear Points

You can use two points to find the slope of the line through the points. In this investigation, you will work more with the slopes of lines. First, you will use the connection between parallel lines and their slopes.

☑ Develop & Understand: A

1. Look back at the graphs of the equations in Group II of Explore. What do you notice about them?

2. If two equations in $y = mx + b$ form have the same m value, what do you know about their graphs? If you are not sure, write a few equations that have the same m value and graph them. Explain why your observation makes sense.

3. Without graphing, decide which of these equations represent parallel lines. Explain.

   a. $y = 2x + 3$ b. $y = 2x^2 + 3$

   c. $y = 2x - 7$ d. $y = 5x + 3$

4. Consider the line $y = 5x + 4$.

   a. A second line is parallel to this line. What do you know about the equation of the second line?

   b. Write an equation for the line parallel to $y = 5x + 4$ that passes through the origin.

   c. Write an equation for the line parallel to $y = 5x + 4$ that crosses the y-axis at the point (0, 3).

   d. Write an equation for the line parallel to $y = 5x + 4$ that passes through the point (2, 11). (Hint: The data pair (2, 11) must satisfy the equation.)

5. A line has the equation $y = 3x - 1$. Find an equation for the line parallel to it and passing through the point (3, 4).

6. A line has the equation $y = 4$.

   a. Graph the line.

   b. Find an equation for the line parallel to this line and passing through the point (3, 6).

Math Link

The origin is the point (0, 0).

Points *A, B,* and *C* below are **collinear**. In other words, they all lie on the same line. Points *D, E,* and *F* are not collinear.

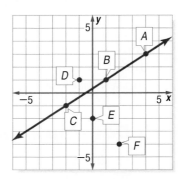

If you are given three points, how can you tell whether they are collinear? You could plot the points to see whether they look like they are on the same line, but you would not know for sure. In the next exercise set, you will develop a method for determining whether three points are on the same line without making a graph.

✅ Develop & Understand: B

7. Without graphing, find a way to determine whether the three points below are collinear. Explain your method.

$$(3, 5) \qquad (10, 26) \qquad (8, 20)$$

8. The points in one of the two sets below are collinear. The points in the other set are not. Which set is which? Test your method on both sets to be sure it works.

Set A	Set B
$(-4, -3)$	$(-1, 1)$
$(1, 2)$	$(0, -1)$
$(7, 7)$	$(2, -5)$

9. Determine whether the three points in each set are collinear.

a. $(-3, -2), (0, 4), (1.5, 7)$

b. $(1.25, 1.37), (1.28, 1.48), (1.36, 1.70)$

1. How would you find an equation for the line through point C and parallel to the line through points A and B? Assume you know the coordinates of each point and that point C is not on the line through points A and B.

2. Describe a method for determining whether point D is on the line through points A and B, if you know the coordinates of each point.

Investigation 2 Recognize Linear Equations

Materials

• graph paper

Equations can be complicated. Sometimes it is not obvious whether a relationship is linear. A graph can tell you whether a relationship *looks* linear, but it still might not be.

Think & Discuss

You know by its form that the equation $y = 3x + 2$ and the equation $4x + 3y = 12$ both represent lines. The equation below is not in that form.

$$y = \frac{4x + 2(3 + x) - 2}{2}$$

How can you determine whether this equation is linear? Explain your reasoning.

If you can write an equation in the form $y = mx + b$, you know it is linear. As you recall, $y = mx + b$ is often called the slope-intercept form of a linear equation.

✓ Develop & Understand: A

Determine whether each equation is linear. If an equation is linear, identify the values of m and b. If an equation is nonlinear, explain how you know.

1. $3y = \frac{x}{2} - 8$

2. $y = 3$

3. $y = \frac{2}{x}$

4. $5y - 7x = 10$

5. $2y = 10 - 2(x + 3)$

6. $y = x(x - 1) - 2(1 - x)$

7. $y = 2x + \frac{1}{2}(3x + 1) + \frac{1}{4}(2x + 8)$

✅ *Develop & Understand: B*

8. Which of these equations describe the same relationship?

    a. $p = 2q + 4$ **b.** $p - 2q = 4$

    c. $p - 2q + 4 = 0$ **d.** $0.5p = q + 2$

    e. $p - 4 = 2q$ **f.** $2p = 8 + 4q$

Write each equation in slope-intercept form.

9. $y - 1 = 2x$

10. $2y - 4x = 3$

11. $2x + 4y = 3$

12. $6y - 12x = 0$

13. $x = 2y - 3$

14. $y + 4 = -2$

15. Choose the equations in Exercises 9–14 that would have graphs that are parallel lines. Sketch them on the same coordinate axes.

16. Group these equations into sets of parallel lines.

    a. $y = (2x - 7) - 3$ **b.** $y + 5 + 2x = 5$

    c. $y = 30 + 4(x - 7)$ **d.** $4y - 5x = 3x - 2$

    e. $y + 3(10 - x) = x$ **f.** $y = 5x + 3(10 - x)$

    g. $y = 8x - \frac{1}{3}(12x - 30)$ **h.** $y = 1 + \frac{1}{2}(2 - 4x)$

    i. $y - 3 = -2x$ **j.** $2y = -4(3 - x)$

17. If two linear equations are in standard form, how can you tell if their graphs are parallel?

Share & Summarize

What are some of the strategies that you used to simplify the equations in this investigation?

Inquiry

Investigation Estimate Solutions With Graphs

Materials

- graphing calculator
- graph paper

In this investigation, you will work with graph paper and a graphing calculator to estimate the solutions to equations. You know that linear relationships can be expressed symbolically, using an equation in the form $y = m + b$.

You can use the standard viewing window $[-10, 10]$ scl: 1 by $[-10, 10]$ scl: 1 or set your own minimum and maximum values for the axes and the scale factor by using the WINDOW option.

The tick marks on the x scale and on the y scale are 1 unit apart.

$[-10, 10]$ scl: 1 by $[-10, 10]$ scl: 1

The x-axis goes from -10 to 10.

The y-axis goes from -10 to 10.

Graph $y = 3x + 4$ and $y = 3x - 2$ in the standard viewing window, and describe how the graphs are related.

1. Graph $y = 3x + 4$ in the standard viewing window.

   - Clear any existing equations from the Y = list. [Y=] [CLEAR]
   - Enter the equation for Y1 and graph.
     [Y=] 3 [X,T,θ,*n*] [+] 4
   - Graph the equation in the standard viewing window. [ZOOM] 6

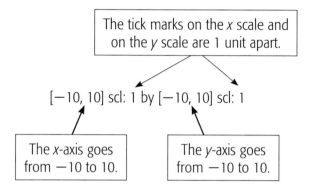
$y = 3x + 4$

2. Graph $y = 3x - 2$

   - The equation $y = 3x + 4$ is already entered as Y1. Enter the equation $y = 3x - 2$ as Y2. [Y=] 3 [X,T,θ,*n*] [−] 2
   - Graph both equations in the standard viewing window. [ZOOM] 6

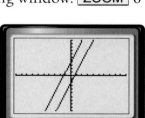

The first equation graphed is Y1 or $y = 3x + 4$. The second equation graphed is Y2 or $y = 3x - 2$. Press [TRACE]. Move along each equation using the right and left arrow keys. Move from one equation to another using the up and down arrow keys. In Chapter 1, you found the slope and the y-intercepts. What is the slope of the two graphs? What are the y-intercepts?

Go on ▶

Set It Up

Although it is true that you cannot rely on a graph for an exact solution to an equation, you can get reliable approximations.

This little-known formula allows you to estimate the air temperature by measuring the speed that ants crawl.

$$t = 15s + 3$$

where t is temperature in °C and s is the ants' speed in cm/s.

You will now represent the equation using a table and a graph.

1. Find the value of t for various values of s. Choose five values from 0.5 to 4.0. Record your results in a table.

2. Graph the values from your table. Put s values on the x-axis and t values on the y-axis. Before you start, decide on a reasonable scale.

Make a Prediction

3. Suppose an ant crawls 1.5 cm/s, what would be the temperature?

Try It Out

4. Andrea timed some ants and estimated their speed to be 2.5 cm/s. Use your graph to estimate the temperature.

5. Early in the morning, Andrea estimated the ants' speed to be about 1.2 cm/s. But by late afternoon, it was 2.0 cm/s. Estimate the change in temperature during that time.

6. Write an equation you could solve to find an ants' speed when the temperature is 25°C. Use your graph to estimate the solution.

7. Use your graph to estimate how much the ants' speed would change if the temperature increased from 10°C to 20°C.

8. What equation would you use if you were trying to figure out how fast the ant would crawl in 24.9°C temperature? How fast would the ant crawl?

9. What equation would you use if you were trying to figure out how fast an ant would crawl in 27°C temperature? How fast would the ant crawl?

Math Link

When using a graphing calculator, your y value is always your dependent variable. Your x value is always your independent variable. This means that if x is given, then y occurs.

Real-World Link

Ants can live in temperatures around 25°C to 27°C. If the temperature goes above 27°C, the queen can become infertile and the colony cannot sustain life.

What Did You Learn?

10. The currency in much of Europe is the euro. One day in January 2007, the rule for converting euros to U.S. dollars was $D = 0.69E$, where D was the price in U.S. dollars and E was the price in euros.

a. Graph this relationship. Put E on the horizontal axis, and show the amounts in euros up to 150.

b. On that day, a French collector advertised a set of miniature trains for 125 euros, including postage. Write an equation you could solve to find how much this was in U.S. dollars. Solve your equation. Use your graph to check your solution.

c. On that same day in January, Megan exchanged $120 she had saved towards a vacation in Germany. Write and solve an equation to find how many euros she received. Use your graph to check your solution.

d. What are the advantages of using either the graph or the rule to solve the equations you wrote?

Vocabulary

line of best fit

outlier

Materials

- graph paper
- transparent ruler or piece of dry spaghetti
- graphing calculator

Sometimes when you graph data, the points will lie close to, but not exactly on, a line.

Think & Discuss

Students in a Science class were measuring how far a cart traveled each time the wheels rotated once. They measured the distance for one rotation, two rotations, three rotations, and so on. The wheels of the cart are 2.5 feet in circumference.

Would you expect the students' data to lie on a line? If so, explain why. Give an equation of that line.

The graph shows a plot of the students' data. Notice that the points do not appear to lie exactly on a line. Why might the data they collected not fit precisely on a line?

Cart Experiment Results

Using mathematics to describe something, such as a set of data from an experiment, is called *modeling*. Modeling is important in many professions, especially in the fields of science and statistics.

People who gather data are often uncertain what kind of relationship the data will show. When data are graphed, they sometimes are close to, but not exactly on, a line. It may be that the variables are linearly related but that there were inaccuracies in the measurements. Or, the relationship between the variables may not be *exactly* linear but close enough to use a line as a reasonable model.

In cases like these, you can use the data to help find a **line of best fit**, a line that fits all the data points as closely as possible. You can then use a line of best fit to make predictions or to solve equations.

There are several ways to find such a line. Some of the techniques are sophisticated, but you can make reasonably good estimates by using more simple techniques.

✅ Develop & Understand: A

Breaths per minute and heartbeats per minute were measured for 16 people after they had each walked for 20 minutes. The data are in the table.

Breaths per Minute	16	16	19	20	20	23	24	26	27	28	28	30	34	36	41	44
Heartbeats per Minute	57	59	66	68	71	70	72	84	82	80	83	91	94	105	116	120

Here is a graph of the data.

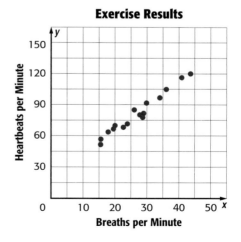

Exercise Results

It seems that the more rapid the breathing, the more rapid the heart rate. The relationship is not exactly linear since it is not possible to draw a single line that passes through every point on the graph. However, if you can find a line that fits the points reasonably well, you can use it to predict the value of one variable, heartbeats per minute or breaths per minute, from the other.

1. With your partner, plot the points in the table on a grid.

   a. Draw a straight line that fits the points as well as possible. Try to draw the line so that about the same number of points lie on either side of it. A transparent ruler or a piece of dry spaghetti will help with this.

   b. Write an equation of the line you drew. To do this, find the slope and the *y*-intercept of the line, or the slope and any point on the line, or two points on the line. Use that information to write an equation.

Math Link

The *mean*, or average, is the sum of all the values divided by the number of values.

2. One technique for improving the "fit" of your line is to use the means of the data.

   a. Find the mean number of breaths per minute and the mean number of heartbeats per minute from the data in the table.

   b. Plot the point that has these two means as its coordinates. Adjust the line you made in Exercise 1 to go through this point.

3. Use your graph to write an equation for your new line. Why might your line be different from someone else's line?

4. Use your equation for Exercise 3 to predict the heart rate for a person who takes 35 breaths per minute after walking for 20 minutes. Compare your prediction with the predictions of other students.

5. Use your equation to predict the heart rate for a person who takes 100 breaths per minute after walking for 20 minutes. Do you think your prediction is reasonable? Explain.

Real-World Link

The winning time for the 400-m relay has decreased steadily since 1928. In the 2004 Olympics, the winner of the gold medal, on the Jamaica team, completed the relay in 41.73 seconds.

✅ *Develop & Understand: B*

The table shows winning times, in seconds, for the women's 400-meter relay in the Olympic Games from 1928 to 2004. No games were held in 1940 or 1944 because of World War II.

Year	Country	Time (s)
1928	Canada	48.4
1932	United States	46.9
1936	United States	46.9
1948	Netherlands	47.5
1952	United States	45.9
1956	Australia	44.5
1960	United States	44.5
1964	Poland	43.6
1968	United States	42.8

Year	Country	Time (s)
1972	West Germany	42.81
1976	East Germany	42.55
1980	East Germany	41.60
1984	United States	41.65
1988	United States	41.98
1992	United States	42.11
1996	United States	41.95
2000	Bahamas	41.95
2004	Jamaica	41.73

The data are plotted below. The points are quite close to falling on a line, with one exception. When a data point seems very different from the others, it is often called an **outlier**. Outliers are typically given less emphasis when analyzing general trends, or patterns.

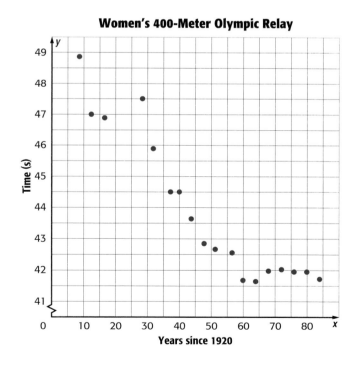

Women's 400-Meter Olympic Relay

6. Which point is the outlier?

7. Can you suggest a reason why that point is so far from the trend of the other values?

8. Copy the data points on a graph of your own. Using the techniques you learned on pages 75 and 76, find an equation for a line that fits the data.

✓ Develop & Understand: C

Graphing calculators use a sophisticated mathematical technique to find lines to fit a set of data. You will now use your graphing calculator to find a line of best fit.

9. Enter the 400-meter-relay data from the previous page into your calculator in two lists.

    a. Use your calculator to determine an equation of a line of best fit.

    b. How close is the calculator's equation to the equation you found on the previous page?

10. The winning times in the table on the previous page are decreasing fairly steadily over the years. Do you think they will always continue to do so? Why or why not?

11. Use your equation from Exercise 9 to predict the winning times for 1984, 1988, 1992, and 1996.

12. The real winning times for these years are given below. How do your predictions compare?

Year	1984	1988	1992	1996
Time (s)	41.65	41.98	42.11	41.95

13. Assume that the winning times actually continue to decrease after 1980 at the same rate as your linear model from Exercise 9 predicts they will.

    a. Write an equation you can use to find when the winning time will be 0 seconds. Solve your equation.

    b. According to your model, in which year do you predict this impossible winning time of 0 seconds will occur?

Share & Summarize

You have just examined two situations that can be modeled with linear equations. Draw a graph containing ten points for which a linear equation would *not* be a good model. Describe why you would have difficulty determining a line of best fit for your graph.

Practice &**Apply**

For each set of equations, tell what the graphs of all four relationships have in common *without* drawing the graphs. Explain your answers.

1. $y = -1.1x + 1.5$

$y = -1.1x - 4$

$y = -1.1x + 7$

$y = -1.1x$

2. $y = 2x$

$y = -2x$

$y = 3x$

$y = -3x$

3. $y = 2x$

$y - 1 = 2x$

$y = 2x + 4$

$y = 2x + 7$

4. $y = 1 - x$

$y = 1 - 2x$

$y = 1 - 3x$

$y = 1 - 4x$

5. In this exercise, you will apply what you have learned about writing equations for parallel lines.

a. Write three equations whose graphs are parallel lines with positive slopes. Write the equations so that the graphs are equally spaced.

b. Graph the lines. Verify that they are parallel.

c. Write three equations whose graphs are parallel lines with negative slopes and are equally spaced.

d. Graph the lines. Verify that they are parallel.

6. You can tell whether a particular point might be on a line by graphing it and seeing whether it seems to lie on the line. But to know for certain whether a particular point is on a line and not just *close* to it, you must test whether its coordinates satisfy the equation for that line.

a. Graph the equation $y = \frac{13}{8}x - 3$.

b. Using the graph alone, decide which points below look like they might be on the line. You may want to plot the points.

$(0, -3)$ $(3, 2)$ $(4, 4)$ $(5, 5)$ $(8, 10)$

c. For each point, substitute the coordinates into the equation and evaluate to determine whether the point satisfies the equation. Which points, if any, are on the line?

If possible, write each equation in the form $y = mx + b$. Then identify the slope and the y-intercept.

7. $y = 5x + \frac{1}{3}(6x + 12)$

8. $y = \frac{1}{5}(10x + 5) - 5 + 7x$

9. $3x + 2(x + 1) = -\frac{1}{2}(4x + 6) + y$

10. $3x^2 - y = 3x + 5$

11. $y - 19 = -2(x - 3)$

12. Within these equations are five pairs of parallel lines. Identify the parallel lines, and give the slope for each pair.

a. $y = 3x - 5(x + 3)$

b. $\dfrac{x - 2y}{2} = 7$

c. $y = 17 - 3(3 + x) + x$

d. $3x + 2y = 4$

e. $y = -\dfrac{x}{2} + 2\left(6 - \dfrac{x}{2}\right)$

f. $x - y + 3 = 0$

g. $4x - 2y - 17 = 20$

h. $4\left(\dfrac{y}{2} - x\right) = 10$

i. $y + x = 4x + 5 - 2(4 + x)$

j. $y = \dfrac{3x + 4}{2} - \dfrac{7 + 2x}{2}$

13. Give the slope-intercept form of each equation. Tell which of these eight equations describe the same relationship.

a. $4x - 2y = 4$

b. $2y - 4x = 4$

c. $2x - y = 2$

d. $y - 2x = 2$

e. $y - 2x = -2$

f. $y = 2x - 2$

g. $y = 2x + 2$

h. $4x + 2y = 4$

14. **Life Science** The average length of gestation, or pregnancy, and the average life span for several animals are listed below.

Animal	Average Gestation (days)	Average Life Span (years)
Cat	63	11
Chicken	22	8
Cow	280	11
Dog	63	11
Duck	28	10
Elephant	624	35
Goat	151	12
Guinea pig	68	3
Hamster	16	2
Horse	336	23
Mule	365	19

a. Plot the points for each animal on one set of axes. Put gestation on the horizontal axis and life span on the vertical axis.

b. Draw a line that fits the data reasonably well. Write an equation for your line.

c. Use your equation to predict the life span of a hog, which has a gestation of about 114 days.

15. **Fine Arts** Wolfgang Amadeus Mozart composed music for most of his very short life. His compositions were numbered in the order that he wrote them.

These data relate the total number of compositions *K* to Mozart's age when they were written *a*.

Age (years), *a*	8	12	16	20	24	27	32	35
Total Number of Compositions, *K*	16	45	133	250	338	425	551	626

Real-World Link
The letter *K* is used for the composition number variable to honor the Austrian scientist Ludwig von Köchel. He sorted the 626 Mozart compositions in the mid-nineteenth century.

a. Plot the points from the table with *a* on the horizontal axis and *K* on the vertical axis.

b. Do the data suggest a general rate at which Mozart wrote new compositions? What is that rate?

c. Draw a line that fits the data points as well as possible. Use the technique of finding a line that passes through the mean age and mean composition number for the given data points.

d. Use your graph to find a linear equation for predicting the number of compositions Mozart wrote at a given age.

e. Mozart died at the age of 35. Would it be reasonable to use your equation to predict the number of compositions Mozart would have produced if he had lived to age 70?

f. What is the value of K for $a = 0$? Does this data point make sense? What does this tell you about your linear model?

Connect & Extend   **16.** The lines for these three equations all pass through a common point.

$$y = \frac{x}{2} - 1 \qquad y = -\frac{2x}{3} + 6 \qquad y = -\frac{x}{6} + 3$$

a. Draw graphs for the three equations. Find the common point.

b. Verify that the point you found satisfies all three equations by substituting the x- and y-coordinates into each equation.

Each table describes a linear relationship. For each relationship, find the slope of the line and the y-intercept. Then write an equation for the relationship in the form $y = mx + b$.

17.

x	2	4	6	8	10
y	8	12	16	20	24

18.

x	−8	−3	3	5	10
y	26	11	−7	−13	−28

19.

x	9	7	5	3	1
y	5	4	3	2	1

20. Hoshi drew graphs for $y = x$ and $y = -x$ and noticed that the lines crossed at right angles at the point $(0, 0)$. Then he drew graphs for $y = x + 4$ and $y = -x + 4$. He noticed that the lines crossed at right angles again, this time at the point $(0, 4)$. He tried one more pair, $y = x - 4$ and $y = -x - 4$. Once again, the lines crossed at right angles, at the point $(-4, 0)$.

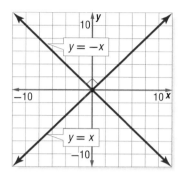 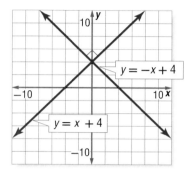 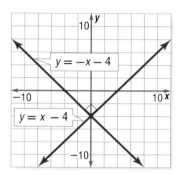

Hoshi made this conjecture, "When you graph two linear equations and one has a slope that is the negative of the other, you always get a right angle."

a. Do you agree with Hoshi's conjecture? Why or why not?

b. Draw several more pairs of lines that fit the conditions of Hoshi's conjecture with different slope values. Do your drawings prove or disprove Hoshi's conjecture?

c. If you think Hoshi's conjecture is false, where do you think he made his mistake?

21. Review Exercises 1–7 on page 69. Some of the equations given there are nonlinear. Write some guidelines for quickly identifying linear and nonlinear equations.

Just as the y-intercept of a line is the y value at which the line crosses the y-axis, the x-intercept is the x value at which the line crosses the x-axis. In Exercises 22–25, find an equation of a line with the given x-intercept and slope.

22. x-intercept 3, slope 2

23. x-intercept -2, slope $-\dfrac{1}{2}$

24. x-intercept 1, slope -6

25. Challenge x-intercept 3, no slope (Hint: If slope is $\dfrac{\text{rise}}{\text{run}}$, when would there be no slope?)

26. You have been using the form $y = mx + b$ to represent linear equations. You saw in Chapter 1 that linear equations are sometimes represented in standard form $Ax + By = C$, where A, B, and C are constants.

   a. Rewrite the equation $Ax + By = C$ in the $y = mx + b$ form. To do this, you will need to express m and b in terms of A, B, and C.

   b. What is the slope of a line with an equation in the form $Ax + By = C$? What is the y-intercept?

27. Social Studies Below is world population data for the years 1950 through 2000.

World Population

Year	Population (billions)
1950	2.52
1960	3.02
1970	3.70
1980	4.45
1990	5.29
2000	6.07

   a. Plot the points on a graph with "Years since 1900" on the horizontal axis and "Population (billions)" on the vertical axis. Try to fit a line to the data.

   b. Write an equation to fit your line.

   c. Use your equation to project the world population for the year 2010, which is 110 years after 1900.

   d. What does your equation tell you about world population in 1900? Does this make sense? Explain.

   e. According to United Nations figures, the world population in 1900 was 1.65 billion. The UN has predicted that world population in the year 2010 will be 6.79 billion.

   Are the 1900 data and the prediction for 2010 different from your predictions? How do you explain your answer?

28. Consider this data set.

x	0	2	4	6	8
y	2	20	6	8	10

   a. Graph the data set.

   b. One point is an outlier. Which point is it?

   c. Find the mean of the x values and the mean of the y values.

   d. Try to find a line that is a good fit for the data and goes through the point (mean of x values, mean of y values). Write an equation for your line.

   e. Now find the means of the variables, *ignoring the outlier.* In other words, do not include the values for the outlier in your calculations.

   f. Try to find a new line that is a good fit for the data, using the means you calculated in Part e for the (mean of x values, mean of y values) point. Write an equation for your line.

   g. Do you think either line should be considered the best fit for the data? Explain.

29. In Your Own Words Describe why determining a line of best fit can be useful in working with data.

Mixed Review

Find the slope of the line through the given points.

30. $(-2, 3)$ and $(5, 8)$

31. $(0, -6)$ and $(-8, 0)$

32. $(-3.5, 1.5)$ and $(0.5, 2)$

33. $(-7, -2)$ and $(-9, -2)$

34. Jonas and Julieta conducted an experiment in which they dropped a rubber ball from various heights and measured how high the ball bounced.

Here are the data they collected.

Drop Height (in.)	6	7	10	14	16	18	20	24	26	30
Bounce Height (in.)	2	2.5	4	5	6	6.5	7	8.5	10	12

   a. Make a graph of the data with drop height on the horizontal axis.

   b. Draw a line that goes through most of the data points. Use your graph to write an equation of the line.

   c. Now find the mean of the drop heights and the mean of the bounce heights from the data in the table.

   d. Plot the point that has these two means as its coordinates. Adjust the line you drew in Part d to go through this point. Write an equation for your new line.

Determine whether the points in each set are collinear. Explain how you know.

35. $(-2, 13), (1.5, -4.5), (3, -12)$

36. $(-1, -4.2), (3, 0.6), (4, 1.6)$

Angle Relationships

Many careers, from engineers and architects to landscape designers and pilots, use geometry and angles everyday. In geometry, angles have special properties based on their angle measures. This lesson is about the relationship between certain special angles.

Investigation 1   Supplementary, Complementary, and Vertical Angles

Vocabulary

complementary angles

supplementary angles

vertical angles

Materials

• protractors

In previous grades, you have seen different angles and learned about some special angles. For example, you learned that the sum of the interior angles of a triangle is 180°. You learned that a right angle measures 90°. In this investigation, you will explore other relationships between angles that can help you determine the measures of angles.

Think & Discuss

In earlier grades, you explored *vertical angles.* In each drawing below, ∠*a* and ∠*b* are vertical angles.

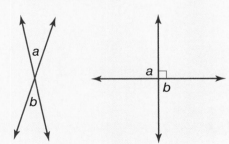

Do you see another pair of vertical angles in each of the figures above?

Math Link

Remember that certain angles have specific measurements. Right angles measure 90°. Acute angles measure < 90°. Obtuse angles measure > 90°.

In these drawings, ∠*a* and ∠*b* are *not* vertical angles.

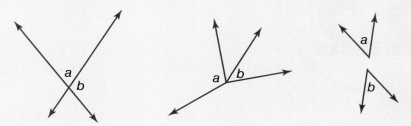

From these examples and non-examples, what do you think is the definition of vertical angles?

What do you think is true about the measures of vertical angles?

Can you prove your conjecture about the measures?

When two lines intersect, two angles that are opposite one another are **vertical angles**. You may have noticed in the Think & Discuss that vertical angles are equal as well.

✅ Develop & Understand: A

For the following exercises, assume all lines that appear to be lines are, unless otherwise indicated.

1. In the figure to the right, can you determine the measurement of *t* without a protractor? If so, what is its measure?

2. In the figure below, which angles can you determine the measurement of without measuring? If you can determine the measurements, find them.

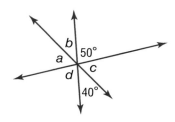

Real-World Link

Architects calculate angle measurements among parts of a building to ensure walls and ceilings align.

3. Look at the figure in Exercise 2. If there were angles you were not able to find, what additional information might help?

Look back at the figure in Exercise 1. You were able to figure out ∠t because the two lines form vertical angles. You also were able to calculate the other missing angles because a straight angle is 180°. You may recall that the degree measures that are opposite from one another on the protractor have a special relationship, their sum is 180°. Angles that add up to a straight angle, or 180°, are **supplementary angles**.

Example

∠1 and ∠2 are examples of supplementary angles, as are ∠3 and ∠4.

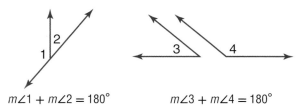

$$m\angle 1 + m\angle 2 = 180° \qquad\qquad m\angle 3 + m\angle 4 = 180°$$

Angles that add up to 90° or a right angle, also have a special name. They are **complementary angles**.

These are examples of pairs of complementary angles.

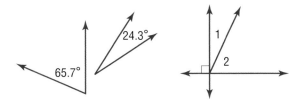

✓ Develop & Understand: B

Math Link

The measure of an angle is indicated by m∠. You add angles just like other addition sentences, m∠A + m∠B.

In the figures below, find the measures of the missing angles indicated by letters.

4.

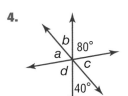

5.

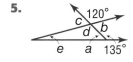

6.

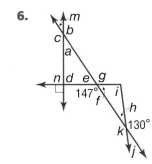

7. What is incorrect in this figure?

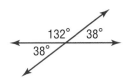

8. Two angles are complementary. If you know one angle is twice the size of the other, can you determine the measures of the two angles? If so, what are the measurements?

9. Can two angles be obtuse and complementary? Obtuse and supplementary? Explain.

10. State whether each of the following statements is always, sometimes, or never true.

   a. If two angles are vertical angles, then they are equal.

   b. If two angles are equal, then they are complementary.

   c. If two angles are obtuse, then they are supplementary.

   d. If two lines intersect, two pairs of vertical angles are formed.

   e. Two angles are both complementary and supplementary.

   f. Two angles in a triangle are supplementary.

   g. If two angles in a triangle are complementary, the triangle is a right triangle.

☑ Develop & Understand: C

11. Write three pairs of complementary angles.

For Exercises 12 and 13, find the complementary angle for the angles shown.

12.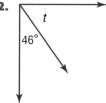
t
$46°$

13.
$67.3°$

14. If you know that two angles are complementary and that one angle measures $m°$, what is the measure of the other angle?

Share & Summarize

Draw the following items.
- vertical angles c and d
- supplementary angles t and r with one angle measuring approximately $20°$
- complementary angles m and o that are congruent

What must be true of the measures of two supplementary angles if they are equal? Two complementary angles?

Investigation ② Parallel Lines Cut By a Transversal

Vocabulary

alternate exterior angles

alternate interior angles

exterior angles

interior angles

transversal

Materials

• protractors

• spaghetti

• pipe cleaners, or other straight material to create lines

In Investigation 1, you explored the special relationships in angles formed by the intersection of two lines. Other special angle relationships are formed when a set of parallel lines are intersected by a third line called the **transversal**.

Explore

Position two pieces of spaghetti to form parallel lines. Lay the third piece of spaghetti so that it crosses the two parallel pieces.

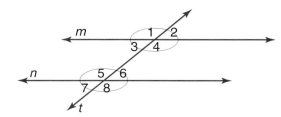

What relationships do you already know among the angles? What else do you observe?

If the placement of the transversal changes, do the observations you made change?

You may have noticed that some angle relationships are *invariant* given two parallel lines and a transversal. In other words, the placement of the transversal has no effect on those relationships.

Which angle measures are *invariant* if you move one of the parallel lines but keep it parallel to the other line?

The angles you have created with the spaghetti have specific names to make it easier to refer to them. The two angles formed inside the parallel lines are called **interior angles**. The angles on the outside of the two parallel lines are called **exterior angles**.

• Name the interior angles in the above figure.

• Name the exterior angles in the above figure.

• What do you think is meant by alternate interior angles? Alternate exterior angles? Name as many pairs as you can.

✓ Develop & Understand: A

When a transversal intersects parallel lines, there is a special relationship among the angles.

Look at the figure to the right. The transversal t intersects both line *m* and line *n*.

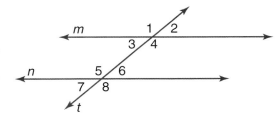

1. Using a protractor, measure ∠1, ∠3, ∠5, and ∠7. Which pairs of angles appear to be equal? Which pairs of angles appear to be supplementary?

2. Why does ∠3 equal ∠2? Does ∠7 equal ∠2?

3. Use the same line of reasoning as in Exercise 2 to show why ∠1 equals ∠8.

The angles that are created on the opposite sides of the transversal inside the parallel lines are called **alternate interior angles**. **Alternate exterior angles** are the angles on the outside of the parallel lines.

4. What can you say about the measures of alternate exterior angles when two parallel lines are cut by a transversal?

5. In Exercise 1, you should have found that ∠1 and ∠5 appear to be equal. What conjecture would you make about ∠4 and ∠5? Use your knowledge of vertical angles to prove your conjecture.

6. Angles 4 and 5 are alternate interior angles. Name another pair of alternate interior angles. Can you show why they are equal?

7. **Challenge** Look at a pair of angles described as *interior angles on the same side of the transversal*. What do you think the relationship is between these angles? Explain why you think this is true.

In the previous exercises, you may have conjectured that when a transversal intersects two parallel lines, the alternate interior angles are equal and the alternate exterior angles are equal. In the next exercises, you will use what you know about angle relationships to find the measures of angles in given situations.

✓ Develop & Understand: B

In Exercises 8–11, find the measures of the missing angles. If you cannot find an angle's measure, explain why not.

8. Given: $l \parallel m$

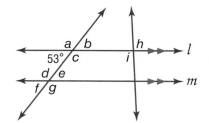

9. Given: $p \parallel q$

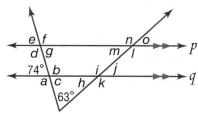

10. Given: $r \parallel s$

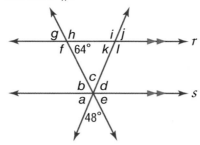

11. Given the figure below where $m \parallel n$. Which of the following statements are impossible? Explain what makes them impossible.

a. ∠4 and ∠5 are supplementary.

b. ∠4 and ∠5 are equal.

c. ∠4 and ∠6 are equal.

d. ∠4 and ∠6 are complementary.

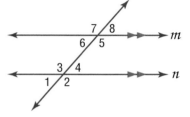

e. The sum of the measures of ∠3 and ∠6 is greater than the sum of the measures of ∠7 and ∠1.

12. Graph the lines $y = x$, $y = x + 3$, and $x = 3$ on the same set of axes.

a. Which lines are parallel?

b. Which line is the transversal?

c. Find the measures of all of the interior and exterior angles formed by the two parallel lines and the transversal. Explain how you found the angles.

13. State whether each of the statements below is always, sometimes, or never true.

a. If a transversal intersects two parallel lines, the alternate interior angles are supplementary.

b. If a transversal intersects two parallel lines, the alternate exterior angles are equal.

c. If a transversal intersects two parallel lines, the alternate interior angles are not equal.

Share & Summarize

Explain to a new classmate the relationship of the interior angles, exterior angles, and the alternate interior and exterior angles. Using a drawing, discuss their position and angle measurements.

Is it possible for a transversal to cross parallel lines to form only obtuse angles? Explain your reasoning.

Practice & Apply

Identify the angle relationship or measurement that is described by the given situation.

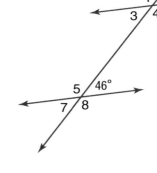

1. ∠1 and ∠4

2. ∠3 and ∠7

3. ∠1 and ∠3

4. $m\angle 2 + m\angle 4$

5. Is it possible to have congruent angles created by complementary angles? Explain your reasoning.

6. A pair of vertical angles measure a total of 78°. What is the total measure of the two angle supplements?

Use the figure from Exercises 1–4 to identify the angle relationship or measurement that is described by the given situation.

7. ∠5 and ∠4

8. $m\angle 2$

9. ∠3 and ∠5

10. ∠7 and ∠8

11. In the figure to the right, can you determine the measurement of W without a protractor? If so, what is the angle measurement? Assume lines that look straight are straight.

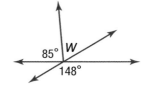

12. Amiri has to calculate the missing angle measure for a pair of parallel lines cut by a transveral. He knows that one angle measure is 56°. Sketch a picture of Amiri's drawing. Label the missing angle measures.

Connect & Extend

Rey is recreating a stained-glass window drawing. In Exercises 13–15, identify the angle relationships formed in the design.

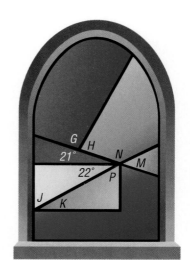

13. What type of angles are formed by the diagonal on the rectangle ∠J and ∠K?

14. Identify the relationship between ∠G and ∠H on Rey's stained-glass window design.

15. What is the angle measure of ∠M? How did you determine the measure of ∠M?

16. Is it possible to have two obtuse angles formed by supplementary angles? Draw a sketch. Explain your reasoning.

Brandi drew the star on the right. Use the best description to identify the angle relationships formed on Brandi's star.

17. $\angle A$ and $\angle E$

18. $\angle G$ and $\angle 3$

19. If $\angle C$ measures $44.5°$, determine the measure of the other angle measures.

20. What angles would be supplements of $\angle 4$?

21. Name three complemetary angle pairs.

22. Angle 4 measures $45.5°$. Determine its supplement.

Mixed Review

23. Identify the three pairs of equivalent equations.

    a. $p = 2q - 4$ **b.** $p - 2q = 4$

    c. $p - 2q - 4 = 0$ **d.** $-2p = 8 + 4q$

    e. $-p - 4 = 2q$ **f.** $0.5p = q - 2$

24. List four numbers that are greater than -1 and less than 1.

Write an equation for a line that would pass through the given points.

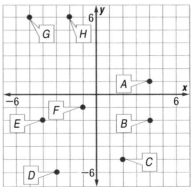

25. Points A and C

26. Points A and B

27. Points D and E

28. Points A and F

29. Point F and the origin

30. Points G and H

LESSON 2.3

Constructions

In this lesson, you will learn to construct some geometrical figures using a compass and a straightedge. Think about how difficult it is to draw a perfect circle with nothing but a pencil.

Vocabulary

arc

straightedge

Materials

- compass
- straightedge
- protractor
- ruler

Explore

Sometimes it is interesting to think about how to construct figures without any measurement tools. In Parts a–c, describe how you made the constructions. Remember, you cannot use a ruler or protractor.

a. Draw a line segment. Find its midpoint.

b. Draw an angle. Draw its bisector.

c. Construct a square from the largest square you can make from an $8\frac{1}{2}$-in. by 11-in. piece of paper. Construct a square with exactly one quarter of the area of your original square. Can you construct a square with exactly half of the area of your original square?

A compass lets you draw arcs and circles precisely. To use a compass, hold the compass point fixed. Move the pencil. You will notice that the only way it can move is around in a circle. You can draw a full circle this way, or an **arc**, which is a curve that is part of a circle. To make a circle with a larger or smaller radius, adjust the distance between the fixed point and the pencil.

A **straightedge** is anything that has a hard, straight edge that you can use to draw a straight line. For example, a ruler is a straightedge that happens to have measurement units marked on it.

Investigation 1 Construct Line Segments

Vocabulary

- **bisect**
- **perpendicular**
- **perpendicular bisector**

Before computers were invented, architects and engineers designed buildings, boats, machines, and other complicated structures by drawing each part by hand. They used compasses and straightedges, along with other tools, to make the drawings accurate. Many historic buildings that are still standing today were designed and built with simple tools like the compass.

You can use a compass and straightedge to draw a line segment that is congruent to this one.

Step 1. Put the fixed point of the compass at one of the endpoints of the segment.

Step 2. Set the radius of the compass so that the pencil's point is at the segment's second endpoint.

Step 3. Without changing the radius, move the compass to another piece of paper. Mark the positions of the fixed point and the pencil point.

Step 4. Use the straightedge to trace a straight line between the two marks.

✓ Develop & Understand: A

With a compass and straightedge, draw a line segment congruent to each of the following segments.

1. •————————•

2.

A **perpendicular bisector** of a line segment has two important characteristics. It intersects the segment at its midpoint and it is perpendicular to the segment. A **bisector** of a line segment is any line that cuts, or *bisects,* the line segment exactly in half. Two line segments that are **perpendicular** form a right angle where they intersect.

Think & Discuss

In this picture, all of the lines bisect $\overline{AB}$. Are any of them perpendicular bisectors? Can you tell for sure?

How could you use a compass and straightedge to draw a line segment and then bisect it with a second line or line segment.

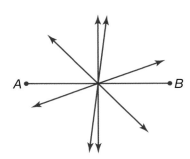

✅ Develop & Understand: B

In this picture, *A* is the center of one circle and *B* is the center of the other. The two circles have the same radius, and they intersect at a single point, *C*.

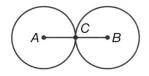

3. Is *C* necessarily exactly halfway between *A* and *B*? Why or why not?

4. Using only a straightedge, could you add a line segment to the drawing that would bisect $\overline{AB}$ perpendicularly? Explain how you would do it, or why it would not be possible.

5. Try to construct a picture like this one using only a compass and a straightedge. The distance between *A* and *B* can be any length.

   a. Explain how you constructed the figure.

   b. Which of the requirements was difficult to meet? Can you be sure you have met them all exactly?

✅ Develop & Understand: C

6. Follow the instructions to construct a figure like the one below.

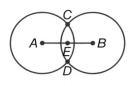

   a. Use a compass to draw two circles, each with the same radius. Draw the circles so that they overlap. Mark the center of each circle, the point where the fixed point of the compass was. Label one *A* and the other *B*. Mark the two points where the circles intersect. Label those points *C* and *D*.

   b. Connect the center points of the two circles with a straight line segment.

   c. Connect *C* and *D* with a straight line segment. Mark the point where $\overline{CD}$ intersects $\overline{AB}$. Label that point *E*.

   d. Use a ruler to measure the distance from *A* to *E*, and from *B* to *E*. Record both distances.

   e. Use a protractor to measure the angle where $\overline{AB}$ and $\overline{CD}$ intersect. Measure the widest angle. Record the angle.

7. Make a conjecture as to what happens to *C* and *D* as the radius of the circles get larger.

8. Imagine that you set the radius of the compass to be the length of $\overline{AB}$ when you are following the procedure from Exercise 4.

   a. What happens to the circles?

   b. What happens to C and D? Where should C and D be logically?

9. Look back over your data from this exploration.

   a. Was $\overline{CD}$ ever a perpendicular bisector of $\overline{AB}$?

   b. Share your results with other students. What do the constructions where $\overline{CD}$ was a perpendicular bisector of $\overline{AB}$ have in common?

   c. Based on the data you have seen, make a hypothesis about when $\overline{CD}$ will be a perpendicular bisector of $\overline{AB}$.

Follow these steps to bisect a line segment using a straightedge and compass.

Step 1. Put the fixed point of the compass at one of the endpoints of the line segment and draw a circle.

Step 2. Keeping the compass set to the same radius, draw a second circle with its center at the other endpoint of the line segment. Make sure the radius is big enough that the two circles intersect.

Step 3. Mark the two points where the circles intersect. Draw a straight line segment connecting these two points. This line segment bisects the original line segment perpendicularly. The point where the two lines intersect is the midpoint of the original line segment.

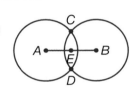

Math Link

Vernier (VUR nee uhr) is an instrument used in measuring lengths and angles. It is named for Pierre Vernier, a French mathematician who invented it in the 1631.

✅ Develop & Understand: D

Use a straightedge and compass to bisect each of these lines.

10. Y •————————————• Z

11.

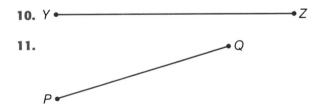

12. Use a straightedge and compass to construct a right triangle. Explain how you did it.

> ### Share & Summarize
>
> Explain the difference between a bisector of a line segment and a perpendicular bisector of the segment.
>
> Give an example of something you can do using both a compass and straightedge that you could not do with just a compass or just a straightedge.

Investigation 2 Constructions Involving Angles

Materials

- compass
- straightedge

Think & Discuss

Look at the triangle below.

If you constructed a triangle that was *similar* to this one, what would have to be true about it?

If you constructed a triangle that was *congruent* to this one, what would have to be true about it?

At times, you may be asked to construct an angle congruent to a given angle or to copy an angle. Using a compass and straightedge, follow the steps to draw a copy of the original angle shown below.

Step 1. Draw the original angle. Draw a straight line segment to the right of the angle.

Step 2. Put the fixed point of the compass at the vertex of the original angle. Draw an arc that intersects the two rays that form the angle. Label the intersection points *A* and *B*.

Step 3. Keep the radius of the compass the same. Put the fixed point at one endpoint of the line segment that you drew. Label this endpoint *C*. Draw an arc that intersects the line segment. Mark the intersection point and label it *D*.

Step 4. Put the fixed point of the compass on point *A*, on the original angle. Adjust the radius until the pencil point is on point *B*.

Step 5. Keep the radius of the compass the same. Put the fixed point of the compass at point *D*, on the angle you are constructing. Draw an arc that intersects the first arc that you drew. Label the intersection point *E*.

Step 6. Draw a straight line connecting *E* to *C*. Your new angle *ECD* is congruent to the original angle.

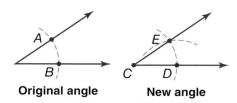

Original angle   New angle

✅ *Develop & Understand: A*

Use a compass and straightedge to construct angles congruent to the angles given.

1.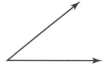

2.

3. Use a compass and straightedge to construct a triangle *similar* to this one. Remember that similar triangles have the same three angles but may have sides of different lengths. Explain how you constructed it.

4. Use a compass and straightedge to construct a second triangle *congruent* to the one that you drew in Exercise 4. Explain how you constructed it.

In Investigation 1, you learned how to bisect a line segment. Now you will learn how to bisect an angle.

Follow these steps to bisect an angle using a compass and straightedge.

Step 1. Draw an original angle.

Step 2. Put the fixed point of the compass at the vertex of the angle. Draw an arc that intersects the two rays that make the angle. Label the intersections *A* and *B*.

Step 3. Put the fixed point of the compass at *A* and draw a circle.

Step 4. Without changing the radius of the compass, put the fixed point at *B* and draw a second circle. Make sure that you choose a radius large enough that the two circles intersect.

Step 5. Mark the point where the circles intersect inside the angle. Label it *D*.

Step 6. Draw a straight line through *A* and *D*. This line bisects the angle.

> ### Think & Discuss
>
> Does this method of bisecting an angle work for a 180° angle?
>
> Does it work for an angle larger than 180°?

✓ Develop & Understand: B

Using a compass and straightedge, construct a congruent angle in Exercises 5–7. Then bisect the angle that you drew.

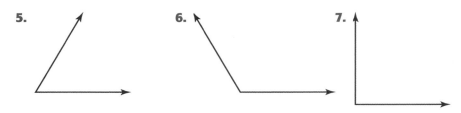

5.   **6.**   **7.**

Using a compass, a straightedge and the angles listed below, construct each of the angles in Exercises 8–11.

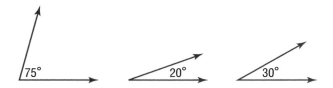

8. 15°

9. 95°

10. 45°

11. 90°

12. Construct a triangle with three 60° angles. Explain how you constructed it.

> ### Share & Summarize
>
> Explain the procedures for constructing an angle congruent to a given angle and for bisecting an angle, so that someone who missed class can understand.

Practice & Apply   Construct a line segment congruent to each of the given segments.

1. •————————————•   2.

Bisect each of the following lines.

3. •——————————•   4.

Construct an angle congruent to each of the given angles.

5. 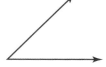   6.

Bisect each angle.

7.   8.

Using a compass and straightedge and the angles listed below, construct each of the angles in Exercises 9–12.

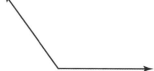

45°   15°   80°

9. 90°

10. 40°

11. 30°

12. Construct a 30° angle in a different way.

Use only a straightedge and a compass to construct the figures.

13. Construct a triangle congruent to this one.

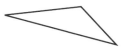

14. Construct a triangle similar to this one but with sides twice as long.

15. Construct a square. Make sure you know that all the angles are 90°. Explain how you did it.

Connect & Extend

16. Cassandra said, "With just a compass and straightedge, I can construct a 45° angle, and I do not need to see any angle for reference."

 a. Explain how she could do this.

 b. Cassandra found a different method to do the same thing. Explain a second way you can construct a 45° angle with compass and straightedge, without using a reference angle.

 c. Give an example of another angle you could construct without using a reference angle. Explain how you would construct it.

 d. Cassandra said: "I can construct a 60° angle without using a reference angle. If I construct three congruent line segments and make them the sides of a triangle, each of the angles will be 60°." Do you think Cassandra's method will work? Explain why or why not.

17. In this picture, the two circles have the same radius. Points *A* and *B* are centers of the circles.

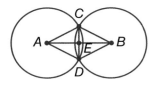

 a. How are the lengths of $\overline{AC}$, $\overline{AD}$, $\overline{BC}$, and $\overline{BD}$ related?

 b. What happens to these lengths when you draw this picture using a larger radius?

 c. What happens to the angles of the quadrilateral *ADBC* if you draw the same picture using a larger radius?

 d. What happens to those angles if you use a smaller radius?

 e. Do you think the quadrilateral *ACBD* is *always, sometimes,* or *never* a square? Explain your reasoning.

18. This is a 60° angle. Use this reference angle and line segment, a compass, and a straightedge to construct the following figures.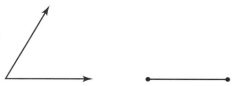

    a. two different triangles, each with a 60° angle

    b. a triangle with two 60° angles and the side between them congruent to the reference line segment

    c. a triangle with a 60° angle, a 30° angle, and the side between them congruent to the reference line segment

    d. Could you construct a different triangle that fits the description in Part c? If so, construct an example. If not, explain why you think it is impossible.

    e. Could you construct a different triangle that fits the description in Part d? If so, construct an example. If not, explain why you think it is impossible.

19.

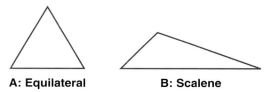

A: Equilateral   B: Scalene

    a. Construct three copies of each triangle.

    b. On one copy of triangle *A,* pick a side and construct its perpendicular bisector. Then construct the line that bisects the angle opposite to that side.

    c. Repeat Step b for the other two sides of triangle A.

    d. Now repeat Steps b and c for triangle B.

    e. What differences do you notice between the results for triangle A and the results for triangle B?

    f. In this triangle, which of the angles do you think might be bisected by the perpendicular bisector of the side opposite it? Explain your reasoning. Then use your compass and straightedge to test your hypothesis.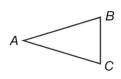

Mixed Review

Rewrite each expression as simply as you can.

20. $x \cdot 4 \cdot x \cdot x$

21. $b + b + b + b + b$

22. $8 \cdot d \cdot d \cdot f \cdot f$

23. $12s + s + s + s$

Find the value of each expression for $y = 2$

24. $y^2 - \dfrac{1}{3}$

25. $12y \div 8$

Find the solution to each equation.

26. $\dfrac{3(b + 5)}{5} = 9$

27. $6(n - 2) - 11 = 1$

Expand each expression.

28. $3(3a - 7)$

29. $2b(8b - 0.5)$

30. $9c(8 + 7c)$

31. $d(d + 6)$

32. $4(e - 6)$

33. $3f(9f - 1)$

34. $g(4 + 7)$

35. $9(3h - 1)$

36. $j(2 + 2)$

37. $km(3 - 2)$

38. Find the slope of each segment.

    a. Segment a

    b. Segment b

    c. Segment c

    d. Segment d

    e. Segment e

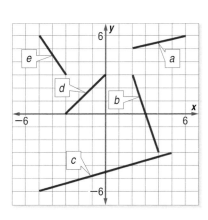

Review & Self-Assessment

Chapter Summary

You learned to find an equation of a line from a written description, a table of values, a graph, a slope and a point on the line, or two points on the line. You also saw that if plotted data show a linear trend, you can fit a line to the data and use the line or its equation to make predictions.

You have learned how to identify and use vertical, complementary, and supplementary angles. You also discovered the relationships among the angles formed when parallel lines are cut by a transversal.

Strategies and Applications

The questions in this section will help you review and apply the important ideas and strategies developed in this chapter.

Using a linear graph to gather information or to make predictions

1. There is a linear relationship between temperature in degrees Fahrenheit and temperature in degrees Celsius. Two temperature equivalents are $0°C = 32°F$ and $30°C = 86°F$.

   a. Make a four-quadrant graph of this relationship, and plot the points (0, 32) and (30, 86). Connect the points with a line.

   b. Use the line joining the two points to convert these temperatures.

      i. $5°F$ **ii.** $20°C$ **iii.** $-30°C$

Fitting a line to data

2. Everyday for a week, Lorena practiced doing pull-ups. Each day, she timed how many pull-ups she could do in a certain number of seconds. She added ten seconds to the time each day.

Seconds	10	20	30	40	50	60	70
Pull-ups	15	25	44	35	42	50	55

   a. Graph the data. Do the data appear to be approximately linear?

   b. Are there any outliers in the data? That is, are there any points that do not seem to fit the general trend of the data? If so, which point or points?

   c. Draw a line that fits the data as well as possible. Find an equation of your line.

   d. Use your equation or graph to predict how many pull-ups Lorena will do when she reaches a time of two minutes.

Vocabulary

alternate exterior angles

alternate interior angles

arc

bisect

collinear

complimentary angles

exterior angles

interior angles

line of best fit

outlier

perpendicular

perpendicular bisector

straightedge

supplementary angles

transversal

vertical angles

Identifying supplementary, complementary, and vertical angles

Identify the relationship or measurement that is described by each situation.

3. $\angle a$ to $\angle b$

4. $\angle a$ to $\angle b$

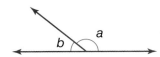

5. $\angle 1$ to $\angle 3$

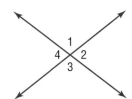

6. $m\angle 1 + m\angle 2$

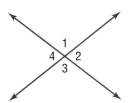

7. A pair of supplementary angles has one angle that measures 78°. What is the measure of the other angle?

8. Find the value of x in the situation shown to the right.

9. An angle has a measure of 95.5°. Find the measure of the supplementary angle.

10. Identify the equations that represent parallel lines.

    a. $y = -x + 2$ **b.** $y = 2x + (1 - x)$ **c.** $3y = 1 - 3x$

Rewrite each equation in slope-intercept form.

11. $y - x - 1 = 2x + 1$ **12.** $2(y - 1) = 3x + 1$

Identify the relationship or measurement that is described by the given situation.

13. $\angle 5$ and $\angle 3$

14. $\angle 2$ and $\angle 8$

15. $\angle 1$ and $\angle 7$

16. $\angle 4$ and $\angle 8$

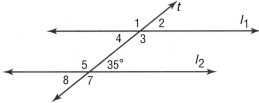

17. Will any of the angles 1 through 8 be anything other than 35° or 145°?

Perform constructions

18. Use a straightedge and a compass to construct a right triangle.

19. Using a compass and a straightedge, construct a line and its perpendicular bisector.

Bisect each angle.

20.

21.

For each of the following angles, use a compass and a straightedge to construct a congruent angle. Then bisect the angle that you drew.

22.

23.

Test-Taking Practice

SHORT RESPONSE

1 In the figure to the right, $n \parallel m$. If $m\angle b = 63°$, find the measures of $\angle a$, $\angle f$, and $\angle h$. State the relationship of each angle to $\angle b$ on the lines to the right.

Answer: $m\angle a$ _____ $m\angle f$ _____ $m\angle h$ _____

_____ _____ _____

MULTIPLE CHOICE

2 The following table shows the costs for tickets for a concert series, depending on the number of tickets purchased.

Tickets	Cost
1	$18
2	$34
3	$50
4	$66

Choose the equation that relates the input to the output.

A $y = x + 16$

B $y = 16x + 2$

C $y = x + 32$

D $y = 2x + 18$

3 In the figure, if $m\angle x = 35°$, which of the following is true?

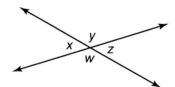

F $m\angle w = 35°$

G $m\angle y = 35°$

H $m\angle z = 35°$

J $m\angle z = 145°$

4 Supplementary angles a and b have measures of $(6x + 17)°$ and $(3x + 1)°$. What is the value of the larger angle?

A 18 **C** 65

B 55 **D** 125

Percents and Proportions

Real-Life Math

Percents and Sports In many sports, individual athletic performances are rated using percents. Baseball players are ranked according to their batting averages and their on-base percentages. In hockey, goalies are evaluated based upon their save percentage. Tennis players keep track of their service ace percentages. Basketball coaches record the two-point and three-point field goal shooting percentages of their players.

Team performances can also be evaluated using percents. A football team that has a season record of eight wins and two losses has won 80% of its games.

Think About It Suppose a football team plays eight games per year. The team's winning percentages for the last five years are shown in the table below.

Year 1	Year 2	Year 3	Year 4	Year 5
50%	75%	75%	25%	75%

How many total games did the football team win during the five-year span? What is the team's five-year win-loss percentage?

Math Online
Take the **Chapter Readiness Quiz** at glencoe.com.

Dear Family,

Percents, which are also referred to as percentages, are used by department stores to advertise sale prices, by sports announcers to evaluate athletic performances, and by food manufacturers to provide nutrition facts. In this chapter, your student will explore percents and apply them to real-world situations.

Key Concept—Percents

Students will use percent as a common scale to make estimates and use the percent proportion to find exact percentages. Students will also use diagrams to calculate percent increase and decrease.

Percent means "out of 100." A percent represents a number as a part out of 100.

37%
Thirty-seven Percent

Percent Increase

The diagram to the right compares the number of births in Colorado in 1999 and 2006. The diagram shows that there was an increase in births because the right bar is higher than the left bar.

It appears that the *percent increase* is about 15%. The exact increase can be found by dividing the difference in the two birth numbers by the number of births in 1999.

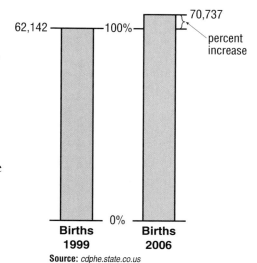

Source: *cdphe.state.co.us*

Chapter Vocabulary

percent decrease   **percent increase**

Home Activities

- When shopping, challenge your student to find the sale price of discounted items.
- Use statistics in the sports section of the newspaper to practice finding percents.
- Convert your student's quiz and test scores to percents.

Understand Percents

In previous courses, you have used percents to represent parts of a whole. You should remember that percent is a ratio that compares a number to 100.

In this lesson, you will build on that knowledge to estimate and calculate percents involving real-world situations. Understanding percents can help you quickly estimate the percent of something, such as a tip in a restaurant or the price of a sale item.

Investigation 1 Estimate Percents

In this investigation, you will estimate percents when presented with real-world contexts.

Think & Discuss

Mateo was asked to find 140% of 80 on a multiple choice test. The answers he had to choose from are shown below.

a. 11.3   **b.** 60

c. 96   **d.** 112

Without doing any computations, Mateo said he could immediately eliminate two of the answer choices. How do you think he was able to do this?

☑ Develop & Understand: A

Washington Middle School has 750 students, and Kennedy Middle School has 600 students.

1. A survey finds that 375 Washington students participate in club activities, while 350 Kennedy students participate in club activities.

   a. What percent of Washington students participate in club activities?

   b. About what percent of Kennedy students participate in club activities?

   c. Do a greater percent of Washington students or a greater percent of Kennedy students participate in club activities?

2. At each school, 500 students play a musical instrument. Which school has a greater percent of students who play a musical instrument? Explain.

3. At each school, about 70% of the students volunteer in the community.

   a. About how many students at Washington volunteer?

   b. About how many students at Kennedy volunteer?

4. The drama teachers at each school selected 50 students to represent the school at a theater workshop.

   a. About what percent of Washington students attended the theater workshop?

   b. About what percent of Kennedy students attended the theater workshop?

5. About 43% of each school's students attended the spring musical.

   a. About how many students did *not* attend the musical?

   b. If each ticket costs $5, what were the approximate ticket sales for each school?

✅ *Develop & Understand: B*

Two neighboring colleges hold an annual blood drive competition the week of their rivalry football game. This year, Smithville University wanted to collect 1,200 pints of blood. Jonesboro College has fewer students and hoped to collect 800 pints.

With one day left, Smithville University had collected 900 pints, and Jonesboro College had collected 720 pints.

6. Suppose each college had reached 30% of its drive goal at the end of the first day.

   a. Determine the number of pints collected by Smithville.

   b. Determine the number pints collected by Jonesboro.

7. Suppose each college had reached 145% of its drive goal by the end of the drive.

   a. Estimate the number of pints collected by Smithville.

   b. Estimate the number of pints collected by Jonesboro.

   c. Explain how you found your estimates.

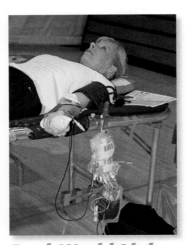

Real-World Link

One pint of blood can save three lives.

8. Halfway through the blood drive, Smithville University had collected 600 pints, and Jonesboro College had collected 550 pints. Malik used his calculator to express the ratios of collected pints to collection goals for schools as percents.

   How could you use your calculator to find $\frac{550}{800}$?

9. Suppose each school was 25 pints short of meeting its goal.

   a. How many pints would Smithville University have collected?

   b. How many pints would Jonesboro College have collected?

   c. Calculate the percentages collected by each school. Comparing percentages, which school came closer to its goal?

10. By the time the blood drive ended, each school had collected exactly 950 pints.

    a. What percent of its goal did Smithville University collect?

    b. What percent of its goal did Jonesboro College collect?

11. Determine which ratio in each pair is greater, or if the ratios are equivalent, by finding the percent each represents.

    a. $\dfrac{56}{69}$ or $\dfrac{76}{89}$

    b. $\dfrac{106}{210}$ or $\dfrac{206}{310}$

    c. $\dfrac{50}{45}$ or $\dfrac{210}{205}$

12. Order the following ratios from least to greatest by finding the percent each represents.

$$\frac{56}{71} \qquad \frac{71}{56} \qquad \frac{102}{203}$$

$$\frac{17}{60} \qquad \frac{203}{102} \qquad \frac{69}{91}$$

$$\frac{91}{69} \qquad \frac{43}{80} \qquad \frac{80}{43}$$

13. Find four ratios equivalent to 25%.

14. Find four ratios equivalent to 120%.

Share & Summarize

1. Suppose you are calculating P% of 1,000 and P% of 700. Which would be the greater number? Explain your answer.

2. Which represents a greater percent, t out of 1,000 or t out of 700? Explain your answer.

Investigation ② Percent and Proportions

Materials
- graph paper (optional)

In the last investigation, you saw that you can use percents as a common scale to compare ratios. Proportions involving percents can be used to solve many problems.

✓ Develop & Understand: A

Work with your group to solve these exercises. Be ready to explain your reasoning.

1. Find the following percents.

   a. What percent of $\frac{2}{3}$ is $\frac{5}{6}$?

   b. What percent of $\frac{5}{6}$ is $\frac{2}{3}$?

2. The circles shown represent 30% of the figure.

   a. Draw 100% of the figure.

   b. Draw 150% of the figure.

   c. What percent of the figure is one circle?

3. Anna is making lemonade for a party. The lemonade will contain water and at least 75% juice concentrate. Anna has six gallons of concentrate. What is the greatest amount of lemonade Anna can make?

One way to work with percents is to set up and solve a proportion. In most situations involving percents, you have values for two things. You want to know the value of the third, so your proportion might look like the proportion shown below.

$$\frac{b}{a} = \frac{n}{100}$$

For example, in Exercise 3 above, you knew the lemonade should contain 75% juice concentrate. You knew that you had six gallons of juice concentrate. The ratio 6 : a compares gallons of juice concentrate to total gallons of liquid.

You could have set up this proportion.

$$\frac{6}{a} = \frac{75}{100}$$

Think & Discuss

This type of percent diagram is used to help compare two numbers. State in words the question represented by this percent diagram.

Write a proportion that is illustrated by the diagram and solve it. How did you think about setting up the proportion?

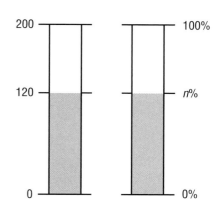

✓ Develop & Understand: B

4. Consider this question: *What percent of 150 is 6?*

   a. Express the question as a proportion.

   b. Solve the proportion in Part a.

5. Now consider this question: *What is 155% of 20?*

   a. Express the question as a proportion.

   b. Solve the proportion in Part a.

6. At an aquarium, 15% of the fish in the tank are angelfish. There are 21 angelfish in the tank. How many fish in all are there in the aquarium tank? Show how you found your answer.

7. Of 125 students studying a foreign language at Liberty Middle School, 45 students are studying Chinese. What percent of the 125 students are studying Chinese? Show how you found your answer.

8. The largest land animal, the African bush elephant, may weigh as much as eight tons. However, that is only about 3.9% of the weight of the largest animal of all, the blue whale. How heavy can a blue whale be? Show how you found your answer.

Real-World Link

The enormous blue whale's primary source of food is plankton, tiny animals and plants that float near the surface of the water.

9. In 2006, Americans spent $282 billion on entertainment, including amusement parks, audio and visual services, toys, and hobbies. They spent 257% of that amount on food. How much did Americans spend on food in 2006?

10. In 1950, the estimated world population was 2.6 billion. It was 6.3 billion in 2003 and 6.6 billion in 2007.

   a. What percent of the 2003 population is the 1950 population?

   b. What percent of the 2007 population is the 2003 population?

   c. What percent of the 2003 population is the 2007 population?

   d. In 2007, about 301 million people lived in the United States. What percent of the world's population lived in the United States in 2007?

Share & Summarize

1. Write a proportion that expresses this statement: 32% of 25 is b. Solve your proportion for b.

2. Write a proportion that expresses this statement, n% of a is b. Explain why your proportion is useful for solving percents.

Practice & Apply

1. There are 145 girls and 160 boys who attend Franklin Middle School. Based on a survey of Franklin Middle School students, 120 girls and 140 boys participate in extracurricular activities.

   a. Which group of students at Franklin Middle School, girls or boys, had a greater percent of extracurricular activity participation?

   In a recent survey of local middle schools, 80% percent of all students participate in extracurricular activities.

   b. How do the two groups compare to the survey of local middle schools?

For each pair of ratios, determine which is greater by finding the percents they represent.

2. $\frac{41}{51}$ or $\frac{23}{28}$

3. $\frac{323}{467}$ or $\frac{16}{23}$

4. $\frac{12}{13}$ or $\frac{60}{65}$

5. Find four ratios equivalent to 60%.

6. Find four ratios equivalent to 75%.

Decide which is a greater number.

7. P% of 10   or P% of 20

8. P% of 312   or P% of 450

9. P% of 2.7   or P% of 3.2

Decide which represents a greater percent.

10. t out of 50   or t out of 75

11. t out of 99   or t out of 81

12. t out of 1.2   or t out of 1.5

13. Consider this question: *5% of what number is 15?*

    a. Express the statement as a proportion.

    b. Solve the proportion in Part a.

14. 48% of what number is 50?

15. 117% of what number is 2.6?

16. $6\frac{1}{2}$ is what percent of 9?

17. 20 is what percent of 15?

18. Josh works 12 hours a week. He recently received a 6% raise, and his new weekly pay is $117.02.

    a. What was Josh's weekly pay before the raise?

    b. What was his hourly pay before the raise?

    c. What is his new hourly rate?

19. Social Studies In 2006, about 25,500,000 children between the ages of 12 and 17 lived in the United States. That was about 8.3% of the entire U.S. population. Find the approximate 2006 U.S. population.

Connect & Extend

20. Social Studies The bar graphs show information about education levels in the United States in a recent year. The left bar shows the percentages of unemployed people at each level. The right bar shows the percentages for the employed.

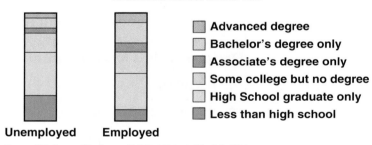

Education Levels in the U.S.

| Unemployed | Employed |

- ▨ **Advanced degree**
- ▢ **Bachelor's degree only**
- ▨ **Associate's degree only**
- ▢ **Some college but no degree**
- ▢ **High School graduate only**
- ▨ **Less than high school**

Source: U.S. Bureau of the Census. *Statistical Abstract of the United States.*

    a. Measure the heights of the bars in centimeters. This height represents 100% of the people classified in each graph.

    b. Now measure the height of the portion for high school graduates who are employed. Use that height to calculate the approximate percent of employed people who only have a high school diploma.

    c. Use a similar procedure to calculate the approximate percentage of unemployed people who have a high school education or less.

    d. Choose two more education levels from the illustration shown above. Repeat Parts b and c for each of them.

21. **World Cultures** The table below shows data about communications in several countries.

Subscribers per 100 People in 2006

Country	Main Phone Lines	Mobile Cellular	Internet
Argentina	24	81	7
Australia	49	97	33
Bosnia	25	48	6
China	28	35	10
Iran	31	19	16
Japan	43	79	27
Mexico	18	53	4
South Africa	10	83	9
Sweden	60	106	38
Thailand	11	63	4
United States	57	77	21

a. The entries in the third column are "per 100 people." Desta said, "The entry for United States is 21. That means 21% of people subscribed to the Internet in 2006." Nicolás convinced her that was not true. Explain why Nicolás was correct.

b. The entries in the second column are "per 100 people." The entry for United States is 77. What would the entry be if the scale was "per 10 people"?

c. Choose one category. Write a paragraph comparing the given countries in terms of that category. You might include information about which country has the highest entry, which has the lowest entry, which have entries close to each other, and how the United States compares to the other countries.

d. Choose one country other than the United States. Compare its data to that for the United States.

22. Make up a reasonable real-world situation that requires finding 65% of 120. Show how to find your solution.

23. The students at four local colleges held a blood drive competition. The local newspaper reported that one of the colleges collected the most pints of blood, 520. The other colleges each raised 20% of the total number of pints collected. However, the paper did not report the total number of pints collected by all four colleges.

Use the information to find the total amount collected by all four colleges. Then find the number of pints collected by each of the other three colleges.

24. Challenge Suppose $S > 0$ and $S = 2T$.

a. Which is greater, 80% of T or 40% of S? Explain your answer.

b. Does your answer to Part a hold true if $S < 0$? Explain.

25. In Your Own Words On a recent quiz, students were asked the following question.

What percent of 30 is 40?

One student wrote an answer of 75%. Do you agree with the student's answer? Why or why not?

Mixed Review

The *x*-intercept is the *x* value at which a line crosses the *x*-axis. Find an equation of the line with the given *x*-intercept and slope.

26. *x*-intercept 5, slope -2

27. *x*-intercept -2.5, slope 0.5

Tell whether the relationship in each table could be linear.

28.

x	0	1	2	3	4
y	2.2	0	-2.2	-4.4	-6.6

29.

x	0	1	2	3	4
y	-2	2	-4	4	-6

30.

x	0	1	2	3	4
y	4.1	13.1	22.1	31.1	40.1

Tell whether the points in each set are collinear.

31. $(3, 1)$; $(8, 12)$; $(-1, -10)$

32. $(-2, 9)$; $(2, 2)$; $(4, -1.5)$

33. $(15, 22)$; $(0, 1)$; $(5, -6)$

Work with Percents

In Lesson 3.1, you learned how to estimate percents, compare ratios, and use a percent proportion to find missing values. In this lesson, you will apply your knowledge of percents to solve problems that involve percent comparisons and percent increase and decrease.

Think & Discuss

A teacher gives weekly quizzes to her math students. The last four weeks, she has determined the median quiz score. The results are shown in the table.

Weekly Quiz Results

Week Number	Total Possible Points	Median Score
1	20	17
2	20	19
3	25	19
4	25	21

Write a description of the students' performance during the four weeks.

Investigation ① Percent Increase and Decrease

Vocabulary

percent decrease

percent increase

Materials

• graph paper (optional)

You have used percent diagrams to compare two ratios and to calculate percentages. You can use similar diagrams to show how a quantity changes in relation to its original size. These diagrams do not have a percent scale because one of the two bars represents 100%.

For example, the number of births in Colorado increased from 62,142 in 1999 to 70,737 in 2006. In the diagram, the **percent increase** from 1999 to 2006 is represented by the portion of the "Births 2006" bar above the 100% line.

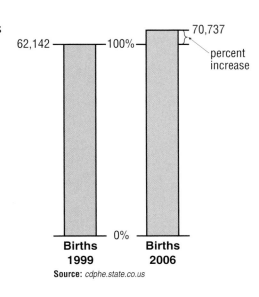

Source: cdphe.state.co.us

We can make a visual estimate by comparing the height of the "percent increase" portion of the bar with the portion representing 100%. In the Colorado birth diagram, the height of the "percent increase" portion is about 15% of the height of the 100% portion.

This diagram shows the change in U.S. military personnel on active duty from 1995 to 2005. The **percent decrease** is represented by the unshaded portion of the bar on the right. In this example, the "percent decrease" portion of the bar looks like about 5% to 10% of the 100% portion.

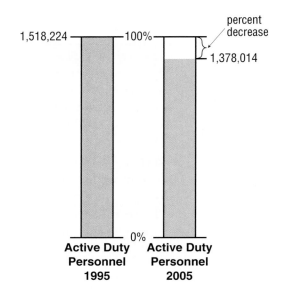

✓ Develop & Understand: A

1. As people grow older, they often say that time seems to pass more quickly. Think about whether two years seems longer to you now than it did when you were six years old.

   a. Make a diagram that shows the percent increase in your age from when you were six years old to when you were eight years old. Draw your diagram so that your age at six years old represents 100%. Use your diagram to estimate the percent increase in your age.

   b. Now make a diagram that shows the percent increase in your age from two years ago to now. Draw your diagram so that your age two years ago represents 100%. Use your diagram to estimate the percent increase in your age.

   c. Does it make sense that time would seem to pass more quickly as you age? Explain.

2. Consider the diagram shown at the right. Estimate the percent change from July to August. Specify whether it is a percent increase or decrease.

If you were to exercise and then measure your heart rate, or pulse, it would probably be quite different from your resting heart rate.

To *calculate* the change in your heart rate, you first need to measure your resting heart rate. You will use this rate as a baseline. Then you need to measure your rate after exercise. A comparison of the two rates can be written as a percent increase.

✅ Develop & Understand: B

Find your pulse in your neck just below your jaw or in your wrist.

3. Count the number of pulses in 20 seconds. Use that value to find your resting heart rate in beats per minute.

4. Following your teacher's directions, do some physical activity for one minute. As soon as you stop, count the pulses in 20 seconds. Find your heart rate after exercise, in beats per minute.

5. If your resting heart rate in beats per minute is considered 100%, what percentage represents your heart rate after exercise?

6. How might you describe the *change* in your heart rate from a resting to an active state as a percentage?

Think & *Discuss*

Zoe and Luis both had a resting heart rate of 84 beats per minute. After jumping rope for a short period, they both had a heart rate of 105 beats per minute. Here is how they each calculated the percent increase.

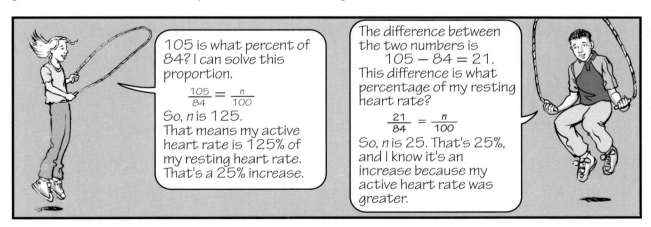

105 is what percent of 84? I can solve this proportion.

$$\frac{105}{84} = \frac{n}{100}$$

So, *n* is 125.
That means my active heart rate is 125% of my resting heart rate. That's a 25% increase.

The difference between the two numbers is
$$105 - 84 = 21.$$
This difference is what percentage of my resting heart rate?

$$\frac{21}{84} = \frac{n}{100}$$

So, *n* is 25. That's 25%, and I know it's an increase because my active heart rate was greater.

Will these two methods always give you the same answer? Explain how you know.

✓ Develop & Understand: C

Use either Zoe's or Luis' method to solve these exercises. Round answers to the nearest tenth of a percent, if needed.

7. In 1985, there were about 39,900,000 U.S. households with cable TV. In 2006, about 65,500,000 households received cable TV. What is the percent increase of households with cable TV from 1985 to 2006?

8. Shoes regularly priced at $55 are on sale for $41.25. What is the percent decrease from the original price to the sale price?

9. In 1987, the average hourly wage in the United States was $9.02. In 1997, it was $12.29. In 2007, it was $17.10.

    a. What is the percent change of the hourly wage from 1987 to 1997? Tell whether it is a percent increase or decrease.

    b. What is the percent increase of hourly wage from 1997 to 2007?

Share & Summarize

A particular tree grew from 125 cm to 133 cm in one year. What was the percent increase in the height that year? Explain how you found your answer.

Investigation 2 Percents of Percents

To figure a tip for a restaurant bill, some people take a percent of the total, after taxes. Since the tax is a percent of the subtotal, these diners are taking a percent of a percent.

There are several other situations in which this might happen.

Think & Discuss

Increase 100 by 20%. Decrease the result by 20%. Is the result the original number, 100? Explain.

✅ Develop & Understand: A

Barry's Bargain Basement and Steve's Super Savings Store are located in the Los Angeles garment district.

1. Rashonda went to Barry's Bargain Basement to buy a pair of jeans. The jeans had been in the store for one month. The original price was $40.

   a. What was the sale price?

   b. What percent of the original price was the sale price?

2. Julie was a smart shopper. The pair of shoes she wanted had an original price of $50 at both Barry's and Steve's. The shoes had been in both stores for two months.

   a. How much would Julie pay for the shoes at Barry's? At Steve's?

   b. Explain why these sale prices are different.

3. Jasmine, a frequent shopper at Barry's, has her eye on a hat with an original price of $20.

   a. The hat has been in the store for one month. How much will Jasmine save?

   b. Jasmine is considering waiting another month to get a better price. How much will she save after two months?

   c. Is the amount Jasmine would save with the 10% discount double the amount she would save with the 20% discount?

   d. How can you easily determine how much Jasmine would save after three months?

4. Omari found a sweatshirt at Steve's with an original price of $30.

   a. If the sweatshirt has been in the store long enough to earn the 10% discount, how much will Omari save?

   b. If the sweatshirt has been there long enough to earn the 20% discount, how much will he save from the original price?

   c. Is the dollar amount of the second discount twice the dollar amount of the first discount? Explain why or why not.

   d. Omari calculated the final price this way. "First I take 10% off, and then I take 20% off. So all together, I'm taking 30% off. That's $9, so the final price is $21." Is Omari correct? Explain.

   e. If Omari buys the sweatshirt after it has been in the store for two months, what is the total percent discount from the original price? Show how you found your answer.

5. Oscar found an item originally marked at $100. It had been in both stores for more than four months. Where would he get the better buy? Explain.

6. Is the ad for Steve's, which claims it "won't be undersold," true when Steve's prices are compared with Barry's prices? Explain.

✅ *Develop & Understand: B*

With most photocopy machines, you can reduce or enlarge your original by entering a percent for the copy. If you enter 200%, the copy machine will double the dimensions of your original.

For example, if you copy a 3-inch-by-5-inch photograph at the 200% setting, the copy of the photograph will measure 6 inches by 10 inches.

7. Jacy wanted to enlarge a 2-cm-by-4-cm drawing. He set a photocopy machine for 150% and copied his drawing.

   a. What will be the dimensions of the copy of the drawing?

   b. The drawing still seemed too small to Jacy. He put his copy in the machine and enlarged it 150%. What are the new dimensions?

   c. Jacy likes this size. He now wants to enlarge another 2-cm-by-4-cm drawing to this same size. What percent enlargement should he enter to make the copy in just one stage?

   d. Is the percentage you calculated in Part c the sum of the two 150% enlargements? Why or why not?

8. Challenge Suppose you put a print in a copy machine and enlarge it by selecting 180%. What percent reduction would you have to take to return the image to its original size?

Share & Summarize

Suppose you reduce an 8-inch-by-10-inch picture by 90% and then increase the copy by 90%.

1. Will you get the same size picture with which you started? Explain.

2. What is the percent change from the original size to the final size?

Investigation 3 — Interpret Comparisons

You probably hear comparisons that use ratios and percentages everyday. News reports and advertisements, such as TV commercials, use such comparisons all the time. People who write advertisements try to state comparisons in a way that will make you want to buy their products.

✓ *Develop & Understand: A*

In a recent election by a seventh-grade class with 90 students, Razi received 36 votes and Jorge received 54 votes.

1. Tell whether each of the following statements presents this information accurately. For each statement, explain or give a calculation to show why it is accurate.

    a. The ratio of the number of students who voted for Razi to the number who voted for Jorge is 2:3.

    b. Forty percent of the voters preferred Razi.

    c. Jorge received 20% more of the total vote than Razi received.

    d. Fifty percent more people voted for Jorge than for Razi.

    e. Eighteen more people voted for Jorge than for Razi.

    f. Jorge received $\frac{3}{5}$ of the votes.

    g. The number of people who voted for Jorge is 1.5 times the number who voted for Razi.

2. Which of the *accurate* statements in Exercise 1 seems to give the best impression of the class' preference for Jorge? Why?

3. Which of the *accurate* statements in Exercise 1 seems to minimize the class' preference for Jorge? Explain.

4. Do some of the statements seem more informative than others?

5. Some ratios compare a part of some group to the whole group, like the 36 students who voted for Razi compared to the total number of voters, 90. Such ratios are called *part-to-whole ratios*.

    Other ratios compare a part of a group to another part of the same or a different group, like the 36 students who voted for Razi compared to the 54 who voted for Jorge. These kinds of ratios are called *part-to-part ratios*.

    a. Which statements in Exercise 1 make part-to-whole comparisons?

    b. Which statements in Exercise 1 make part-to-part comparisons?

In the next exercise set, you will see how different impressions can be created by presenting the same data in different ways.

✓ Develop & Understand: B

The organizers of a basketball tournament had one team slot left to fill but two teams they wanted to invite. They decided to look at the two teams' win-loss records.

The Wayland Lions had won 24 games and lost 16. The Midway Barkers played more games, and their record was 32 wins and 22 losses.

6. Find the ratio of wins to the total number of games for the Lions and for the Barkers. Are these part-to-part or part-to-whole ratios?

7. Find the ratio of losses to the total number of games for each team.

8. What percent of all their games did the Lions win? What percent of all their games did the Barkers win?

9. What percent of all their games did the Lions lose? What percent of all their games did the Barkers lose?

Now you will compare the teams using a different method of comparison.

10. Use ratios to compare each team's wins to losses. Are these part-to-part or part-to-whole ratios?

11. For each team, the number of wins is what percent of the number of losses?

12. Use a percentage to report how many more wins than losses each team had.

13. One of the organizers thinks the Lions is definitely the team to invite. What statements, using comparisons and other information, might this organizer make so that the Lions look like the best choice?

14. Another organizer thinks the Barkers is the team to invite. What statements, using comparisons and other information, might this organizer make so that the Barkers look like the best choice?

Share & Summarize

In 2006, California had about 32,440,000 registered motor vehicles. About 19,640,000 were cars. The other 12,800,000 were buses and trucks. That same year, New York had about 11,860,000 registered motor vehicles. About 8,970,000 were cars, and about 2,890,000 were buses and trucks.

Write at least five comparison statements using these statistics.

Inquiry

Investigation 4   Find the Better Deal

Materials

- advertising circulars
- coupon pages for several local stores
- links to several Web sites for retail stores or online stores
- computer with spreadsheet software or graphing calculators (optional)

You and your partners are in charge of sales for the Cool Stuff Company. You are planning a clearance sale to reduce inventory before you introduce a new line of products. You still want to make a good profit on the items you sell.

Research Sales Strategies

1. Look through some advertising circulars or browse the Web sites of some stores online. Notice what kind of information is presented for items that are on sale or discounted. Can you tell what the item's price was before it went on sale? Can you tell what the new price is? Can you tell the percent or dollar amount of the discount?

2. Find examples of advertisements for these sale formats.
   - the discount is given as a percent off the original price
   - the discount is given as a dollar amount off the original price
   - the item's price after discount is given

3. Make a prediction of which type of discount is more attractive to customers. Explain your answer.

Plan Your Sale

Here are the products the Cool Stuff Company sells. The current price of each product is 100% more than its production cost. Determine the production cost of each item.

Company policy is that you must make at least a $5 profit on any item that costs less than $20 per unit to manufacture. You must make at least a $20 profit on any item that costs $20 or more per unit to manufacture.

Item	Current Price
Widget	$86
Gizmo	$50
Whatsit	$48
Thingummy	$38
Doohickey	$24
Jobbie	$12

4. Work with a group to advertise Cool Stuff Company products using the three sale formats. Each group member should pick a different strategy.
   - Express the discount by giving the old price and the new price.
   - Express the discount as a dollar amount off the original price.
   - Express the discount as a percentage off the original price.

5. Decide what will be the discount on each item. Make sure that your profits are high enough to meet company standards.

Math Link

When an item costs someone $50 to make or buy and the person then sells it for 125% of that original cost, a 25% *profit* is made. The item's price has been *marked up* 25% from its original value. You can say that the price is a 25% *markup* from the original cost. Similarly, a 25% decrease in price is a 25% *markdown*.

6. You will need to create a spreadsheet or a table to record your calculations. What columns will you need to implement your sales strategy?

7. Write an advertisement announcing your sale. Describe the deal that your customers will receive.

Compare Results

8. Compare the advertisements from all of the team members. Are all three discounted amounts equivalent? Would a customer receive a better deal at one sale than at another?

What Did You Learn?

9. Your supervisors at the company are worried that you have slashed prices too much. Write a memo using percents to argue that the discounts you have proposed are not unreasonably large. Compare the way that you present the numerical information in this memo to the way that you have presented the information in your advertisement to the customers. What did you do differently? What was similar?

Practice & Apply

In Exercises 1 and 2, estimate the percent change. Be sure to indicate whether it is an increase or a decrease.

1. the price of lettuce from one week to the next

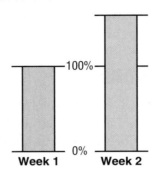

2. the price of a DVD player from one month to the next

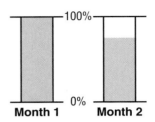

3. **Social Studies** The table lists U.S. crime rates for 1990, 1995, 2000, and 2005.

Crime in the United States

Year	Crimes per 100,000 Inhabitants
1990	5,088.5
1995	4,591.3
2000	3,618.3
2005	3,900.5

Source: Disaster Center

a. To the nearest tenth, what is the percent change in the crime rate from 1990 to 1995? Indicate whether it is a percent increase or decrease.

b. To the nearest tenth, what is the percent change in the crime rate from 2000 to 2005? Indicate whether it is a percent increase or decrease.

4. A used minivan priced at $20,000 went on sale for 5% off.

a. What is the van's reduced price?

b. One month later, the van was discounted 5% off the sale price. What are the dollar amounts of the first discount and the second discount?

c. At the end of the year, the van was discounted 5% off the last sale price. What was the price of the minivan after the third discount?

d. To the nearest tenth, what is the total percent decrease after all three discounts?

5. A family bought a house valued at $350,000. An office park was then constructed next to the house. Five years after the family had bought the house, its value had dropped 30%.

   a. What was the house's value after five years?

   b. What percent increase is now needed for the house to be sold for what the family paid for it? Explain.

   c. Another family updated its house with a new roof and new flooring. These changes increased the value of the house by 6%. It is now worth $265,000. What was the original value of the house?

Connect & Extend

Real-World Link

The DTP-3 immunization protects infants from diptheria, tetanus, and pertussis. It is a strong indicator of access to basic health services.

6. Social Studies The table gives DTP3-immunization data for several countries.

   a. Write a ratio statement comparing Afghanistan and Cambodia.

   b. Write a percent statement comparing Eritrea and France.

   c. Compare Haiti and the United States using a statement that is not a ratio or percent statement.

Infants Immunized for DTP3

Country	Percent of Infants
Afghanistan	76
Cambodia	82
Eritrea	83
France	98
Haiti	43
United States	96

Source: childinfo.org

7. The table shows the resting and active heart rates of several students. The active rates were taken just after different types of exercises.

Complete the table.

Name	Resting Heart Rate	Active Heart Rate	Difference (Active–Resting)	Percent Increase
Juanita	91	138		
Kris	92	190		
Nathan	82	157		
Aaliyah	88	176		
Nieve	85	131		
Tyree	79	147		
Ramón	77	119		

8. Warren typed an English paper on his computer. Using a font size of 12, the first page contained 250 words. When Warren changed the font size to 10, the first page held 330 words.

  a. What is the percent increase in the number of words on the page from the larger font (size 12) to the smaller font (size 10)?

  b. Warren's paper was long enough to fill several pages. What percent increase in the number of words on three pages would you expect when he changes from font size 12 to font size 10? Explain.

  c. When Warren printed his paper using a font size of 12, it was exactly five pages long. About how many pages would you expect it to be if he uses a font size of 10?

  d. Write a formula for calculating the approximate number of pages in a font size of 10 if you know the number of pages in a font size of 12.

9. Food advertisements often contain claims worded to give the best impression.

  a. Explain why the claim "20% fewer calories" is, by itself, not very informative.

  b. Some reduced-fat milks are labeled "2% milk" or "98% fat free." Below are nutrition data for one cup of milk from a carton of whole milk and a carton of 2% milk. How much of a reduction of fat is "98% fat free"?

Nutrition Facts	
Whole Milk	
Calories	150
Total Fat	8g
Total Carbohydrates	13g
Protein	8g

Nutrition Facts	
2% Milk	
Calories	140
Total Fat	5g
Total Carbohydrates	15g
Protein	10g

  c. Explain why advertisers might prefer "98% fat free" over, for example, stating the percent change in the quantity of fat.

10. Karla's Discount Madness store does not like to keep merchandise in the store for long. Any item that has been in the store for a month is marked down 20%. For each additional month an item stays in the store, Karla's staff marks it down another 20% from the previous month's price.

The "Ever-Whiny Baby" doll did not generate much interest. It began as a $30 doll. Several months later, Karla's still had numerous dolls in stock.

a. Complete the table, showing the price of the doll each month.

Doll Price

Month	0	1	2	3	4	5	6	7
Price ($)	30							

b. Graph the data in your table. Connect the points with a smooth, dashed curve.

c. What kind of relationship does this appear to be? Look back through your book if you need help remembering the relationships you have studied.

11. There are 864 students and 32 teachers at Scott School. There are 1,428 students and 42 teachers at King School.

a. Find the ratios of students to teachers at each school. Can you form a proportion with these ratios? Why or why not?

b. Suppose King School wants to make its student-teacher ratio equal to that at Scott School. How many new teachers would King have to hire?

12. Annette recorded how she spent her time each day for a week.

Daily Activities

Activity	One Week, in Hours	School Year (39 Weeks), in Days	75 Years (Age 5 to 80), in Years
Sleeping	54.6		
At school	35.0		
Socializing	14.1		
Watching TV	11.2		
Doing homework	9.0		
Talking on phone	8.4		
Reading	7.0		
Eating	7.0		
Bathing, grooming	6.3		
Other activities	15.4		

a. What assumption would you have to make to complete the table?

b. Is the assumption in Part a reasonable for the "School Year" column? If not, would it be reasonable for *some* of the activities? If so, which ones?

c. Is the assumption in Part a reasonable for the "75 Years" column? If not, would it be reasonable for *some* of the activities? If so, which ones?

d. Complete those entries in the table for which your assumption seems reasonable.

e. Calculate the percent of time Annette spent in each activity during the week she made her records.

f. Make a circle graph of the percentages from Part e.

13. In Your Own Words Describe several ways percentages are used to compare different quantities.

14. Preview In the next chapter, you will extend the laws of exponents to negative exponents. Identify the missing numbers in the following sequences.

2^3	2^2	2^1	2^0	2^{-1}	2^{-2}	$2^?$	$2^?$
8	4	2	1	$\frac{1}{2}$	$\frac{1}{2^2}$	$\frac{1}{2^?}$	$\frac{1}{2^?}$

Mixed Review

15. A 24-oz box of corn flakes costs $2.99. A 32-oz box costs $4.19. Which costs the least per ounce? Why?

16. Janine wants to wallpaper a 6 meter by 9 meter room with 3-meter-high ceilings. The room has two 1.5-meter square windows and one entranceway, 80 cm wide by 2.4 meters high. How many square meters of wallpaper will she need, including a little extra for practice?

Write each linear equation in slope-intercept form, $y = mx + b$.

17. $\frac{2}{3}x - \frac{5}{8} = 2y$

18. $3y - 4x - 1 = 8x - 2y$

19. $4(x + y) - 6(5 - x) = 6$

20. $\frac{7(y + 5)}{x + 2} = 10$

21. Consider the line $y = -5x - 7$.

   a. A second line is parallel to this line. What do you know about the equation of the second line?

   b. Write an equation for the line parallel to $y = -5x - 7$ that passes through the origin.

   c. Write an equation for the line parallel to $y = -5x - 7$ that crosses the y-axis at the point $(0, -2)$.

   d. Write an equation for the line parallel to $y = -5x - 7$ that passes through the point $(3, 0)$.

Find each sum or product. Give your answers in lowest terms.

22. $\frac{1}{4} + \frac{2}{3}$

23. $\frac{5}{9} + \frac{7}{12}$

24. $\frac{5}{6} + \frac{8}{9}$

25. $\frac{1}{2} \cdot \frac{1}{3}$

26. $\frac{4}{7} \cdot \frac{5}{6}$

27. $\frac{3}{5} \cdot \frac{20}{9}$

Review & Self-Assessment

Chapter Summary

You estimated percents, and you combined difference with percentages when you worked with percent change. You also learned how to use the percent proportion to solve real-world situations involving ratios and percentages.

Strategies and Applications

The questions in this section will help you review and apply the important ideas and strategies developed in this chapter.

Estimating percents

1. A local college is supported by two fan clubs, the Best Fans in the Land fan club and the We Have Spirit fan club.

   A survey reported the following findings.

   Fan Club Survey Results

	Number of Members	Number Attending Games	Percent Viewing Televised Games
Best Fans in the Land	1,249	370	75%
We Have Spirit	1,521	600	90%

   a. About what percent of Best Fans in the Land fan club members attended football games?

   b. About how many We Have Spirit fan club members viewed televised football games?

   c. Explain how solving Part a was different from solving Part b.

Understanding percent change

2. In 1999, about 48.5 million people visited the United States. In 2000, there were 50.9 million visitors. In 2001, there were 45.5 million visitors.

   a. Describe two ways to find the percent change from 1999 to 2000. Explain how to tell whether the change is an increase or a decrease.

   b. Find the percent change from 1999 to 2000 and from 2000 to 2001. Be sure to specify if each change is an increase or a decrease.

Writing and interpreting comparisons

3. A brand of shampoo was sold in 12-ounce bottles and priced at $3.60. When the company switched to 15-ounce bottles, it kept the pricing the same to be more competitive. To get attention on the store shelf, the company printed "25% more FREE!" on the new bottles.

a. Explain what is meant by the statement "25% more FREE!"

b. Write as many statements as you can comparing the old size and price to the new size and price.

Calculating percents of percents

4. At Denim World, all jeans are on sale for 20% off the regular price of $45. The store is also running an unadvertised special on a few select pairs of jeans. These jeans are an additional 10% off the already discounted price.

a. What would you pay for a pair of jeans included in the unadvertised special?

b. Explain why an additional 10% off an item receiving an initial discount of 20% is not the same as a 30% savings.

Demonstrating Skills

Find the value of the variable in each proportion.

5. $\dfrac{12}{5} = \dfrac{x}{9}$

6. $\dfrac{3.2}{y} = \dfrac{4}{7}$

7. $\dfrac{92}{36} = \dfrac{23}{w}$

8. $\dfrac{a}{4.7} = \dfrac{13}{61.1}$

Set up and solve a proportion to answer each question.

9. 17 is 5% of what number?

10. What percent of 450 is 25?

11. What is 32% of 85?

In a recent survey of two new cold medicines given to 120 doctors, Cold Away received 45 votes and Sick Free received 75 votes.

12. Tell whether each of the following statements provide information accurately. For each accurate statement, explain or give a calculation to show why it is accurate.

a. Cold Away received $\dfrac{3}{8}$ of the votes.

b. The ratio of doctors who voted for Cold Away to the doctors who voted for Sick Free is 5:8.

c. Sick Free received 62.5% of the doctors' total votes.

13. Which *accurate* statement in Exercise 12 would the makers of Sick Free most likely want to use in an advertising campaign? Explain your reasoning.

14. Are the accurate statements from Exercise 12 part-to-whole comparisons, part-to-part comparisons, or a mixture of both? Explain your reasoning.

Use the following information to answer Questions 15–19.

Below are the win-loss records for two baseball teams.

	Win	Loss
Bearcats	56	14
Falcons	72	8

15. Find the ratio of wins to total number of games for the Bearcats and the Falcons.

16. Find the ratio of losses to the total number of games for each team.

17. What percent of games did the Falcons lose?

18. What percent of games did the Bearcats win?

19. Use ratios to compare each team's wins to losses.

The following items are set to be discounted by 50% on New Year's Day. Determine the best way to advertise each item using one of the three strategies listed below. Explain how the advertisement could be written.

- *Strategy A*: Express the discount as a dollar amount off the original price.
- *Strategy B*: Express the discount as a percent off of the original price.
- *Strategy C*: Express the discount by giving the old price and the new price.

20. $1,500 diamond earrings

21. $1,000 opal necklace

22. $12 cuff links

23. $500 ruby bracelet

Test-Taking Practice

SHORT RESPONSE

1 Daniel put his money into a savings account. He initially deposited $350. After 15 years, he had $533.75 in the bank. What was the percent increase in his savings?

Show your work.

Answer _____

MULTIPLE CHOICE

2 Alano leaves a $5.00 tip for his server after dinner. If he is leaving a 15% tip, find the amount of the dinner.
   A $28.33
   B $30.00
   C $33.33
   D $75.00

3 Solve the following proportion.
$$\frac{9}{21} = \frac{6}{p}$$
   F 14
   G 16
   H 18
   J 27

4 What percent of 780 is 273?
   A 0.35%
   B 29%
   C 35%
   D 65%

5 Of a group of 65 eighth graders, 20% of them said they had never been to summer camp. How many students said they had never been to summer camp?
   F 13 students
   G 15 students
   H 22 students
   J 52 students

6 A biologist was growing bacteria in a dish. At hour 1, there were 800 bacteria in the dish. At hour 2, there were 600 bacteria. What percent decrease occurred between hour 1 and hour 2?
   A 20%
   B 25%
   C 33%
   D 75%

7 What number is 135% of 120?
   F 42
   G 89
   H 162
   J 198

Exponents and Exponential Variation

Contents in Brief

Real-Life Math

Computer Viruses People who use e-mail often receive warnings about computer viruses from friends. Suppose one person sent an e-mail about a virus to ten friends. An hour later, those recipients sent the e-mail to ten friends, and so on. In about seven hours, 10,000,000 households with e-mail might have received the message. This is why some people say the virus warning *is* the virus.

In this chapter, you will learn that the spread of computer viruses, rumors, and other situations can often be modeled using an exponential relationship.

Think About It How many households do you think have e-mail in the U.S.? Do you think that after seven hours most of them would have received the e-mail virus?

Math Online
Take the **Chapter Readiness Quiz** at glencoe.com.

Dear Family,

The class is continuing its study of exponents, exponent laws, and scientific notation. Exponents can be thought of as a shortcut for expressing repeated multiplication. For example, 5^2 equals $5 \cdot 5$, or 25. We'll deepen our understanding of exponents by exploring exponential growth and decay.

Key Concept—Exponential Growth

In exponential relationships, the amount of change gets larger or smaller each time. Compound interest and population growth are examples of exponential growth.

Year	Interest Earned	Account Balance
1	$7.00	$107.00
2	$7.49	$114.49
3	$8.01	$122.50
4	$8.58	$131.08
5	$9.18	$140.26
6	$9.81	$150.07

Suppose you deposit $100 in an account earning 7% interest annually. The balance will grow exponentially. The table on the right shows that the earned interest amount increases from year to year, even though the interest percentage is the same. This growth can be modeled by the exponential equation $y = 100(1.07)^x$.

Chapter Vocabulary

exponential decay

exponential growth

nth root

radical sign

scientific notation

square root

Home Activities

- Notice real-world references to extremely large amounts, such as the national debt on a news report or in the newspaper. Ask your student to express these numbers in both standard and scientific notation.
- Discuss everyday occurrences of exponential growth and decay, such as compound interest and car depreciation.

LESSON 4.1

Exponents

In your mathematics studies in earlier years, you probably encountered exponents. In this lesson, you will take another look at exponents and how to work with expressions that involve them.

Explore

Consider this list of numbers.

$$2, 4, 8, 16, 32, 64, 128, \dots$$

What could be the next two numbers in this list?

The ellipsis "..." means the list continues in the same pattern.

How would you describe this list?

Choose any number in the list and double it. Is the result another number in the list? Would this be true for *any* number in the list? Why or not?

Choose two numbers in the list and multiply them. Is the product also a number in the list? Would this be true for *any* two numbers in the list? Why or why not?

Investigation 1 Positive Integer Exponents

Vocabulary

scientific notation

All the numbers in the list from Explore are powers of 2. That means they can all be found by multiplying 2 by itself a certain number of times. Recall that you can use positive integer *exponents* to show that a number, called the *base,* is multiplied by itself.

2^3 exponent →   base ↗

$$2^3 = 2 \cdot 2 \cdot 2$$

$$5^4 = 5 \cdot 5 \cdot 5 \cdot 5$$

$$120^1 = 120$$

✅ Develop & Understand: A

1. Look again at the list from Explore.

$$2, 4, 8, 16, 32, 64, 128, \dots$$

   a. Use an exponent to write 128 as a power of 2.

   b. Write an expression for the nth number in the list.

   c. Is 6,002 in this list? How do you know?

   d. Is 16,384 in this list? How do you know?

2. Shanequa started another list of numbers using positive integer powers with 4 as the base.

   a. What are the first ten numbers in Shanequa list?

   b. Is 2,048,296 in her list? How do you know?

   c. The number 262,144 is in her list. Is it also a number in the list in Exercise 1? How do you know?

   d. The numbers 4,096 and 8,192 are in the list in Exercise 1. Are they also in Shanequa's list?

   e. **Prove It!** Shanequa made the conjecture that every number in her list is also in the list in Exercise 1. Explain why her conjecture is true, or give a counterexample.

> **Math Link**
>
> A counterexample is an example for which a conjecture does not hold. A counterexample proves a conjecture wrong.

✅ Develop & Understand: B

Tell whether each statement is *sometimes true, always true,* or *never true* for positive integer values of n. If a statement is sometimes true, state for what values it is true.

3. $2^n = 2{,}048$

4. $4^n = 2{,}048$

5. 3^n is less than 1,000,000 (that is, $3^n < 1{,}000{,}000$)

6. 0.5^n is between 0 and 1 (that is, $0 < 0.5^n < 1$)

Use this poem to answer Exercises 7–11. Use powers to write your answers.

> *I pried the side from the crate;*
> *Inside, six drawers lay in wait.*
> *The drawers each held six yellow boxes*
> *Whose six faces each showed six foxes.*
> *The foxes each had six green eyes.*
> *Alien foxes? That's a surprise!*

7. How many drawers were in the crate?

8. How many boxes were in the crate?

9. What was the total number of faces on the boxes in the crate?

10. How many pictures of foxes were in the crate?

11. What was the total number of eyes on the foxes in the crate?

Math Link

The exponent laws also apply to negative integers.

$(-2)^2 = (-2)(-2) = 4$

$(-2)^3 = (-2)(-2)(-2)$
$\quad = -8$

$-2^2 = -(2^2) = -4$

Without computing the values of the numbers in each pair, determine which number is greater. Explain your answers.

12. 3^{18} or 3^{20} **13.** $(-2)^8$ or $(-2)^{19}$ **14.** -3^{500} or -3^{800}

15. Now consider powers of $\frac{1}{2}$.

   a. Without using your calculator, find three positive integer values of n that make $\left(\frac{1}{2}\right)^n$ less than $\left(\frac{1}{2}\right)^3$.

   b. Describe all the positive integers that would make $\left(\frac{1}{2}\right)^n$ less than $\left(\frac{1}{2}\right)^3$.

   c. How do you know your answer to Part b is correct?

16. Without computing, determine which number is greater, $\left(-\frac{1}{3}\right)^{34}$ or $\left(-\frac{1}{3}\right)^{510}$. Explain how you know your answer is correct.

You may recall that scientific notation makes use of powers of 10. A number is written in **scientific notation** when it is expressed as the product of a power of 10 and a number greater than or equal to 1 but less than 10. Two examples are shown below.

$$3{,}456 = 3.456 \times 10^3 \qquad 10{,}000{,}000 = 1 \times 10^7$$

✓ Develop & Understand: C

17. Copy the chart. Complete it without using your calculator.

Description	Number in Standard Form (approximate)	Number in Scientific Notation
Time since dinosaurs began roaming Earth (years)	225,000,000	
Projected World population in 2010		6.8×10^9
Distance from Earth to Andromeda galaxy (miles)		1.5×10^{19}
Mass of the sun (kg)	2,000,000,000,000,000,000,000,000,000,000	

For each pair of numbers, indicate which is greater.

18. 2.34×10^5 or 1.35×10^6

19. 3.83312×10^{31} or 8.1×10^{32}

Investigation 2 Negative Integer Exponents

In Investigation 1, you worked with exponents that were positive integers. You may remember from other mathematics courses that exponents can also be negative integers.

Think & Discuss

Consider this list of numbers, which continues in both directions.

$$\ldots, \frac{1}{27}, \frac{1}{9}, \frac{1}{3}, 1, 3, 9, 27, \ldots$$

What number could follow 27? What number could precede $\frac{1}{27}$?

Write 27, 9, and 3 using integer exponents and the same base.

Assume the pattern in the exponential forms that you just wrote continues with the other numbers in the above list. Use the pattern to write the exponential forms of 1, $\frac{1}{3}$, $\frac{1}{9}$, and $\frac{1}{27}$.

Think & Discuss may have reminded you how to use zero and negative integers as exponents.

Example

For any number a not equal to zero and any integer b,

$$a^0 = 1 \qquad \text{and} \qquad a^{-b} = \frac{1}{a^b} = \left(\frac{1}{a}\right)^b.$$

Some examples follow.

$$2^0 = 1 \qquad\qquad 8^{-3} = \left(\frac{1}{8}\right)^3 = \frac{1}{512}$$

$$0.25^0 = 1 \qquad\qquad (-5)^{-5} = \frac{1}{(-5)^5} = -\frac{1}{3{,}125}$$

$$\left(\frac{3}{7}\right)^{-2} = \frac{1}{\left(\frac{3}{7}\right)^2} = \frac{1}{\frac{9}{49}} = 1 \div \frac{9}{49} = 1 \cdot \frac{49}{9} = \frac{49}{9}$$

✅ Develop & Understand: A

For the following exercises, do not use your calculator.

1. Write each number without using an exponent.

   a. 4^{-1}

   b. -5^{-3}

   c. 1.43536326^0

2. Consider these numbers.

   $$3^2 \qquad 3^{-2} \qquad \left(\frac{1}{3}\right)^2 \qquad \frac{1}{3^2} \qquad \frac{1}{9} \qquad 9$$

   a. Sort the numbers into two groups so that all the numbers in each group are equal to one another.

   b. In which group does $\left(\frac{1}{3}\right)^{-2}$ belong?

3. Write each number without using an exponent.

   a. $\left(\frac{4}{5}\right)^{-2}$ **b.** 0.5^{-3}

4. Sort these numbers into two groups so that all the numbers in each group are equal to one another.

   $$\left(\frac{2}{3}\right)^2 \quad \left(\frac{2}{3}\right)^{-2} \quad \left(\frac{3}{2}\right)^2 \quad \left(\frac{3}{2}\right)^{-2} \quad \frac{3^2}{2^2} \quad \frac{9}{4} \quad \frac{4}{9}$$

5. Sort these numbers into four groups so that all the numbers in each group are equal to one another.

   $$10^3 \quad 10^{-3} \quad (-10)^{-3} \quad \frac{1}{(-10)^3} \quad \frac{1}{1,000} \quad -1,000 \quad \left(\frac{1}{10}\right)^{-3}$$

   $$\left(\frac{1}{10}\right)^3 \quad \left(-\frac{1}{10}\right)^{-3} \quad \frac{1}{10^3} \quad 1,000 \quad \left(-\frac{1}{10}\right)^3 \quad \frac{-1}{1,000} \quad (-10)^3$$

6. Which of these are equivalent to a^{-n}?

   $$\frac{1}{a^n} \qquad -a^n \qquad \left(\frac{1}{a}\right)^n \qquad 1 \div a^n$$

7. Which of these are equivalent to $\left(\frac{1}{a}\right)^{-n}$?

   $$\frac{1}{a^{-n}} \qquad a^n \qquad -\frac{1}{a}^n \qquad 1 \div a^n$$

8. Which of these are equivalent to $\left(\frac{a}{b}\right)^{-n}$?

   $$\frac{a}{b^n} \qquad \left(\frac{b}{a}\right)^n \qquad \frac{a^{-n}}{b^{-n}} \qquad \frac{b^n}{a^n}$$

9. **Challenge** Sort these numbers into two groups so that all the numbers in each group are equal to one another.

   $$\left(\frac{a}{b}\right)^3 \quad \left(\frac{b}{a}\right)^{-3} \quad \left(\frac{a}{b}\right)^{-3} \quad \left(\frac{b}{a}\right)^3 \quad \frac{b^3}{a^3} \quad b^3 \div a^3 \quad a^3 \div b^3$$

Math Link

A reciprocal is a multiplicative inverse. The numbers 3 and $\frac{1}{3}$ are reciprocals, as are $\frac{4}{5}$ and $\frac{5}{4}$. Dividing by a number is equivalent to multiplying by its reciprocal.

When a fraction is raised to a negative integer power, how can you find an equivalent fraction using a positive integer power? For example, how do you find an equivalent fraction for $\left(\frac{5}{7}\right)^{-3}$?

✅ Develop & Understand: B

Tell whether each statement is *sometimes true*, *always true*, or *never true* for integer values, positive, negative, or 0, of n. If a statement is sometimes true, state for what values it is true.

10. $2^n = \dfrac{1}{2,048}$

11. 3^n is less than 1 (that is, $3^n < 1$)

12. 5^n is between 0 and 1 or is equal to 0 or 1 (that is, $0 \leq 5^n \leq 1$)

Without computing the values of the numbers in each pair, determine which number is greater. Explain how you know your answer is correct.

13. 7^{-89} or 7^{-90}

14. 3^{-15} or 6^{-15}

15. 0.4^{-5} or 0.4^{-78}

16. -9^{-4} or -0.5^{-4}

17. $(-2)^{-280}$ or $(-2)^{-282}$

18. $(-50)^{-45}$ or $(-50)^{-51}$

19. 0.3^{-50} or 1.3^{-50}

As you know, you can use scientific notation with positive integer powers of 10 to express very large numbers. In the same way, you can use negative integer powers of 10 to express very small numbers. Two examples are shown below.

$$0.003456 = 3.456 \cdot 10^{-3}$$

$$0.0000001 = 1 \cdot 10^{-7}$$

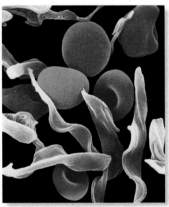

Real-World Link
These red blood cells have been invaded by the protozoa Trypanosoma, the cause of West African sleeping sickness. Red blood cells measure about 0.005 cm, or 5×10^{-3} cm.

☑ Develop & Understand: C

20. Complete the chart without using your calculator.

Description	Number in Standard Form (approximate)	Number in Scientific Notation
Average mass of a hydrogen atom (grams)	0.0000000000000000000000016735	
Diameter of the body of a Purkinje cell (meters)		8×10^{-5}
Diameter of some fats in the body (meters)		5×10^{-10}
Average mass of an oxygen atom (grams)	0.000000000000000000000026566	

For each pair of numbers, indicate which is greater.

21. 2.34×10^{-5} or 1.35×10^{-6}

22. 1.92×10^{-3} or 0.21×10^{-2}

23. 6.391×10^{-9} or 7.814×10^{-10}

24. 5.1×10^{-19} or 6.92×10^{-20}

25. 9.384×10^{-23} or 7.6×10^{-24}

26. 3.83312×10^{-31} or 8.1×10^{-32}

Share & Summarize

1. Explain what a^{-b} means, assuming b is a positive integer.

2. How would you decide whether a^{-5} is greater than or less than a^{-7}? Assume a is positive and not equal to 1.

Investigation ③ Laws of Exponents

You may recall from earlier mathematics classes that the following *exponent laws* can make calculations with exponents much simpler. In these laws, the bases a and b cannot be zero if they are in a denominator or if they are raised to a negative exponent or to zero.

Product Laws

$$a^b \cdot a^c = a^{b+c}$$

$$a^c \cdot b^c = (ab)^c$$

Quotient Laws

$$\frac{a^b}{a^c} = a^{b-c}$$

$$\frac{a^c}{b^c} = \left(\frac{a}{b}\right)^c$$

Power of a Power Law

$$(a^b)^c = a^{bc}$$

Example

Ben explains how he remembers the first product law.

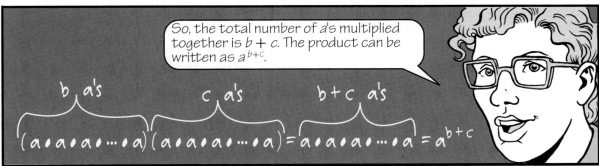

In Exercises 1 and 2, think about ways to explain why some of the other exponent laws are true. Although these laws are true for all integer values of the exponents, focus on either positive or negative integers to make your job easier. Use Ben's line of reasoning as a guide.

✅ Develop & Understand: A

1. **Prove It!** In this exercise, you will show that the first quotient law, $\frac{a^b}{a^c} = a^{b-c}$, works for positive integer exponents. Assume $a \neq 0$.

   a. First, show that the law works when $b > c$.

   b. Now, show that the law works when $b < c$.

2. **Prove It!** Show that the second product law, $a^c \cdot b^c = (ab)^c$, works for negative integer exponents. Hint: Let $c = -x$ for a positive integer x.

✅ Develop & Understand: B

3. Copy this multiplication chart. Without using your calculator, find the missing expressions. Write all entries as powers.

×	3^{-2}	3^x	3^4
3^4	3^2	3^{x+4}	3^8
3^a			
-3^2			

4. Copy this division chart. Without using your calculator, find the missing expressions by dividing the expressions in each row by the expressions in each column. For example, the first beige cell represents the quotient $2a^4 \div a^5$, or $\frac{2a^4}{a^5}$, which is equivalent to $2a^{-1}$. Write all entries as powers or products of powers.

÷	a^5	a^{-2}	$(a^3)^2$
$2a^4$	$2a^{-1}$	$2a^6$	$2a^{-2}$
a^{-3}			
$(2a)^5$			

5. Copy this chart. Without using your calculator, fill in the missing expressions. Write all entries as powers or products of powers.

×	?	2a	?	?
b^{-4}	b^4	$2ab^{-4}$		
a^8			$a^{10}b^{-4}$	a^8b^{-8}
?			b	
?		$(2ab)^4$		

✓ Develop & Understand: C

For the following exercises, do not use your calculator.

6. Rewrite each expression using a single base.

   a. $(a^{-m})^0$ (Assume a ≠ 0.)

   b. $[(-d)^3]^4$

   c. Challenge $(-10^{-4})^{-5}$

7. Rewrite each expression using a single base.

   a. $(2^3 \cdot 2)^2$

   b. $(a^m)^n \div (a^{-m})^n$

   c. $4^3 \cdot n^3 \div (-16n)^3$

8. Find at least two ways to write each expression as a product of two expressions.

   a. $32n^{10}$ **b.** m^7b^{-7}

Share & Summarize

Without using a calculator, rewrite this expression as simply as you can. Show each step of your work. Record which exponent law, if any, you used for each step.

$$\frac{2^6n^3}{(16n^2)^3}$$

Investigation 4 · Exponent Laws and Scientific Notation

Lucita and Tala were discussing how to multiply 4.1×10^4 by 3×10^6.

We can start by rearranging things a little.

4.1 times 3 is 12.3, and we can use a product rule for the powers of 10.

$(4.1 \times 10^4)(3 \times 10^6) = (4.1 \times 3)(10^4 \times 10^6)$
$= 12.3 \times 10^{4+6}$
$= 12.3 \times 10^{10}$

But the answer's not in scientific notation. The first number is greater than 10.

$10^4)(3 \times 10^6) = (4.1 \times 3)(10^4 \times 10^6)$
$= 12.3 \times 10^{4+6}$
$= 12.3 \times 10^{10}$

Rewrite it as 1.23×10^{11}.

Think & Discuss

How would you divide two numbers written in scientific notation? For example, what is $(2.12 \times 10^{14}) \div (5.3 \times 10^6)$? Write your answer in scientific notation.

✅ Develop & Understand: A

For the following exercises, do not use your calculator unless otherwise indicated.

1. There are about 4×10^{11} stars in our galaxy and about 10^{11} galaxies in the observable universe.

   a. Suppose every galaxy has as many stars as ours. How many stars are there in the observable universe? Show how you found your answer.

   b. Suppose only 1 in every 1,000 stars in the observable universe has a planetary system. How many planetary systems are there? Show how you found your answer.

   c. Suppose 1 in every 1,000 of those planetary systems has at least one planet with conditions suitable for life as we know it. How many such systems are there? Show how you found your answer.

   d. At the end of the 20th century, the world population was estimated at about 6 billion people. Compare this number to your answer in Part c. What does your answer mean in terms of the situation?

Math Link
You may recall that 1 billion is 1,000,000,000. What is its equivalent in scientific notation?

Real-World Link
There are 88 recognized constellations, or groupings of stars, that form easily identified patterns. This is the constellation Orion, "the Hunter."

2. *Escherichia coli* is a type of bacterium that is sometimes found in swimming pools. Each *E. coli* bacterium has a mass of 2×10^{-12} gram. The number of bacteria increase so that after 30 hours, one bacterium has been replaced by a population of 4.8×10^8 bacteria.

  a. Suppose a pool begins with a population of only one bacterium. What would be the mass of the population after 30 hours?

  b. A small paper clip has a mass of about 1 gram. How does the mass of the paper clip compare to the mass of the 4.8×10^8 *E. coli* bacteria? Show how you found your answer.

3. The speed of light is about 2×10^5 miles per second.

  a. On average, it takes light about 500 seconds to travel from the Sun to Earth. What is the average distance from Earth to the Sun? Write your answer in scientific notation.

  b. The star Alpha Centauri is approximately 2.5×10^{13} miles from Earth. How many seconds does it take light to travel between Alpha Centauri and Earth?

  c. Use your answer to Part b to estimate how many years it takes for light to travel between Alpha Centauri and Earth. You may use your calculator.

4. These data show current estimates of the energy released by the three largest earthquakes recorded on Earth. The *joule* is a unit for measuring energy, named after British physicist James Prescott Joule.

Largest Recorded Earthquakes

Location	Date	Energy Released (joules)
Tambora, Indonesia	April 1815	8×10^{19}
Santorini, Greece	about 1470 B.C.	3×10^{19}
Krakatoa, Indonesia	August 1883	6×10^{18}

  a. The Santorini earthquake was how many times as powerful as the Krakatoa earthquake?

  b. The Tambora earthquake was how many times as powerful as the Krakatoa earthquake?

5. A scientist is growing a culture of cells. The culture currently contains 2×10^{12} cells.

    a. The number of cells doubles every day. If the scientist does not use any of the cells for an experiment today, how many cells will she have tomorrow?

    b. Suppose she uses 2×10^9 of the 2×10^{12} cells for an experiment. How many will she have left? Show how you found your answer. Be careful. This exercise is different from the others you have done.

✅ Develop & Understand: B

Copy these multiplication and division charts. Without using your calculator, find the missing expressions. Write all entries using scientific notation. For the division chart, divide the row label by the column label.

6.

×	4×10^{28}	?
-2×10^{12}		4×10^5
6×10^{-20}		
8×10^a		

7.

÷	2×10^6	?
-4×10^{12}		2×10^5
8×10^{-10}		
8×10^a		

Share & Summarize

Jordan has written his calculations for three expressions involving scientific notation. Check his work on each exercise. If his work is correct, write "correct." If it is incorrect, write a note explaining his mistake and how to correctly solve the exercise.

1. $(2 \times 10^5) \cdot 2 = 2 \cdot 2 \times 10^5 = 4 \times 10^5$

2. $(6 \times 10^{-5}) \cdot (2 \times 10^{-7}) = 6 \cdot 2 \times 10^{-5} \cdot 10^{-7} = 1.2 \times 10^{-12}$

3. $(3 \times 10^{12}) - (3 \times 10^{10}) = 3 \times 10^{12-10} = 3 \times 10^2$

Inquiry

Investigation **5** Model the Solar System

Materials

- masking tape
- ruler

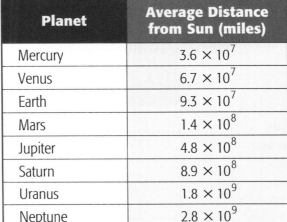

Real-World Link

For seventy-five years, Pluto was classified as a planet. However, in 2006, the International Astronomical Union revised the definition of *planet*. Pluto was reclassified as a *dwarf planet*.

In this investigation, you will examine the relative distances between objects in the solar system.

Make a Prediction

The average distances of planets from the Sun are often written in scientific notation.

Planet	Average Distance from Sun (miles)
Mercury	3.6×10^7
Venus	6.7×10^7
Earth	9.3×10^7
Mars	1.4×10^8
Jupiter	4.8×10^8
Saturn	8.9×10^8
Uranus	1.8×10^9
Neptune	2.8×10^9

Imagine lining up the planets with the Sun on one end and Neptune on the other so that each planet is at its average distance from the Sun. How would the planets be spaced? Here is one student's prediction.

1. Now make your own prediction. Without using your calculator, sketch a scale version of the planets lined up in a straight line from the sun. Do not worry about the sizes of the planets.

Create a Model

To check your prediction, some of the members of your class will represent parts of the solar system in a large-scale model. The scale model will allow you to compare the planets' average distances from the sun. However, it will not model the relative sizes of the planets.

In a large space, use masking tape to mark a line along which you will make your model. The sun will be at one end of the line and Neptune will be at the other. As a class, answer the following questions to determine the locations of the planets in your model.

Go on ▶

2. Measure the length of the line. This will be the distance between the sun and Neptune in your model.

3. What is the actual distance between the sun and Neptune in miles?

4. What number would you multiply distances, in feet, in your model by to find the actual distance, in miles?

5. How can you calculate the distances of the planets from the Sun in your model?

6. Use your answer to Question 5 to estimate the distance between each planet and the Sun in your scale model. Copy the table, fill in the proper unit in the last column, and then record the scaled distances.

Planet	Average Distance from Sun (miles)	Average Distance from Sun in Scale Model (___?___)
Mercury	3.6×10^7	
Venus	6.7×10^7	
Earth	9.3×10^7	
Mars	1.4×10^8	
Jupiter	4.8×10^8	
Saturn	8.9×10^8	
Uranus	1.8×10^9	
Neptune	2.8×10^9	

Your teacher will assign your group to a planet or the Sun. Your group should create a sign stating the name of your planet or the Sun and its average distance from the Sun, in miles, in scientific notation.

Decide as a class at which end of the line the Sun will be. Determine where your celestial body belongs in the scale model. Choose one member of your group to represent your celestial body by standing on the line with the sign.

Check Your Prediction

Answer the following questions as a class.

7. Which two planets are closest together? What is the actual distance between them?

8. Which two adjacent planets are farthest apart? *Adjacent* means next to each other. What is the actual distance between them?

9. Compare the two distances in Questions 7 and 8. How many times farther apart are the planets in Question 8 than the planets in Question 7?

10. Did your sketch give a reasonably accurate picture of the distances?

11. Was there anything about the spacing of the planets that surprised you? If so, what?

12. Was there anything about the spacing of the planets that did not surprise you? If so, what?

What Did You Learn?

13. Draw a scale version of the planets that shows the relative distances between the planets. Do not worry about representing the sizes of the planets, but do your best to get the distances between planets correct.

14. Below are three number lines marked with numbers in scientific notation. Which number line has numbers in the correct places?

a.

5×10^7

$0 \quad 1 \times 10^1 \quad 1 \times 10^2 \quad 1 \times 10^3 \quad 1 \times 10^4 \quad 1 \times 10^5 \quad 1 \times 10^6 \quad 1 \times 10^7 \quad 1 \times 10^8$

b.

1×10^6

$0 \quad 1 \times 10^7 \quad 2 \times 10^7 \quad 3 \times 10^7 \quad 4 \times 10^7 \quad 5 \times 10^7 \quad 6 \times 10^7 \quad 7 \times 10^7 \quad 8 \times 10^7 \quad 9 \times 10^7 \quad 1 \times 10^8$

c.

$1 \times 10^7 \quad 2 \times 10^7 \quad 3 \times 10^7 \quad 4 \times 10^7 \quad 5 \times 10^7 \quad 6 \times 10^7$

$0 \qquad\qquad\qquad 7 \times 10^7 \quad 8 \times 10^7 \quad\quad 9 \times 10^7 \qquad\quad 1 \times 10^8$

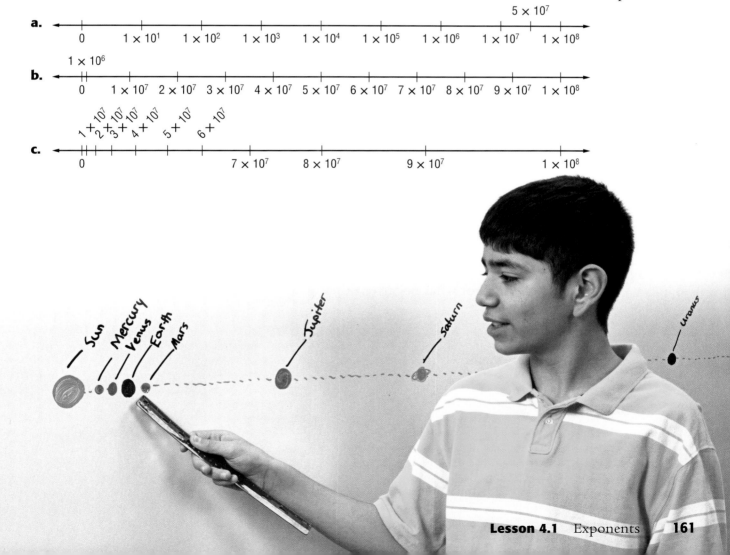

On Your Own Exercises
Lesson 4.1

Practice & Apply

1. **Social Studies** One of these numbers is in standard notation, and one is in scientific notation. One is the world population in 1750, and the other is the world population in 1950.

$$2.56 \times 10^9 \qquad 725{,}000{,}000$$

Which number do you think is the world population in 1750? In 1950? Explain your reasoning.

2. For what values of n, if any, will n^2 be equal to or less than 0?

3. For what values of n, if any, will n^3 be equal to or less than 0?

Given that n represents a positive integer, decide whether each statement is *sometimes true*, *always true*, or *never true*. In Exercises 4–7, if a statement is sometimes true, state for what values it is true.

4. $4^n = 65{,}536$

5. 4^n is less than 1,000,000 (that is, $4^n < 1{,}000{,}000$)

6. n^2 is negative (that is, $n^2 < 0$)

7. 0.9^n is greater than or equal to 0 and, at the same time, 0.9^n is less than or equal to 1. That is, $0 \le 0.9^n \le 1$.

8. For what positive values of x will x^{20} be greater than x^{18}?

9. For what positive values of x will x^{18} be greater than x^{20}?

10. For what positive values of x will x^{18} be equal to x^{20}?

11. For what negative values of x will x^{20} be greater than x^{18}?

12. For what negative values of x will x^{18} be greater than x^{20}?

13. For what negative values of x will x^{18} be equal to x^{20}?

Real-World Link

In 2006, the United States Census Bureau estimated the population of New York City as 8.2×10^6 residents. If each resident produces four pounds of trash per day, that is about 3.28×10^7 pounds of garbage every year.

14. **Challenge** In Investigation 1, you explored positive integer powers of 2 and of 4.

n	1	2	3	4	5	6	7	8	9
2^n	2	4	8	16	32	64	128	256	512
4^n	4	16	64	256	1,024	4,096	16,384	65,536	262,144

Now think about positive integer powers of 8

a. List the first five positive integer powers of 8.

b. Name three numbers that are on all three lists, that is, three numbers that are powers of 2, 4, and 8.

c. List three numbers greater than 16 that are powers of 2 but are not powers of 8.

d. List three numbers greater than 16 that are powers of 4 but are not powers of 8.

e. Describe the powers of 2 that are also powers of 8.

f. Describe the powers of 4 that are also powers of 8.

15. For what positive values of x will x^{-20} be greater than x^{-18}?

16. For what positive values of x will x^{-18} be greater than x^{-20}?

17. For what values, positive or negative, of x will x^{-18} be equal to x^{-20}?

18. The sixth power of 2 is 64, or $2^6 = 64$.

a. Write at least five other expressions, using a single base and a single exponent, that are equivalent to 64.

b. Write the number 64 using scientific notation.

Sort each set of expressions into groups so that the expressions in each group are equal to one another. Do not use your calculator.

19. m^2 $\left(\frac{1}{m}\right)^2$ m^{-2} $\left(\frac{1}{m}\right)^{-2}$ $\frac{1}{m^2}$ $1 \div m^2$

20. 3^x $\left(\frac{1}{3}\right)^x$ $\left(\frac{1}{3}\right)^{-x}$ $\frac{1}{3^x}$ 3^{-x} $1 \div 3^x$

21. **Prove It!** Prove that the second quotient law, $\frac{a^c}{b^c} = \left(\frac{a}{b}\right)^c$, works for positive integer exponents c. Assume b is not equal to 0.

22. Challenge Prove that the power of a power law, $(a^b)^c = a^{bc}$, works for positive integer exponents b and c.

Rewrite each expression using a single base and a single exponent.

23. $2^7 \cdot 2^{-4} \cdot 2^x$ **24.** $(-4^m)^6$ **25.** $m^7 \cdot 28^7$

26. $(-3)^{81} \cdot (-3)^{141}$ **27.** $\dfrac{55^{-8}}{9^{-8}}$ **28.** $\left(\dfrac{m^{84}}{m^{12}}\right)^x$

29. $3^{-5} \cdot 8^5$ **30.** $n^a \div n^{\frac{a}{3}}$ **31.** $(22^2 \cdot 22^5)^0$

Simplify each expression.

32. $4a^4 \cdot 3a^3$ **33.** $m^{-3} \cdot m^4 \cdot b^7$ **34.** $\dfrac{10n^{-15}}{5n^5}$

35. $(4x^{-2})^6$ **36.** $(-m^2 n^3)^4$ **37.** $(a^m)^n \cdot (b^3)^2$

38. $(x^{-2})^3 \cdot x^5$ **39.** $\dfrac{12b^5}{4b^{-2}}$ **40.** $\dfrac{(x^4 y^{-5})^{-3}}{(xy)^2}$

Copy each chart. Without using your calculator, find the missing expressions. Write all entries as powers or products of powers. For the division chart, divide the row label by the column label.

41.

×	2^{10}	2^{-x}	-2^x	?
-2^{-3}				
?		2^{a-x}		$(2n)^a$
2^{2a}				

42.

÷	4^{-2}	4^x	-4^x	n^7
-4^7				
4^a				
4^7				

Real-World Link
Sirius, also called the Dog Star, is a double star orbited by a smaller star called Sirius B, or the Pup.

43. Physical Science The speed of light is about 2×10^5 miles per second. At approximately 5×10^{13} miles from Earth, Sirius appears to be the brightest of the stars. How many seconds does it take light to travel between Sirius and Earth? How many years does it take?

44. Social Studies The population of the world in the year 1 A.D. has been estimated at 200,000,000. By 1850, this estimate had grown to 1 billion. By 2000, the population was close to 6×10^9.

a. The 1850 population was how many times the 1 A.D. population?

b. The 2000 population was how many times the 1850 population?

c. Did the world population grow more during the 1,850 years from 1 A.D. to 1850 or during the 150 years from 1850 to 2000?

45. Copy this division chart. Without using your calculator, find the missing expressions by dividing the row label by the column label. Express all entries in scientific notation.

÷	?	?	3×10^{x}
3×10^{-20}	6×10^{-29}		
6×10^{14}		3×10^{134}	
?			$5 \times 10^{a-x-1}$

Connect & Extend

46. Social Studies According to the 1790 census, the population of the United States was 3,929,214. You can approximate this value with powers of various numbers. For example, 2^{21} is 2,097,152 and 2^{22} is 4,194,304. Using powers of 2, the number 2^{22} is the closest possible approximation to 3,929,214.

What is the closest possible approximation using powers of 3? Powers of 4? Powers of 5?

47. Which of these sets of numbers share numbers with the powers of 2? Explain how you know.

   a. positive integer powers of 6

   b. positive integer powers of 7

   c. positive integer powers of 16

48. Fine Arts A piano has eight C keys. The *frequency* of a note determines how high or low it sounds. Moving from the left of the keyboard to the right, each C note has twice the frequency of the one before it. For example, "middle C" has a frequency of about 261.63 vibrations per second. The next higher C has a frequency of about 523.25 vibrations per second.

If the first C key has a frequency of x, what is the frequency of the last C key?

49. Economics Julián's mother offered him $50 a month in allowance. Julián said he would rather have his mother pay him 1 penny the first day of the month, 2 pennies the second day, 4 the third day, 8 the fourth day, and so on. His mother would simply double the number of pennies she gave him each day until the end of the month. His mother said that sounded fine with her.

  a. Would Julián receive more money with an allowance of $50 a month or using his plan? Explain why.

  b. If Julián's plan produces more money, on what day would he receive more than $50 a month?

  c. With his plan, how much money would Julián receive the last day of June, which has 30 days?

  d. Challenge With his plan, how much would Julián receive in all for the month of June? Filling in a table like the one below might help you answer this question.

Day	Amount Received Each Day	Total Amount
1	$0.01	$0.01
2	$0.02	$0.01 + 0.02 = $0.03
3	$0.04	$0.03 + 0.04 = $0.07
4		

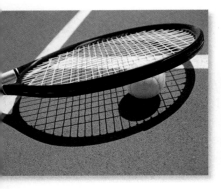

50. A particular tennis tournament begins with 64 players. If a player loses a single match, he or she is knocked out of the tournament. After one round, only 32 players remain. After two rounds, only 16 players remain, and so on.

Six students have conjectured a formula to describe the number of players remaining p after r rounds. Which rule or rules are correct? For each rule you think is correct, show how you know.

- Terrill: $p = \dfrac{64}{2^r}$

- Peter: $p = 64 \cdot \left(\dfrac{1}{2}\right)^r$

- Mi-Yung: $p = 64 \cdot 2^{-r}$

- Damon: $p = 64 \cdot 0.5^r$

- Antonia: $p = 64 \cdot \dfrac{1}{2^r}$

- Tamera: $p = 64 \cdot (-2)^r$

51. This list of numbers continues in the same pattern in both directions.

$$\ldots, \frac{1}{5}, 1, 5, 25, 125, 625, \ldots$$

Ivana wanted to write an expression for this list using n as a variable. To do that, she had to choose a number on the list to be her "starting" point. She decided that when $n = 1$, the number on the list is 5. When $n = 2$, the number is 25.

a. Using Ivana's plan, write an expression that will give any number on the list.

b. What value for n gives you 625? 1? $\frac{1}{5}$?

Without computing the value of each pair of numbers, determine which number is greater. For each exercise, explain why.

52. 2^{80} or 4^{42} **53.** $3^{-1,600}$ or 27^{-500} **54.** 12^{20} or 4^{45}

55. A pastry shop sells a square cake that is 45 cm wide and 10 cm thick. A competitor offers a square cake of the same thickness that is 2 cm wider. The first baker argues that the area of the top of the rival cake is $(45 + 2)^2$ cm^2 and is therefore only 4 cm^2 larger than the one he sells.

What computational mistake did the first baker make? What is the actual difference in areas?

56. Astronomy Earth travels around the sun approximately 6×10^8 miles each year. At approximately what speed must Earth travel in miles per second? Give your answer in scientific notation.

57. Life Science The diameter of the body of a Purkinje cell is 8×10^{-5} m.

a. If a microscope magnifies 1,000 times, what will be the scaled diameter, in meters, as viewed in the microscope?

b. What is the scaled diameter, in centimeters, as viewed in the microscope?

58. In Your Own Words Write a letter to a student who is confused about exponents. Explain how to multiply two numbers when each is written as a base to an exponent. Be sure to address the following.

• The numbers have the same exponent.

• The numbers have the same base.

• The numbers have different exponents and different bases.

Mixed Review

59. Measurement How many days is 1,000,000 seconds?

60. How many years is 1,000,000 hours?

61. How many decades is 3,500 years?

62. Without graphing, decide which of these equations represent parallel lines. Assume that q is on the horizontal axis. Explain.

a. $2p = 3q + 5$

b. $p = 3q^2 + 5$

c. $p = 1.5q - 7.1$

d. $p = 3q + 3$

Determine whether the three points in each set are collinear.

63. $(1, -1.6), (5, 0), (6, 0.4)$

64. $(7, -4), (3, -1), (-2, 2.75)$

Each table represents a linear relationship. Write an equation to represent each relationship.

65.

a	-4	-3	-2	-1	0	1
b	-8.8	-6.6	-4.4	-2.2	0	2.2

66.

c	-4	-3	-2	-1	0	1
d	-0.5	0.5	1.5	2.5	3.5	4.5

67.

e	-4	-3	-2	-1	0	1
f	-15	-13.75	-12.5	-11.25	-10	-8.75

LESSON 4.2

Exponential Relationships

Have you ever heard someone say, "It's growing exponentially"? People use this expression to describe things that grow very rapidly.

In this lesson, you will develop a more precise meaning of *exponential growth*. You will also investigate what it means to decrease, or decay, exponentially.

Explore

Suppose you make up a funny joke one evening. The next day, you tell it to two classmates. The day after that, your two classmates each tell it to two people. Everyone who hears the joke tells two new people the next day.

On the fourth day after you made it up, how many new people will hear the joke?

On which day will more than 1,000 new people hear the joke?

If the joke was told only to people in your school, how many days would it take for everyone in your school to hear it?

The number of students who hear the joke for the first time doubles each day. *Repeated doubling* is one type of exponential growth. In the next two investigations, you will look at other situations in which things grow exponentially.

Investigation 1 Compare Exponents

The expressions 2^x, $2x$, and x^2 look similar. They all include the number 2 and the variable x, but they have very different meanings. The same is true of 3^x, $3x$, and x^3. Looking at tables and graphs for these expressions will help you discover just how different they are.

✅ Develop & Understand: A

Real-World Link

The fruit fly is one of the world's most studied animals. Because it reproduces so rapidly, it is ideal for many scientific experiments.

Mr. Brooks brought two fruit flies to his biology class. He told his students that the flies would reproduce and that the population of flies would double every week. Lois and Roberto wanted to figure out how many flies there would be each week.

Lois made this table.

Week	1	2	3	4	5
Flies	2	4	6	8	10

Roberto made this table.

Week	1	2	3	4	5
Flies	2	4	8	16	32

1. Whose table is correct? That is, whose table shows the fruit fly population doubling every week?

2. How do the fly-population values in the incorrect table change each week?

3. Complete this table to compare the expressions $2x$, 2^x, and x^2.

x	0	1	2	3	4	5
2x	0	2				
2^x	1	2				
x^2	0	1				

4. As x increases from 0 to 5, which expression's values increase most quickly?

5. Which expression describes Lois' table? Which describes Roberto's table?

6. How many flies does the expression for Lois' table predict there will be in week 10? How many flies will there actually be, assuming all of the flies live?

7. Use your table from Exercise 3 to plot the points for each set of ordered pairs, $(x, 2x)$, $(x, 2^x)$, and (x, x^2), on one set of axes. If you can, use a different color for each set of points to help you remember which is which. How are the graphs similar? How are they different?

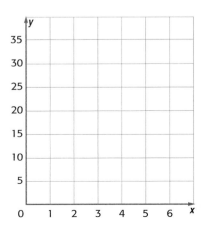

✅ *Develop & Understand: B*

Rebecca and India made tables to show the relationship between the edge length and the volume of a cube.

Math Link

The formula for the volume of a cube with edge length x is $V = x^3$.

Rebecca made this table.

Edge Length	1	2	3	4	5
Volume	3	9	27	81	243

India made this table.

Edge Length	1	2	3	4	5
Volume	1	8	27	64	125

8. Whose table shows the correct volumes?

9. Complete this table to compare the expressions $3x$, 3^x, and x^3.

x	0	1	2	3	4	5
3x	0					
3ˣ	1					
x³	0					

10. As x increases from 0 to 5, which expression's values increase most quickly?

11. Which expression describes Rebecca's table? Which describes India's table?

12. Look at the values in the table for 3^x. Describe in words how to get from one value to the next.

13. Look at the values in the table for $3x$. Describe in words how to get from one value to the next.

Share & Summarize

1. Compare the way the values of $2x$ change as x increases to the way the values of 2^x change as x increases.

2. Consider the expressions $4x$, 4^x, and x^4. Without making a table, predict which expression's values grow the fastest as x increases.

Investigation ② Exponential Growth and Decay

Vocabulary

exponential decay

exponential decrease

exponential increase

exponential growth

Suppose you buy four comic books each week. The table shows how your comic book collection would grow.

Weeks	1	2	3	4	5	6
Comics	4	8	12	16	20	24

Since you add the same number of comics each week, your collection grows at a *constant rate*. You add 4 comics each week, so the number of comics in your collection after x weeks can be expressed as $4x$.

Now consider the fruit fly population from Exercises 1–7 in Investigation 1.

Weeks	1	2	3	4	5	6
Flies	2	4	8	16	32	64

The fly population grows at an *increasing rate*. That is, the number of flies added is greater each week. In fact, the total number of flies is multiplied by 2 each week.

Quantities that are repeatedly multiplied by a number greater than 1 are said to grow exponentially, or to show **exponential increase**, or **exponential growth**. The number of flies after x weeks can be expressed as 2^x.

Think & Discuss

Look back at your tables from Investigation 1.

The values of $2x$ and $3x$ grow at a constant rate. How do the values of $2x$ and $3x$ change each time x increases by 1? How is the change related to the expression?

The values of the expression 3^x grow exponentially. How do the values of 3^x change each time x increases by 1? How is the change related to the expression?

✓ Develop & Understand: A

In an ancient legend, a ruler offers one of his subjects a reward in return for a favor. For the reward, the subject requests that the ruler place 2 grains of rice on the first square of a chessboard, 4 grains on the second square, 8 on the third, and so on. Each time, the number of grains of rice on each square are doubled.

The ruler thinks this would not take much rice, so he agrees to the request.

1. Copy and complete the table to show the number of grains of rice on the first six squares of the chessboard.

Square	1	2	3	4	5	6
Grains of Rice	2	4	8			

2. Which expression describes the number of grains on square x?

$$x^2 \qquad\qquad 2^x \qquad\qquad 2x$$

3. Use the expression to find the number of grains on square 7, square 10, and square 20.

4. For one type of uncooked, long-grain rice, 250 grains of rice have a volume of about 5 cm^3. What would be the volume of rice on square 5? On square 7? On square 10?

5. When the ruler saw there were about 20 cm^3 of rice on square 10, he assumed there would be 40 cm^2 on square 20. Is he correct? How many cubic centimeters of rice will actually be on square 20?

6. The number of grains of rice on the last square is 2^{64}. Use your calculator to find this value. What do you think your calculator's answer means?

The value of 2^{64} is too large for calculators to display normally. So, calculators use scientific notation.

In Exercises 1–6, the number of grains of rice on each square grows exponentially. It is multiplied by 2 each time. In Exercises 7–14, you will look at a different kind of exponential change.

☑ Develop & Understand: B

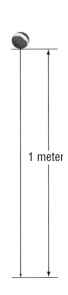

1 meter

Imagine that a superball is dropped onto concrete from a height of 1 meter and is allowed to bounce repeatedly. Suppose the height of each bounce is 0.8 times the height of the previous bounce. To understand how the height changes, you can imagine sending a piece of putty through a ×0.8 machine repeatedly.

Math Link

1 m = 100 cm

7. How high does the ball rise on the first bounce?

8. How high does the ball rise on the second bounce?

9. How high does the ball rise on the third bounce?

10. Complete the table to show how the ball's height in centimeters changes with each bounce.

Bouncing a Superball

Initial Height	100	100
1st Bounce Height	100×0.8	100×0.8^1
2nd Bounce Height	$100 \times 0.8 \times 0.8$	100×0.8^2
3rd Bounce Height		
4th Bounce Height		
10th Bounce Height		

11. How high does the ball rise on the 10th bounce?

12. How many times do you think the ball will bounce before coming to rest?

13. Use the pattern in the right column of your table to write an expression for the ball's height on the nth bounce.

14. Assuming the ball continues to bounce, use your expression to find the height of the ball on the 35th bounce. Does your answer surprise you?

In both the bouncing-ball and the rice exercises, a quantity is repeatedly multiplied by the same factor. For the ball exercise, however, the factor is 0.8, which is between 0 and 1. This means the bounce height gets smaller with each bounce. This type of exponential change is called **exponential decrease**, or **exponential decay**.

Share & Summarize

1. Briefly explain why the amount of rice on each square of the chessboard grows so rapidly.

2. As n increases, what happens to the value of $\left(\frac{1}{5}\right)^n$?

3. As n increases, what happens to the value of 5^n?

Investigation 3 Identify Exponential Growth and Decay

Vocabulary

decay factor

growth factor

Materials

• graphing calculator

The kingdom of Tonga is a group of more than 150 islands in the Pacific Ocean. In 2006, the population of Tonga was estimated to be 101,200 people. Suppose the population has been increasing by about 2% each year since 2006.

Think & Discuss

Make a table that shows the estimated population of Tonga for the five years after 2006.

Can you find an equation that shows what the population p would be n years after 2006?

Years after 2006	Estimated Population
0	101,200
1	
2	
3	
4	
5	

In your equation, 1.02 is called the **growth factor**.

Recall the bouncing superball in Investigation 2. The equation for the ball's height h after the nth bounce was $h = 100 \cdot 0.8^n$. Since this equation represents exponential decay, this 0.8 is called the **decay factor**.

✅ *Develop & Understand: A*

1. For each equation, state whether it represents an exponential relationship. If it does, tell whether that relationship involves growth or decay.

   a. $y = 100 \cdot 3^x$

   b. $y = 100 \cdot 0.3^x$

   c. $y = 32 \cdot x$

   d. $y = 20 \cdot \left(\frac{1}{4}\right)^x$

   e. $y = 0.3 \cdot 1.5^x$

   f. $y = 3x^2$

2. For each table, state whether the relationship described could be exponential. If it could be, tell whether it would involve growth or decay. Explain your reasoning.

   a.

x	y
1	6
2	9
3	12
4	15

   b.

x	y
1	6
2	9
3	13.5
4	20.25

   c.

x	y
111	56
112	28
113	14
114	7

3. Consider the following graphs.

   i.

   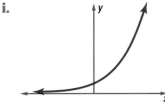

   ii.

   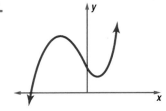

   iii.

   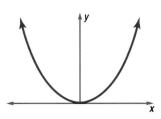

   iv.

   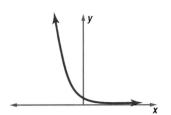

   a. Which of these graphs could *not* describe an exponential relationship? Explain.

   b. Assume that the graphs that *might* describe an exponential relationship actually *do*. For each, would the relationship involve growth or decay?

4. A scientist is growing a culture of amoebas. Amoebas reproduce by splitting in half. One amoeba becomes two. Two amoebas become four, and so on. Suppose that each amoeba in this culture splits twice in a day so that after one day, a single amoeba has become four amoebas. Assume that all of the amoebas live.

Is the relationship between the number of days since the scientist began growing the culture and the number of amoebas exponential? If so, what is the growth factor? If not, explain how you know.

5. Maria is hiking on the Appalachian Trail. She hikes 15 miles every day. Is the relationship between the number of days since Maria began hiking and the number of miles she has hiked exponential? If so, what is the growth factor? If not, explain how you know.

✅ Develop & Understand: B

In the following exercises, you will compare exponential relationships with different growth and decay factors.

Imagine that you have just discovered a table of data from a scientist who is experimenting with four cultures of bacteria. Each culture contains cells that were treated differently.

Next to the table, the scientist noted that all of the populations changed exponentially but with different growth or decay factors. Unfortunately, the scientist spilled coffee on the page, so many of the entries are not legible.

6. Copy the scientist's table. Fill in the missing pieces of data. Remember that each culture changed exponentially.

Bacteria Count

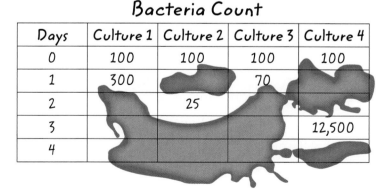

Days	Culture 1	Culture 2	Culture 3	Culture 4
0	100	100	100	100
1	300		70	
2		25		
3				12,500
4				

Real-World Link

The Appalachian Trail, the longest marked hiking trail in the world, runs more than 2,000 miles from Maine to Georgia.

7. Which cultures *grew* exponentially? What is the growth factor?

8. Which cultures *decayed* exponentially? What is the decay factor?

9. Now you will compare equations for the four cultures.

   a. For each culture, write an equation giving the population p of the culture d days after the experiment began.

   b. How are your equations similar?

   c. How are your equations different?

10. Now you will compare graphs for the four cultures.

    a. Start with the cultures that *grew*. Graph the equations for those cultures in the same window of your calculator with the horizontal axis showing Days 0 to 3. Sketch the graphs, and label each with the appropriate equation. Be sure to label the axes as well.

    b. Now consider the cultures that *decayed*. Graph them in the same window with the horizontal axis showing Days 0 to 3. Sketch the graphs.

    c. How are the graphs for the cultures that grew different from those for the cultures that decayed?

    d. How are the graphs for the cultures that grew similar to those for the cultures that decayed?

    e. How can you tell which of two growth factors is greater just by looking at the graphs of the equations? How can you tell from the graphs which of two decay factors is greater?

Real-World Link

Bacteria are single-celled organisms that are visible only under magnification.

. . . .

Share & Summarize

1. How can you tell from a table when the relationship between two quantities might be exponential and not some other type of relationship? Assume the values for one quantity are consecutive integers.

2. How can you tell from a graph when the relationship between two quantities might be exponential?

3. How are graphs and equations representing exponential decay different from those representing exponential growth?

Practice & Apply

1. The rows of this table represent the expressions 4^x, $4x$, and x^4. Copy the table. Fill in the first column with the correct expressions and then fill in the missing entries.

x	0	1	2	3	4	5	6
?	1	4		64			
?	0	4				20	
?	0	1	16				

2. For a science experiment, Jinny put a single bacterium in a dish and placed it in a warm environment. Each day at noon, she counted the number of bacteria in the dish. She made the following table.

Day	0	1	2	3
Bacteria	1	4	16	64

   a. How do the bacteria population values change each day?

   Jinny repeated her experiment, beginning with one bacterium. She wanted to see if she could get the number of bacteria to triple each day by lowering the temperature of the environment. She predicted the number of bacteria in the dish with the following table.

Day, d	0	1	2	3
Bacteria, b	1	3	6	9

   b. Which expression describes her table for $d = 1, 2,$ and 3?

   $$b = 3d \qquad b = d^3 \qquad b = 3^d$$

   c. Is the table correct? If not, which expression should Jinny have used?

3. It is the first day of camp, and 242 campers are outside the gates waiting to enter. The head counselor wants to welcome each camper personally, but she does not have enough time. At 9:00 A.M., she welcomes two campers and leads them into camp. This takes ten minutes.

   Counting the head counselor, the camp now has three people. The three return to the gate, and each welcomes two more campers into the camp. This round of welcomes, for six new campers this time, also takes ten minutes.

   a. How many people are in the camp after this second round of welcomes?

   b. At what time will all of these people be ready to welcome another round of campers?

c. With every round, each person in the camp welcomes two new campers into camp. After the third round of welcomes, how many people will be in the camp? What time will it be then?

d. Copy and complete the table.

Welcome Rounds, n	0	1	2	3	4	5	6	7	8
Time at End of Round	9:00	9:10							
People in Camp, p	1	3							

e. At what time will all 242 campers have been welcomed into camp? Do not forget about the head counselor.

f. Look at the last row in your table. By what number do you multiply to get from one entry to the next?

g. Which expression describes the total number of people in the camp after n rounds, 3^n, $3n$, or n^3? Explain your choice.

h. Use your expression from Part g to find the number of welcome rounds that would be required to bring at least 1,000 campers into camp and to bring at least 10,000 campers into camp.

i. Lunch is served at noon. Could a million campers be welcomed before lunch? Ignore the fact that it would be very chaotic.

4. Match each expression with the situation it describes.

$$3p \qquad\qquad p^2 \qquad\qquad 2^p$$

a. area of the squares in a series with side length p for the square at stage p

b. a population of bacteria that doubles every hour

c. a stamp collection to which 3 stamps are added each week

5. A botanist recorded the number of duckweed plants in a pond at the same time each week.

Week, w	0	1	2	3	4
Plants, d	32	48	72	108	162

a. Look at the second row of the table. By what number do you multiply each value to get the next value?

b. Predict the number of plants in the fifth week.

c. Which expression describes the number of plants in the pond after w weeks?

$$32 \cdot 1.5^w \qquad\qquad 1.5w + 32 \qquad\qquad w^2 + 32$$

6. Start with a scrap sheet of paper. Tear it in half and throw half away. Then tear the piece you have left in half and throw half away. Once again, tear the piece you have left in half and throw half away.

    a. What fraction of the original piece is left after the first tear? After the second tear?

    b. Make a table showing the number of tears and the fraction of the paper left.

    c. Look at the second row of your table. By what number do you multiply each value to get the next value?

    d. Write an expression for the fraction of the paper left after t tears.

    e. If you continue this process, will the paper ever disappear? Why or why not?

For each equation, state whether the relationship between x and y is exponential. If the relationship is exponential, tell whether growth or decay is involved. Name the growth or decay factor.

  7. $y = 4x$ **8.** $y = 4^x$ **9.** $y = 5 \cdot 4^x$

10. $y = 5x^4$ **11.** $y = 5 \cdot 0.25^x$ **12.** $y = 5$

13. Life Science When you take medicine to combat a headache or to lower a fever, the medicine enters your bloodstream. Your body eliminates the medicine in such a way that each hour a particular fraction is removed.

Suppose you have 200 mg of medicine in your bloodstream. Every hour, $\frac{1}{3}$ of the medicine still in your bloodstream is eliminated.

    a. How many milligrams will remain in your bloodstream at the end of the first hour?

    b. How many milligrams will remain at the end of n hours?

    c. How much of the medicine will have been eliminated at the end of n hours?

14. Below are five equations of exponential relationships and five graphs that represent them. Match each equation to a graph.

    a. $y = 2(1.1^x)$

    b. $y = 2(0.2^x)$

    c. $y = 0.1(1.3^x)$

    d. $y = 0.1(1.5^x)$

    e. $y = 2(0.9^x)$

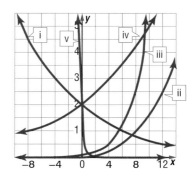

Connect & Extend

15. Preview You can evaluate the expression 2^x for negative values of x. This table is filled in for x values from 0 to 5.

x	−5	−4	−3	−2	−1	0	1	2	3	4	5
2^x						1	2	4	8	16	32

  a. How do the 2^x values change as you move from *right to left*?

  b. Use the pattern you described in Part a to complete the table.

Tell whether each expression represents exponential growth, exponential decay, or neither.

16. 3.2^x **17.** $3x^2$ **18.** $32x$ **19.** $\left(\frac{3}{2}\right)^x$ **20.** $\left(\frac{2}{3}\right)^x$

Tell whether each sentence describes exponential growth, exponential decay, or neither. Explain how you decided.

21. Each time a tennis ball is used, its pressure is $\frac{999}{1,000}$ what it was after the previous use.

22. Each year, the quantity of detergent in a liter of groundwater near an industrial area is expected to increase by 0.05 gram.

23. A star becomes twice as bright every 10 years.

24. It is predicted that the membership in a club will increase by 20 people each month.

. . . .
Math Link
An equilateral triangle has three sides of the same length.
. . . .

25. Triangles This drawing was created by first drawing a large equilateral triangle. A second equilateral triangle was drawn inside the first by connecting the midpoints of its sides. The same process was used to draw subsequent triangles. The sides of each triangle are half as long as the sides of the previous triangle.

  a. If the sides of the smallest triangle are 1 cm long, what are the side lengths of the other triangles?

  b. What is the length of the purple spiral in the drawing?

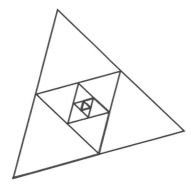

Real-World Link

Prior to the onset of commercial whaling in the 19th century, the worldwide population of humpback whales may have been close to 1.5 million. The current humpback whale population is estimated to be between 20,000 and 30,000.

26. Ecology "Whale Numbers up 12% a Year" was a headline in a 1993 Australian newspaper. A 13-year study had found that the humpback whale population off the coast of Australia was significantly increasing. The actual data suggested the increase was closer to 14%, or 1.14 times the previous year's population.

a. When the study began in 1981, the humpback whale population was 350. If the population grew to 1.14 times this number in the next year, what was the population in 1982? In 1983?

b. Complete the table to show how the population grew each year.

Humpback Whale Population

1981: Year 0	350	350
1982: Year 1	350×1.14	350×1.14^1
1983: Year 2	$350 \times 1.14 \times 1.14$	350×1.14^2
1984: Year 3		
1985: Year 4		
1993: Year 12		

c. Write an expression for the number of whales x years after the study began in 1981.

d. How long did it take the whale population to double from 350?

27. Dion and Katrina are looking at this table of data, which represents the exponential growth of bacteria in a certain swimming pool.

Dion says the equation $p = 10 \cdot 2^h$ describes the table, where p is the population and h is the number of hours.

Hour	Population
0	10
1	20
2	40
3	80
4	160
•	•
•	•
•	•
24	167,772,160
25	335,544,320

a. Show that Dion's equation is correct.

Katrina says that the equation $p = 10 \cdot 16{,}777{,}216^{d}$ describes the table, where p is the population and d is the number of days.

b. Express 1 hour, $h = 1$, as a fraction of a day. Do the same for 2 hours, 3 hours, and 24 hours.

c. Using a calculator, test Katrina's equation. Find the population for the number of days corresponding to 1 hour, 2 hours, 3 hours, and 24 hours. Does her equation seem to be correct?

d. Resolve Dion and Katrina's dilemma. If one or both of the equations is incorrect, explain why. If both of the equations are correct, explain how this can be.

28. **Challenge** Examine this table.

x	y
0.1	−1
1	0
10	1
100	2
1,000	3
10,000	4

    a. Plot the points in the table. Draw a dashed curve through them.

    b. In all of the exponential relationships you have seen, y grew or decayed exponentially as x increased. In this relationship, x is growing exponentially as y increases. What is the growth factor?

29. **In Your Own Words** Describe the difference in the way 4^x and $4x$ grow. In your description, tell which expression grows exponentially. Explain how you could tell from a table of values.

Write an equation for a line that is parallel to the given line.

Mixed Review

30. $y = 2x - 4$

31. $2(y - 3) = 7x + 1$

32. $x = -2$

33. **Architecture** An architect is designing several staircases for a home in which the distance between floors is ten feet. In designing a staircase, she considers these two ratios.

$$\frac{\text{total rise}}{\text{total run}} \qquad \frac{\text{step rise}}{\text{step run}}$$

The diagram shows how these quantities are measured.

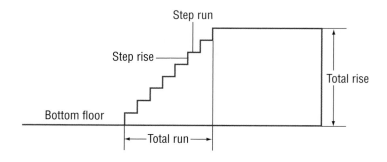

a. One staircase is to have 18 steps with a total run of 14 ft. What is the $\dfrac{\text{total rise}}{\text{total run}}$ ratio for the staircase?

b. Find the step rise and step run, in inches, for each stair in this staircase. What is the ratio $\dfrac{\text{step rise}}{\text{step run}}$?

c. Compare your results for Parts a and b. Explain what you find.

LESSON 4.3

Radicals

You have previously worked with numbers raised to positive and negative integer exponents. You know how to evaluate 2^3, 4^{-3} and even $(-100)^{100}$.

But what if you were asked, "What number squared is 441?" This is the same as knowing the exponent, in this case, 2, because you are squaring, and the result, in this case, 441, but not knowing the base. In this lesson, you will examine situations like this.

Explore

Why do you think these numbers are called "square numbers" or "perfect squares"? What do they have to do with a square?

16　　　64　　　100

Without using the $\boxed{\sqrt{}}$ key on your calculator, try to find the side length of a square whose area is 12.5316.

You can take a number that is not an integer, like 5.5, and make it the side length of a square. A square with a side length of 5.5 has an area of 30.25. Even though 30.25 is not a perfect square, it is still equal to a number multiplied by itself.

Investigation ① Square Roots

Vocabulary

radical sign

square root

To solve an equation like $2x = 6$, you can think about "undoing" the multiplication by 2 that gives the product 6. To undo multiplication, you divide 6 by 2. Since $6 \div 2 = 3$, x must equal 3.

Think again about the question, "What number squared is 441?" This can be written as the equation $x^2 = 441$. To find the answer, you can think about "undoing" the process of squaring a number. That is, you have to find the **square roots** of 441.

Thinking about solutions to an equation gives you a new way to look at square roots, as the students in this class discovered.

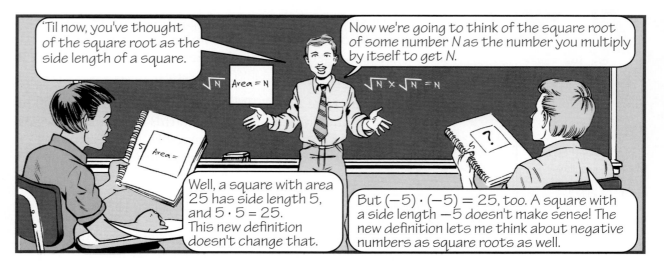

Math Link

The numbers with which you work are called *real numbers*. The set of real numbers is all the numbers on a number line. Mathematicians and engineers also use the set of *complex numbers*. While the square root of a negative number does not exist in the real number system, negative numbers do have square roots in the complex number system.

Both 5 and −5 are square roots of 25. This is important to remember when you want to solve $x^2 = 25$ because both $x = 5$ and $x = -5$ are solutions.

The symbol $\sqrt{}$ is called a **radical sign**. It is used to mean the *positive* square root of a number, so $\sqrt{25} = 5$. Indicate the *negative* square root of a number by writing a negative sign in front. For example, $-\sqrt{25} = -5$.

✅ Develop & Understand: A

Solve these exercises without using your calculator.

1. $\sqrt{49}$

2. What are the square roots of 100?

3. $-\sqrt{225}$

4. What are the square roots of 2.25?

5. $\sqrt{0}$

6. What is the negative square root of 0.01?

7. How many square roots does a positive number have?

8. How many square roots does 0 have?

9. In the set of real numbers, how many square roots does a negative number have?

10. Decide whether the statement below is true or false. Explain your answer.

$$-4 = \sqrt{16}$$

✅ Develop & Understand: B

Solve these exercises without using a calculator.

11. $\left(\sqrt{49}\right)^2$

12. $\left(\sqrt{2.25}\right)^2$

13. $\left(\sqrt{81}\right)^2$

14. $\left(-\sqrt{100}\right)^2$

15. $\left(-\sqrt{4}\right)^2$

16. $-\left(\sqrt{0.04}\right)^2$

17. What happens when you square the square root of a number n? Why does this happen?

Use a calculator to check your answers to these exercises.

18. $\sqrt{100^2}$

19. $\sqrt{(-4)^2}$

20. $\sqrt{0.04^2}$

21. $\sqrt{(-0.04)^2}$

22. Consider the value of $\sqrt{n^2}$.

   a. Assume n is positive. Write an expression equivalent to $\sqrt{n^2}$ that has no radical signs. (Hint: Consider your answer to Exercise 18.)

   b. Assume n is negative. Write an expression equivalent to $\sqrt{n^2}$ that has no radical signs. (Hint: Consider your answers to Exercises 19 and 21.)

   c. If possible, write a single expression equivalent to $\sqrt{n^2}$ that has no radical signs, for any value of n, positive, negative, or 0. (Hint: Consider your answers to Exercises 20 and 21.)

✅ Develop & Understand: C

Solve these equations, if possible, using whatever method you prefer. If an equation has no solution, write "no solution" and explain why.

23. $\sqrt{x} = 7$

24. $\sqrt{x} = 5$

25. $\sqrt{x} + 8 = -6$

26. $0 = 2.3 + \sqrt{x}$

27. $\sqrt{x + 2} = 5$

28. $x = \sqrt{-16}$

29. $x^2 = 64$

30. $3x^2 = 48$

Share & Summarize

1. Juana said, "My older sister loves to tell me about her math class. When she learned about square roots, she showed me some examples, like $\sqrt{4} = 2$. At first, I thought finding a square root meant the same thing as finding half of something."

   Write a sentence or two explaining the difference between finding half of something and finding the square root of something.

2. Are there any numbers for which taking half gives the same result as taking the square root? If so, what are they?

Investigation ② Simplify Radical Expressions

What happens when you try to add, subtract, multiply, or divide numbers that involve radical signs? The next set of exercises will help you think about how and when you can combine terms with radical signs.

✅ Develop & Understand: A

1. Does the square root of a sum equal the sum of the square roots? That is, does $\sqrt{x + y} = \sqrt{x} + \sqrt{y}$?

   Copy and complete the following table, testing several examples. For the last column, choose your own positive x and y values to test. Make a conjecture about whether the two expressions are equivalent.

(x, y)	(0, 0)	(4, 4)	(36, 16)	$(25, \frac{1}{4})$	(___, ___)
$\sqrt{x} + \sqrt{y}$					
$\sqrt{x + y}$					

2. Now do the same for multiplication. Does $\sqrt{x \cdot y} = \sqrt{x} \cdot \sqrt{y}$? Copy and complete the table and then make a conjecture.

(x, y)	(0, 0)	(5, 5)	(9, 25)	(0.64, 100)	(___, ___)
$\sqrt{x} \cdot \sqrt{y}$					
$\sqrt{x \cdot y}$					

3. Copy and complete the table. Make a conjecture about the differences of square roots.

(x, y)	(0, 0)	(4, 4)	(81, 49)	$\left(\frac{25}{9}, \frac{16}{9}\right)$	(___, ___)
$\sqrt{x} - \sqrt{y}$					
$\sqrt{x - y}$					

4. Copy and complete the table. Make a conjecture about dividing square roots.

(x, y)	(0, 2)	(3, 3)	(4, 16)	$\left(\frac{4}{9}, 2.25\right)$	(___, ___)
$\sqrt{x} \div \sqrt{y}$					
$\sqrt{x \div y}$					

Expressions that contain one or more numbers or variables under a radical sign are called *radical expressions*. The following are radical expressions.

$$3\sqrt{x} \qquad 2.3m\sqrt{3m} \qquad x + 3\sqrt{x}$$

Computing with radical expressions can be difficult when the expressions are complicated. Sometimes it helps to *simplify* radical expressions to make computations easier. A radical expression is simplified when:

- numbers under radical signs have no square factors.
- the number of radical signs in the expression is as small as possible.

In Exercise 2, you probably conjectured the following.

$$\sqrt{x \cdot y} = \sqrt{x} \cdot \sqrt{y}.$$

As long as both *x* and *y* are nonnegative (positive or 0) this conjecture is true.

You might also have conjectured the following.

$$\sqrt{x} \div \sqrt{y} = \sqrt{x \div y}.$$

This rule is true for nonnegative *x* and positive *y*. These rules will help you learn how to simplify radical expressions.

To simplify a number under a square root sign, look for factors of the number that are perfect squares. Rewrite the number as a product in which at least one factor is a perfect square. Take the square roots of the perfect squares.

Math Link

The number or quantity under a radical sign is called a *radicand*.

Example

To simplify $\sqrt{24}$, rewrite it as $\sqrt{4 \cdot 6}$, or $\sqrt{4} \cdot \sqrt{6}$. Since $4 = 2 \cdot 2$, $\sqrt{4} \cdot \sqrt{6}$ is equivalent to $2 \cdot \sqrt{6}$ or $2\sqrt{6}$.

To simplify $\sqrt{18x^4}$, rewrite it as $\sqrt{9x^4 \cdot 2}$, or $\sqrt{9x^4} \cdot \sqrt{2}$. Since $9x^4 = 3x^2 \cdot 3x^2$, $\sqrt{9x^4} \cdot \sqrt{2}$ is equivalent to $3x^2\sqrt{2}$.

To simplify $\sqrt{30}$, you could rewrite it as $\sqrt{15 \cdot 2}$, $\sqrt{10 \cdot 3}$, or $\sqrt{5 \cdot 6}$. However, since none of the factors of 30 are perfect squares, $\sqrt{30}$ is already simplified.

Perfect squares can be hard to find. For example, to simplify $\sqrt{48}$, you could rewrite it as $\sqrt{8} \cdot \sqrt{6}$. But since neither 8 or 6 is a perfect square, you might think $\sqrt{48}$ is already simplified.

However, if you rewrite $\sqrt{48}$ as $\sqrt{4} \cdot \sqrt{12}$, you *can* simplify.

$$\sqrt{48} = \sqrt{4} \cdot \sqrt{12}$$
$$= 2 \cdot \sqrt{12}$$

Then simplify $\sqrt{12}$ further.

$$= 2 \cdot \sqrt{4 \cdot 3}$$
$$= 2 \cdot \sqrt{4} \cdot \sqrt{3}$$
$$= 2 \cdot 2 \cdot \sqrt{3}$$
$$= 4 \cdot \sqrt{3}, \text{ or } 4\sqrt{3}$$

✓ Develop & Understand: B

Simplify the following square roots if possible. Assume all variables represent positive numbers.

5. $\sqrt{75}$

6. $\sqrt{60}$

7. $\sqrt{42}$

8. $\sqrt{\dfrac{1}{8}}$

9. $\sqrt{50x^3}$

10. $\sqrt{72a^4b^5}$

Now see whether you can "unsimplify" expressions. Fill each blank with an expression different from the given one. Assume all variables represent nonnegative numbers.

11. $2\sqrt{3}$ is the simplified form of _____.

12. $6\sqrt{2}$ is the simplified form of _____.

13. $5y\sqrt{3}$ is the simplified form of _____.

14. $\dfrac{1}{4}x^2\sqrt{3x}$ is the simplified form of _____.

When adding or subtracting terms with radical signs, the terms with radical signs behave in ways similar to expressions with variables.

Example

To simplify expressions with variables, combine *like terms*.

$$3x + 4y - 2x + 5y = 3x - 2x + 4y + 5y$$
$$= (3 - 2)x + (4 + 5)y$$
$$= 1x + 9y$$
$$= x + 9y$$

To simplify expressions with radicals, combine *like radical terms*.

$$3\sqrt{2} + 4\sqrt{5} - 2\sqrt{2} + 5\sqrt{5} = 3\sqrt{2} - 2\sqrt{2} + 4\sqrt{5} + 5\sqrt{5}$$
$$= (3 - 2)\sqrt{2} + (4 + 5)\sqrt{5}$$
$$= 1\sqrt{2} + 9\sqrt{5}$$
$$= \sqrt{2} + 9\sqrt{5}$$

Sometimes, you must simplify the terms before you can combine them.

$$\sqrt{20} + \sqrt{45} = \sqrt{4 \cdot 5} + \sqrt{9 \cdot 5} = 2\sqrt{5} + 3\sqrt{5}$$
$$= (2 + 3)\sqrt{5}$$
$$= 5\sqrt{5}$$

✓ Develop & Understand: C

For Exercises 15–18, decide whether the two expressions are equivalent. Explain your reasoning or show your work.

15. $\sqrt{2} + \sqrt{2} + \sqrt{2}$ and $\sqrt{6}$

16. $3\sqrt{2} + \sqrt{3}$ and $3\sqrt{5}$

17. $\sqrt{50} + \sqrt{98}$ and $12\sqrt{2}$

18. $-\frac{1}{2}\sqrt{80}$ and $\sqrt{45} + \sqrt{20} - 7\sqrt{5}$

19. Write an expression equivalent to $2\sqrt{3}$ that includes addition or subtraction.

20. Below are four radical expressions.

    i. $-3\sqrt{3} + \sqrt{48}$ **ii.** $2\sqrt{12} - \sqrt{27}$

    iii. $12\sqrt{2} - 7\sqrt{2}$ **iv.** $3\sqrt{32} - \sqrt{98}$

  a. Simplify each expression.

  b. Are any of the expressions equivalent to each other? If so, which?

Share & Summarize

When Susan is trying to simplify radical terms, she gets stuck if she cannot immediately find a factor that is a perfect square. For example, she said, "When I try to simplify $\sqrt{60}$, I come up with $\sqrt{6} \cdot \sqrt{10}$. Neither of those are perfect squares, so I can't figure out how to simplify it."

1. Simplify $\sqrt{60}$.

2. Describe a method that Susan could use to find the right factors. Your method should work for any situation that she tries.

Investigation ③ *n*th Roots

Vocabulary

*n*th root

In Investigation 1, you "undid" the process of squaring numbers. Now you will undo the process of taking numbers to higher powers.

Think & Discuss

Think about why "cubic numbers" or "perfect cubes" have those names. Give an example of a cubic number. Show how it is related to a cube.

Some number cubed is equal to 8. What is that number? How many answers can you find?

Some number cubed is equal to −64. What is that number? How many answers can you find?

Is there a number squared that equals −64? If so, what is it?

Some number cubed is equal to −0.027. What is that number?

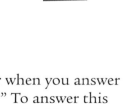

You are undoing the process of cubing a number when you answer questions like "What does N equal, if $N^3 = -64$?" To answer this question, suppose someone has cubed a number and told you the result, and you want to find the original number.

The *cube root* of 64 is 4 because 4 *cubed* is 64. That is, $4^3 = 4 \cdot 4 \cdot 4 = 64$. The radical sign with an index of 3 is used to indicate cube roots.

$$\sqrt[3]{64} = 4$$

For square roots, the radical sign by itself always represents the positive square root. You will find that every real number has one cube root.

The cube root of a positive number is always positive, and the cube root of a negative number is always negative. So, the symbol $\sqrt[3]{}$ represents a number's one cube root.

$$\sqrt[3]{8} = \sqrt[3]{2 \cdot 2 \cdot 2} = 2$$

$$\sqrt[3]{-8} = \sqrt[3]{(-2)(-2)(-2)} = -2$$

✅ Develop & Understand: A

Evaluate without using a calculator.

1. $\sqrt[3]{8}$

2. $\sqrt[3]{-125}$

3. $\sqrt[3]{0.000001}$

4. $\left(\sqrt[3]{8}\right)^3$

5. $\left(\sqrt[3]{-125}\right)^3$

6. $\left(\sqrt[3]{37}\right)^3$

7. What happens when you cube the cube root of any number? Why?

Math Link

Using the set of *complex numbers*, every number except 0 has three cube roots.

Solve each equation using whatever method you choose.

8. $\sqrt[3]{x} = -\dfrac{1}{8}$

9. $6 = \sqrt[3]{2n}$

10. $\sqrt[3]{z + 5} = -2$

You can also take higher roots. The *fourth root* of 625 is 5 because 5 to the fourth power is 625. That is, $5^4 = 5 \cdot 5 \cdot 5 \cdot 5 = 625$. Another fourth root of 625 is -5 because $-5 \cdot -5 \cdot -5 \cdot -5 = 625$.

The radical sign for the *positive* fourth root is $\sqrt[4]{}$, so $\sqrt[4]{625} = 5$. However, if you want the fourth roots of 625, you need to consider both -5 and 5.

Think & Discuss

What do you think a *fifth root* is? Give an example.

What do you think a *sixth root* is? Give an example.

What's another name for a *second root*?

What's another name for a *third root*?

Fifth roots, sixth roots, and so on are similar to square roots and cube roots.

In general, $\sqrt[n]{}$ denotes the **nth root** of x.

- When n is even, $\sqrt[n]{x}$ denotes the positive nth root. The negative nth root is $-\sqrt[n]{x}$. When n is even, x must be nonnegative.
- When n is odd, $\sqrt[n]{x}$ is positive if x is positive and negative if x is negative.

The nth root of 0 is always 0, no matter what n is.

Notice that the square root of x is written $\sqrt{x}$, not $\sqrt[2]{x}$. If there is no number, the radical is assumed to represent a square root.

✅ Develop & Understand: B

Evaluate without using a calculator.

11. $\sqrt[5]{32}$

12. Find the fourth roots of 256.

13. Find the seventh root of -128.

14. $\sqrt[6]{729}$

15. Are there any numbers for which the second, third, fourth, fifth, and sixth roots are the same number? If so, list them.

16. Are there numbers with both a positive and a negative fourth root? If so, give an example. If not, explain why not.

17. Is it possible to find a number with only a positive fourth root? If so, give an example. If not, explain why not.

18. Is it possible to find a number with both a positive and a negative fifth root? If so, give an example. If not, explain why not.

Order each set of numbers from least to greatest.

19. $\sqrt[5]{7}$ $\sqrt[3]{7}$ $\sqrt[4]{7}$ $\sqrt{7}$ $\sqrt[6]{7}$ **20.** $\sqrt[4]{\dfrac{1}{4}}$ $\sqrt[81]{\dfrac{1}{4}}$ $\sqrt{\dfrac{1}{4}}$ $\sqrt[80]{\dfrac{1}{4}}$

Share & Summarize

1. Describe the general relationship between finding an nth root of some number and raising some number to the nth power.

2. Is it possible to find a number with both a positive and a negative ninth root? If so, give an example. If not, explain why not.

Practice & Apply   In Exercises 1–12, find the indicated roots without using a calculator.

1. the square roots of 36

2. $\sqrt{100}$

3. the square roots of 1.21

4. $\sqrt{1.21}$

5. the square roots of $\frac{4}{49}$

6. $\sqrt{\frac{4}{49}}$

7. the square roots of 0.0064

8. $\sqrt{0.0064}$

9. $\left(\sqrt{26}\right)^2$

10. $\left(\sqrt{0.09}\right)^2$

11. $\sqrt{(-3)^2}$

12. $\sqrt{x^2}$

Solve each equation. If it has no solution, write "no solution."

13. $\sqrt{x-3} = 9$

14. $5\sqrt{x} = 25$

15. $\sqrt{\frac{x}{7}} = 3$

16. $\sqrt{x} = 36$

17. $\sqrt{x+2} + 8 = 1$

18. $\sqrt{x-20} = -18$

19. $\sqrt{x} = 4$

20. $\sqrt{x+4} = 10$

Tell whether each computation is correct or incorrect.

21. $\sqrt{2} \cdot \sqrt{3} = \sqrt{6}$

22. $\sqrt{4} + \sqrt{15} = \sqrt{19}$

23. $\sqrt{3} - \sqrt{0} = \sqrt{3}$

24. $\sqrt{20} \div \sqrt{45} = \frac{2}{3}$

Simplify each radical expression. State if it is already simplified.

25. $\sqrt{200}$

26. $\sqrt{125}$

27. $\dfrac{\sqrt{20} + \sqrt{80}}{\sqrt{20}}$

28. $\sqrt{17} - \sqrt{30}$

29. $\sqrt{8} \cdot \sqrt{12}$

30. $\sqrt{32} \cdot \sqrt{48}$

31. Challenge $\sqrt{x+2} + \sqrt{4x+8}$

32. Below are four radical expressions. Assume x is positive.

    i. $\sqrt{800x^3}$ **ii.** $4x\sqrt{50x}$

    iii. $3\sqrt{32x^3}$ **iv.** $5\sqrt{8x^3} + 2x\sqrt{2x} + 2\sqrt{32x^3}$

    a. Simplify each expression.

    b. Which expressions are equivalent to each other?

Decide whether the expressions in each pair are equivalent. Explain.

33. $3\sqrt{5}$ and $\sqrt{5} + \sqrt{5} + \sqrt{5}$ **34.** $\sqrt{32} - \sqrt{18}$ and $\sqrt{14}$

In Exercises 35–42, find the indicated roots without using a calculator.

35. the cube root of -216 **36.** $\sqrt[3]{-216}$

37. the sixth roots of 64 **38.** $\sqrt[6]{64}$

39. the fifth root of -243 **40.** $\sqrt[5]{-243}$

41. Challenge the eighth roots of x^{16}

42. Challenge $\sqrt[8]{x^{16}}$

Order each set of numbers from least to greatest.

43. $\sqrt[9]{-41}$ $\sqrt[3]{-41}$ $\sqrt[11]{-41}$ $\sqrt[5]{-41}$ $\sqrt[7]{-41}$

44. $\sqrt[3]{-\frac{1}{3}}$ $\sqrt[311]{-\frac{1}{3}}$ $\sqrt[5]{-\frac{1}{3}}$ $\sqrt[105]{-\frac{1}{3}}$

Connect & Extend

45. For this exercise, assume x is positive.

    a. Name three values of x for which $\sqrt{x}$ is less than x.

    b. Name three values of x for which $\sqrt{x}$ is greater than x.

    c. In general, how can you tell whether $\sqrt{x}$ is greater than x?

46. Prove It! Ralph said, "If $x^2 = y^2$, then $x = y$."

    a. Try several values for x and y to investigate Ralph's conjecture.

    b. Is Ralph's conjecture true? If it is, explain why. If not, give a counterexample.

47. For what values of x and y is $\sqrt{x} + \sqrt{y} = \sqrt{x + y}$ true?

48. For what values of x and y is $\sqrt{x} - \sqrt{y} = \sqrt{x - y}$ true?

49. Prove It! In Investigation 2, you saw that for nonnegative values of x and y, $\sqrt{x} \cdot \sqrt{y} = \sqrt{x \cdot y}$. In Parts a–d, you will prove this conjecture.

To make things easier, define two new variables, $u = \sqrt{x}$ and $v = \sqrt{y}$.

  a. What is the product uv in terms of x and y?

  b. In terms of x and y, to what are u^2 and v^2 equivalent?

  c. Fill in the blank, using x and y.

$$(uv)^2 = u^2 v^2 = \underline{\hspace{2cm}}$$

  d. Use your answer to Part c to write an expression in terms of x and y that is equivalent to uv.

  e. Parts a and d both asked you to write the product uv in terms of x and y, so your answers to those parts are equivalent. Have you shown that $\sqrt{x} \cdot \sqrt{y} = \sqrt{(x \cdot y)}$?

  f. Challenge Use Parts a–d as a guide to prove that

$$\sqrt{x} \div \sqrt{y} = \sqrt{x \div y}$$

  if x is nonnegative and y is positive.

  If it helps, rewrite both sides as fractions.

$$\frac{\sqrt{x}}{\sqrt{y}} = \sqrt{\frac{x}{y}}.$$

50. Geometry Isosceles right triangles have two legs that are the same length. If the length of one leg of an isosceles right triangle is a, what is the length of the hypotenuse? Simplify your answer.

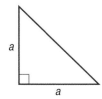

51. **Number Sense** List these seven numbers from least to greatest.

$$0 \qquad 1 \qquad -1 \qquad \sqrt[51]{2}$$

$$\sqrt[51]{-2} \qquad \sqrt[51]{0.2} \qquad \sqrt[51]{-0.2}$$

52. **Number Sense** Consider $\sqrt[8]{n}$.

    a. If n is greater than 1, is $\sqrt[8]{n}$ greater than or less than n?

    b. If n is greater than 0 but less than 1, is $\sqrt[8]{n}$ greater than or less than n?

53. **Fractional Exponents** You have previously worked with integer exponents. Exponents can also be fractions.

    The laws of exponents apply to fractional exponents just as they do to integer exponents.

    When $\frac{1}{n}$ is used as an exponent, it means "take the nth root." These are examples.

    $$81^{\frac{1}{2}} = \sqrt{81} = 9$$
    $$(-27)^{\frac{1}{3}} = \sqrt[3]{-27} = -3$$
    $$64^{\frac{1}{4}} = \sqrt[4]{64} = 4$$

    Evaluate each expression without using a calculator. In Parts e–h, use the laws of exponents to help you.

    a. $1.44^{\frac{1}{2}}$

    b. $125^{\frac{1}{3}}$

    c. $(-32)^{\frac{1}{5}}$

    d. $-32^{\frac{1}{5}}$

    e. $\left(3^{\frac{1}{3}}\right)^3$

    f. $\left(\frac{9}{25}\right)^{\frac{1}{2}}$

    g. $(64)^{-\frac{1}{3}}$

    h. $16^{\frac{3}{4}}$

54. **Challenge** Consider $\sqrt[n]{x}$.

    a. If $\sqrt[n]{x}$ is positive, what can you say for sure about x? Explain.

    b. If $\sqrt[n]{x}$ is negative, what can you say for sure about n and x? Explain.

55. **In Your Own Words** Write a letter to a younger student explaining the difference between rational and irrational numbers. Include the definitions of both kinds of numbers and several examples of each.

56. Preview You have used the distributive property to *expand* expressions. You can also use the distributive property to *factor* expressions.

$$\text{Expand: } 7(c + d) = 7c + 7d$$
$$\text{Factor: } 7c + 7d = 7(c + d)$$

Factor the following expressions.

a. $10a + 10b$

b. $5x + 5y$

c. $3p + 9$

d. $k^2 + 20k$

Mixed Review

Tell whether the points in each set lie on a line.

57. $(0, 0); (9, 0); (0, 2)$

58. $(0.3, -2); (-1, 4.5); (10.1. -51)$

59. $(-3, -14.8); 0, -7.3); (1.4, -3.8)$

60. Write an equation to represent the value of A in terms of t.

t	0	1	2	3	4
A	9	27	81	243	729

61. Write 64, 256, and 1,024 using integer exponents and the same base.

62. Sort these expressions into two groups so that the expressions in each group are equal to one another.

$$m^3 \qquad \left(\frac{1}{m}\right)^3 \qquad m^{-3}$$

$$\left(\frac{1}{m}\right)^{-3} \qquad \frac{1}{m^3} \qquad m \div m^4$$

Review & Self-Assessment

CHAPTER 4

Vocabulary

decay factor

exponential decay

exponential decrease

exponential growth

exponential increase

growth factor

nth root

radical sign

scientific notation

square root

Chapter Summary

This chapter began with a review of positive integer exponents. You expanded the laws of exponents to include negative exponents. You also used negative exponents and scientific notation to compare numbers.

Next, you looked at quantities that grow exponentially and investigated how exponential and constant growths differ. You explored situations in which quantities decrease exponentially. You learned how to recognize exponential growth and decay in tables, expressions, and written descriptions.

You then worked with square roots of numbers. You also generalized the concept of square roots for cube roots, fourth roots, and even higher roots.

Strategies and Applications

The questions in this section will help you review and apply the important ideas and strategies developed in this chapter.

Understanding integer exponents

1. Suppose y is a positive integer.

   a. Explain what x^y means.

   b. Explain how x^y and x^{-y} are related.

2. Suppose r is a number not equal to 0 or 1. Which is greater, r^{11} or r^{21}, in each of the following cases? Explain your answers.

   a. r is greater than 1.   **b.** r is between 0 and 1.

   c. r is between -1 and 0.   **d.** r is less than -1.

Understanding scientific notation

3. Which is greater, 2.3×10^{32} or 3.2×10^{23}? Explain.

4. An atom of an element is composed of protons, neutrons, and electrons. The resting mass of a proton is about 1.7×10^{-24} gram. The resting mass of an electron is about 9.1×10^{-28} gram. Answer the following questions without using your calculator.

   a. A proton has how many more grams of mass than an electron?

   b. A proton has how many times the mass of an electron?

Recognizing and describing exponential relationships

5. Determine which of these equations describe an exponential relationship. For each exponential relationship, indicate whether it is growth or decay.

$$y = 3(x^4) \qquad y = 3 \cdot 4^x \qquad y = 3x + 4$$

$$y = \frac{1}{4x} \qquad y = \left(\frac{1}{4}\right)^x \qquad y = x^4$$

For Questions 6–7, do Parts a–c.

a. Identify whether the relationship described is exponential.

b. Explain your reasoning.

c. If the relationship is exponential, write an equation describing it.

6.

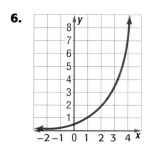

7. *Modems* are devices that allow computers to "talk" to other computers over a phone line. Each time modem speeds increased, David bought a new one. His first modem had a speed of only 1,200 kilobytes per second (kps). The table shows the speed of each of his modems. Consider the relationship between the modem number and the speed of the modem.

Modem Number	Speed (kps)
1	9,600
2	14,400
3	28,800
4	33,600
5	57,600

Understanding the laws of exponents

8. Prove It! By completing Parts a and b, prove that the power to a power law works when one exponent is a positive integer and the other is a negative integer. Assume a is a positive number and b and c are positive integers.

a. Show that $\left(a^b\right)^{-c} = a^{-bc}$.

b. Now show that $\left(a^{-b}\right)^c = a^{-bc}$.

Understanding roots

9. Explain why there are two values of x for which $x^2 = 16$ but only one value of x for which $x = \sqrt{16}$. What are the values of x in each case?

10. Why is -3 the fifth root of -243?

Demonstrating Skills

Evaluate or simplify each number without using your calculator.

11. 0.4^3

12. $\left(\dfrac{2}{3}\right)^4$

13. 8^{-3}

14. $\left(\dfrac{2}{3}\right)^{-4}$

15. $\sqrt{121}$

16. $\sqrt{32}$

17. $\sqrt[3]{-1{,}000}$

18. $\sqrt[3]{0.027}$

19. $\sqrt[5]{32}$

20. $2\sqrt[7]{(-3)^{14}}$

21. $\dfrac{1}{3}\sqrt{27}$

22. $\sqrt[x]{13^x}$

Write each expression using a single exponent and a single base. Assume x is not zero.

23. $a^3 \cdot b^3$

24. $(2x)^4 \cdot (2x)^{-7}$

Simplify each expression as much as possible. When necessary, assume the variable is not negative.

25. $\sqrt{52}$

26. $-\sqrt{70}$

27. $\sqrt{18x^3}$

28. $3m\sqrt{4m^6}$

29. Complete this table to compare the expressions $5x$, 5^x, x^5.

x	0	1	2	3	4	5
5x	0					
5^x	1					
x^5	0					

30. As x increases from 0 to 5, which expression's values increase the most quickly?

31. For the function 5^x, how do you get from one value to the next?

32. Use your table from Question 29 to plot the points for each function on one set of axes.

33. Imagine that a tennis ball is dropped from a height of 10 feet and bounces repeatedly. Suppose the height of each bounce is 0.5 times the height of the previous bounce.

   a. How high does the ball rise on its first bounce?

   b. How high does the ball rise on the second bounce?

   c. Write an expression for the ball's height on the nth bounce.

Test-Taking Practice

SHORT RESPONSE

1 Max is doing his math homework. He simplifies the expression $(3a^2)(2b)^3$ as follows.

$$(3a^2)(2b)^3 = 3a^2 \cdot 8b^3 = 24a^2b^3$$

Did Max make a mistake? Explain why or why not?

Show your work.

Answer _____

MULTIPLE CHOICE

2 Which list shows numbers in order from greatest to least?

   A 6.2×10^{-2}, 2.3×10^4, 1.7×10^6

   B 2.3×10^4, 1.7×10^6, 6.2×10^{-2}

   C 1.7×10^6, 2.3×10^4, 6.2×10^{-2}

   D 6.2×10^{-2}, 2.3×10^4, 1.7×10^6

3 Which of these equations represents an exponential relationship?

   F $y = x^3$

   G $y = \left(\dfrac{1}{3}\right)^x$

   H $y = 4(x^6)$

   J $y = \dfrac{2}{5x}$

Algebraic Expressions

Real-Life Math

Heavy Lifting Do you think you could lift a 10,000-pound elephant? If you could find a lever long enough and strong enough, you could. Using a lever allows you to apply more force to an object than you could with your bare hands.

Suppose the elephant is four feet from the fulcrum of the lever. Then, the amount of force F you would need to apply is $F = \dfrac{40,000}{d}$, where d is the distance from the fulcrum to where you apply the force.

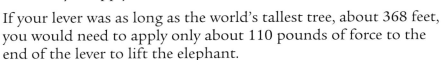

If your lever was as long as the world's tallest tree, about 368 feet, you would need to apply only about 110 pounds of force to the end of the lever to lift the elephant.

Think About It In the equation $F = \dfrac{40,000}{d}$, what do you think the number 40,000 represents?

Math Online
Take the **Chapter Readiness Quiz** at glencoe.com.

Dear Family,

The class is about to begin the topic of *algebraic expressions*. Algebra is one of the most powerful tools of mathematics. Part of its usefulness is the way algebra allows you to state and solve many situations with little effort. Here is an example.

> The cost of a movie is $9.00 for each adult plus half of the adult price for each of four children, less your $3 coupon. How much will you pay?

The answer is $[9.00x + 4.50(4) − 3]$, where x is the number of adults.

Key Concept—Geometric Model

In this chapter, the class will begin by learning to multiply expressions like $x(x + 5)$ and $(x + 1)(x + 5)$ by referring to a *geometric model*. For example, to find the area of the large rectangle below, you can multiply the length by the width, or $x(x + 3)$. However, we can also find the area of the square, $x \cdot x$, or x^2, and add the area of the small rectangle, $x \cdot 3$, or $3x$, to get $x^2 + 3x$. This shows that $x(x + 3) = x^2 + 3x$.

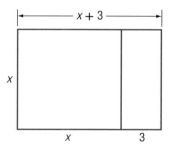

Chapter Vocabulary

binomial	like terms	trinomial
expanding	monomial	

Home Activities

- Have your student explain his or her work to you, using both geometric models and symbols.

- Have your student share newly learned strategies for working with algebraic expressions.

Rearrange Algebraic Expressions

Ideas from geometry can sometimes shed light on certain concepts in algebra. In this investigation, you will look at a geometric model involving rectangles to help you work with and simplify algebraic expressions.

Vocabulary

expanding

Think & Discuss

This rectangle can be thought of as a square with a strip added to one side. The large rectangle's width is one unit longer than its height.

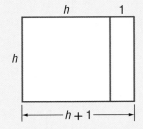

If you cut the large rectangle apart, you get a square and a small rectangle with the dimensions shown below.

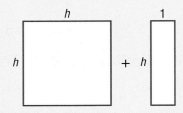

- What is the area of the square?

- What is the area of the small rectangle?

- Using the expressions that you wrote above, write an expression for the area of the large rectangle.

- What does this tell you about $h(h + 1)$ and $h^2 + h$? Why?

Rectangle diagrams, like those above, are geometric models that can help you think about how the distributive property works. That is, they can help you understand why $a(b + c) = ab + ac$.

Using the distributive property to multiply the factors a and $(b + c)$ is called **expanding** the expression. For example, to expand the expression $2(x + 1)$, multiply 2 and $(x + 1)$ to get $2x + 2$. The expanded version of $h(h + 1)$ is $h^2 + h$.

Investigation 1 Geometric Models

In this investigation, you will use rectangle models to represent algebraic expressions.

✅ Develop & Understand: A

1. One of the rectangles in this diagram has an area of $x(x + 3)$.

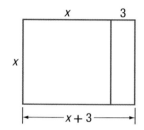

   a. Copy the diagram. Indicate the rectangle that has area $x(x + 3)$.

   b. Use your diagram to expand the expression $x(x + 3)$. Explain what you did.

2. Dante and Héctor are making another rectangle diagram.

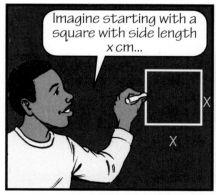

   a. Answer Dante's question by writing an expression.

   b. Is Héctor correct? Explain your thinking.

3. In this diagram, the rectangle that is shaded has been removed from the square.

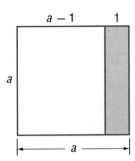

Explain two ways to use the diagram to find an expression for the area of the unshaded rectangle. Give your expressions.

4. Use the distributive property to rewrite each expression. Then draw a rectangle diagram that shows why the two expressions are equivalent. Use shading to indicate when a region's area is being removed.

a. $b(b + 4)$

b. $m(m - 6)$

You will now explore more complex combinations of rectangles and the algebraic expressions they represent.

✅ Develop & Understand: B

5. Start with a square that has side length x. Create a large rectangle by adding another square of the same size and a one-unit strip.

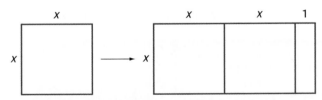

a. What is the height of the large rectangle?

b. Write a simplified expression for the width of the large rectangle.

c. Use the dimensions from Parts a and b to write an expression for the area of the large rectangle.

d. The large rectangle is composed of two squares and a smaller rectangle. Write an expression for the area of each of these parts. Then use the areas to write an expression for the area of the large rectangle, simplifying it if necessary.

e. Your two expressions for the area of the large rectangle, from Parts c and d, are equivalent. Write an equation that states this. Then use the distributive property to verify your equation.

6. Start with a square with side length x, add another square of the same size, and *remove* a strip with width 1.

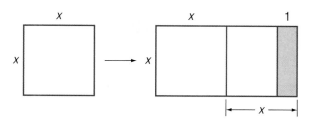

Use the diagram to help explain why $x(2x - 1) = 2x^2 - x$.

7. Now start with a square with side length x and make a rectangle with sides $2x$ and $x + 1$.

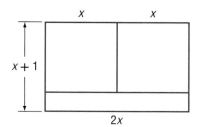

a. Use the distributive property to expand $2x(x + 1)$.

b. Use the diagram to explain why your expansion is equivalent to the original expression.

8. Draw a rectangle diagram that models each expression. Use your diagram to help you write the expression in a different form.

   a. $2a(a - 1)$

   b. $b^2 - 3b$

   c. $c^2 + 2c$

9. Expand each expression.

   a. $3a(a + 4)$

   b. $2m(3m - 2)$

   c. $4x(3 + 2x)$

Share & Summarize

According to the distributive property, $a(b + c) = ab + ac$. Use a rectangle model to explain why the distributive property makes sense.

Investigation 2 Simplify Expressions

Vocabulary

like terms

Materials

• graphing calculator

In Investigation 1, you learned about expanding expressions by removing parentheses. To make expanded expressions easier to use, you will sometimes want to shorten, or simplify, them.

Think & Discuss

This floor plan is for a living room. Imagine that you want to buy new carpeting for the room. All dimensions are in feet.

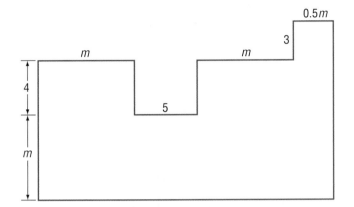

• Write an expression for the area of the floor.

• Tamara wrote this expression for the area of the floor.

$$m^2 + 4m + 5m + m^2 + 4m + 3.5m + 0.5m^2$$

Is her expression correct?

• Evaluate Tamara's expression for $m = 6$.

Tamara's area expression has several terms. You can create an equivalent expression that is written more simply.

For example, consider this expression.

$$k + 4k^2 + 3 - 2k^3 + 2k - 16 - 6k^4 + 3k^2 + 7k^3 + 19k^8$$

Two of the terms are k and $2k$. You can reason that their sum is $3k$, even though you do not know what k represents. A number plus twice that number is three times that number, no matter what the number is.

The parts k and $2k$ are called *like terms*. **Like terms** have the same variable raised to the same power. They can be added or subtracted and then written as a single term, like in the above expression.

$$4k^2 + 3k^2 = 7k^2 \qquad \text{and} \qquad -2k^3 + 7k^3 = 5k^3$$

Similarly, 3 and -16 are like terms because they are both constants, or terms with no variable, and can be combined to give -13. Since the terms $-6k^4$ and $19k^8$ are unlike each other and unlike the other terms, they stand alone.

You can rewrite the expression more simply.

$$19k^8 - 6k^4 + 5k^3 + 7k^2 + 3k - 13$$

Notice that, in the above expression, the terms are ordered by the exponent on the variable k.

✅ Develop & Understand: A

1. Which of these expressions are equivalent to the expression $p + 2p - p + 6 - 3 + 2p$?

$3p$	$7p$	$4p + 3$
$6p + 3$	$2p + 3 + 2p$	$4p - 3$

2. Which of these expressions are equivalent to the expression $y(2y + 3) - 5 + 2y - 2 + 3y^2 + 7$?

10	$2y^2 + 3y^2 + 5y$	
$10y^2$	$12y + 14$	$5y^2 + 5y$

Avery tried to simplify the expressions in Exercises 3–6 but made some errors. For each, tell whether the simpler expression is correct. If it is not, identify Avery's mistake and write the correct expression.

3. $x + x + 7 = 2x + 7$

4. $m^2 + m^2 - 4 = m^4 - 4$

5. $2 + b + b^2 = 2 + b^3$

6. $3 - b^2 + b(b + 2b) = 2b^2 + 3$

7. Copy the expression that you wrote for the area of the floor in Think & Discuss on page 210.

  a. Write the expression as simply as you can.

  b. How many square feet of carpet do you need if m is 6 ft?

  c. In Think & Discuss, you evaluated Tamara's expression for $m = 6$. Which was easier, evaluating Tamara's expression or your expression from Part a?

8. Lana and Keenan simplified $5a^2 + 10 - 4a^2 - 5 + 3a^2$ to $3a^2 + 5$. Lana checked the answer by substituting 0 for a. She found that both expressions equal 5 when a is 0. She concluded that they are equivalent.

Keenan asked, "But what happens when we substitute 2?" Using $a = 2$, he found that the first expression equals 21 and that the simplified expression equals 17.

  a. Did Lana and Keenan simplify correctly? Explain.

  b. Lana and Keenan tested the equivalence of the expressions by substituting the same value into each expression to see whether the results were equal. Do you think this test should work? What did you learn from the results of their tests?

✅ Develop & Understand: B

Math Link

Subtracting a number is equivalent to adding its opposite. For example, $3 - 5 = 3 + (-5)$.

Write each expression as simply as you can.

9. $3(x + 1) + 7(2 - x) - 10(2x - 0.5)$

10. $3a + 2(a - 6) + \frac{1}{2}(8 - 4a)$

11. $3y + 9 - (2y - 9) - y$

12. $(x^2 - 7) - 2(1 - x + x^2)$

In this addition chart, the expression in each tan cell is the sum of the first expressions in that row and column. For example, the sum of a^2 and $a(a-1)$ is $a^2 + a(a-1) = a^2 + a^2 - a = 2a^2 - a$.

+	a	$a(a-1)$
$a-1$	$2a-1$	a^2-1
a^2	a^2+a	$2a^2-a$

13. Copy and complete this chart by finding the missing expressions.

+	?	$2a(a-5)$	$a(2a+1)$
?		$2a^2-9a$	
$a(a+1)$	a^2+2a+3		
?			$3a^2+5a$

14. By completing the expression below, create an expression that simplifies to $4x - 3$.

$$3(x^2 + x - 2) + 2(\underline{})$$

Share & Summarize

Challenge your partner. Write an expression that simplifies to one of these three expressions.

$$3x - 1 \qquad 5x + 2 \qquad 3x^3 - 7x - 2$$

Include at least five terms in your expression. Let no more than two terms be single numbers. Include some terms with variables raised to a power. When you are done, swap with your partner. Figure out which expression above is equivalent to your partner's expression.

Inquiry

Investigation ③ Make the Cut

Materials

- 1-cm graph paper
- scissors

In this H-shape, *a* and *b* are positive numbers and the angles are all right angles.

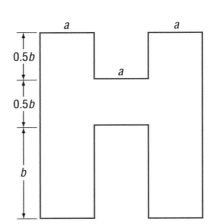

Analyze the H-Shape

1. On graph paper, draw an H-shape in which $a = 1$ cm and $b = 2$ cm.

2. Find the perimeter and the area of your H-shape.

3. Write an algebraic expression for the perimeter of your H-shape in terms of *a* and *b*.

4. Write an expression for the area of your H-shape in terms of *a* and *b*.

5. Use your expression from Question 3 to calculate the perimeter of your H-shape when $a = 1$ cm and $b = 2$ cm. Does it agree with your answer to Question 2?

6. Now use your expression from Question 4 to calculate the area of your H-shape when $a = 1$ cm and $b = 2$ cm. Does it agree with your answer to Question 2?

Transform the H-Shape

7. Create a new figure, different from the H-shape, with an area the same as the H-shape shown above but with a different perimeter.

   a. Draw your figure.

   b. Write an algebraic expression for your figure's area. Is it equivalent to the expression for the area of the H-shape from Question 4?

   c. Write an expression for your figure's perimeter.

   d. The expression for the perimeter of the figure you drew probably looks very different from the perimeter expression for the H-shape. Try to find values of *a* and *b* for which the perimeters of the two figures are the same. If you find such values, does that mean the perimeters of the general shapes are equivalent? Explain.

8. Now draw a figure that has the same perimeter as the H-shape but a different area. Draw such a figure.

    a. Write an expression for your figure's perimeter. Is it equivalent to the expression for the perimeter of the H-shape?

    b. Write an expression for your figure's area.

    c. Are there any values of a and b for which the areas of the two figures are the same? To answer this question, you might want to write and try to solve an equation.

Make a Rectangle

Copy the H-shape from page 214, cut it into pieces, and rearrange the pieces to form a rectangle. Keep track of the lengths of the sides of your pieces in terms of a and b. You will need this information later.

9. Draw the rectangle you formed from the H-shape. Label the lengths of the sides.

Make a Prediction

10. Without doing any calculations, think about how the perimeter of the original H-shape compares to the perimeter of your rectangle. Are they the same or different?

11. Without doing any calculations, think about how the area of the original H-shape compares to the area of your rectangle. Are they the same or different?

Check Your Prediction

12. Write an expression for the perimeter of your rectangle in terms of a and b. Check your prediction from Question 10. Are there specific values of a and b that would make the perimeters the same? Different?

13. Write an expression for the area of your rectangle in terms of a and b. Check your prediction from Question 11. Are there specific values of a and b that would make the areas the same? Different?

What Did You Learn?

14. Jenny thinks that if you increase the perimeter of a figure, the area must also increase. Write a letter to her explaining whether she is correct and why. Include examples or illustrations.

Investigation 4 Find Angle Measures Using Algebra

In the previous investigation, you simplified expressions by combining like terms. In this investigation, you will use this skill to simplify one side of an equation to make the equation easier to solve. The equations you solve, along with angle relationships you know, will help you find missing angle measures for geometric figures.

Think & Discuss

What do you recall about the angle relationships in figures like these?

✅ Develop & Understand: A

These two mathematical phrases both give the same result.

$$35 - 7 \qquad \text{and} \qquad 4 \cdot 7$$

If you use the variable h to represent 7, then the phrases can be written as expressions in terms of h.

$$35 - h \qquad \text{and} \qquad 4h$$

1. Using the number 10, give three different expressions that result in 20.

2. Using the variable n to represent the 10, write an expression for each of the mathematical phrases you wrote for Exercise 1.

3. Jaime drew a triangle with angle measures 50°, 25°, and 105°.

   a. Let the variable d represent the 25° angle measure. Write an expression in terms of d to represent each of the other angles.

   b. Write an equation showing the sum of the three angles is 180°. Use the expressions in terms of d to write the equation.

   c. Solve your equation for d. Did you get $d = 25$? Why or why not?

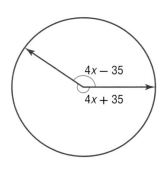

4. Write an equation for the sum of the angles in the figure. Solve for x. Determine the measure of each angle in the figure.

$4x - 35$

$4x + 35$

5. Matt drew a parallelogram that was not rectangular. He labeled the acute angles a and the obtuse angles b. He measured the angles and found that the measure of b was twice the measure of a.

    a. Draw a sketch to represent Matt's parallelogram. Label its angles a and b as appropriate.

    b. Write an expression to rename b in terms of a.

    c. Describe how you would find the measure of each angle in Matt's parallelogram.

Think & Discuss

Create an exercise for another triangle or quadrilateral similar to Exercise 3. Draw a sketch of your figure. Label the angles with the expressions you created. Trade figures with your neighbor, write the equation, and solve to find each angle measure.

✓ Develop & Understand: B

6. Write an equation to show the sum of the angles in each of these figures. Solve each equation to find each missing angle measure.

    a.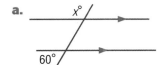

    b.

    c.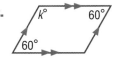

7. Write an equation to show the sum of the angles for this triangle.

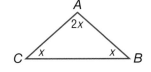

    a. Solve the equation to find the value of x.

    b. Use the value of x to determine the measure of each angle.

8. Write an equation to show the sum of the angles for this angle pair. Solve the equation to find the value of t.

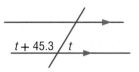

Use the value of t to determine the measure of each angle.

9. Write an equation to show the sum of the angles for this quadrilateral. Solve the equation to find the value of x.

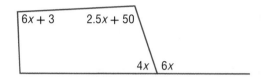

Use the value of x to determine the measure of each angle.

10. If you draw a right triangle where one acute angle is twice the measure of the other, what will be the measurements of each of the three angles of the triangle?

Share & Summarize

How can you use expressions and equations to find missing angle measures in geometric figures?

Draw a polygon with more than three sides and label the angles using algebraic expressions. Trade figures with a partner and determine the measure of each angle.

Practice & Apply

1. Start with a square with side length x cm. Imagine extending the length of one side to 7 cm.

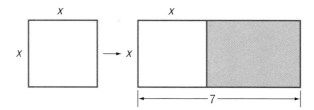

   a. What is the area of the original square? What is the area of the new, large rectangle?

   b. Use the areas you found to write an expression for the area of the shaded rectangle.

   c. Write expressions for the length and width of the shaded rectangle.

   d. Use the dimensions from Part c to write an expression for the area of the shaded rectangle.

   e. What do your answers to Parts b and d suggest about the expansion of $x(7 - x)$?

2. Use the distributive property to expand $3x(x - 2)$. Draw a rectangle diagram and use it to help explain why the two expressions are equivalent.

Use the distributive property to expand each expression.

3. $3z(z + 1)$

4. $\frac{1}{2}x(x - 2)$

Draw a rectangle diagram to match each expression. Use it to write the expression in factored form.

5. $2x^2 + x$

6. $2x^2 - x$

7. This diagram shows the grassy area between two buildings and the rectangular walkway through the middle of the area. The length of the grassy area is four times its width. The edges of the walkway are 2.5 meters from the sides of the rectangle.

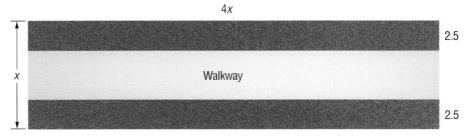

   Write an expression for the area of the walkway.

8. In Your Own Words Describe some of the steps you take when you simplify an algebraic expression. Explain how you know when an expression is simplified as much as possible.

Write each expression as simply as you can.

9. $2a(a + 2) - 4a + 3$

10. $n(n + 1) - n$

11. $x(3 - 2x) + 2(x^2 - 4)$

12. $p(3p - 4) - 2(3 - 5p)$

13. Complete this expression to create an expression that simplifies to $x + 2$.

$$-4(x + x^2 - 1) + \underline{\hspace{3cm}}$$

14. Sort these expressions into groups of equivalent expressions.

a. $5(x^4 - 1) - 10 - 2x^4 + 5 - 5x^2 + 2x^4 + 2x^2 + 7$

b. $3x^5 + 2x^4 + 3x^2$

c. $5x^5 + 2(x^4 - x^5) - 10 - 2x + 3x^2 + 2x + 3 + 7$

d. $-3x^2 - 3 + 5x^4$

e. $x + 4 + 5(x + x^2) - 8 + 2x - 10x + 5 - 2x - 5x^2$

f. $3(x^5 + x^2) + 2(x^4 + 3x) - 6x$

In Exercises 15–19, set up equations to find the measure of each angle.

15. Quadrilateral *ABCD* is a square. Triangle *EFG* is isosceles. The measure of angle *E* is twice the measure of angle *F*.

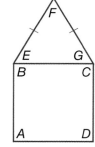

16.

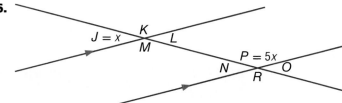

17.

18. Figure *HIJLK* is a regular pentagon. The sum of the angles in a regular pentagon is 540°.

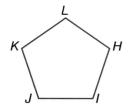

19. Line *l* is perpendicular to line *m*. Solve for *x* and determine the value of angles *A* and *B*.

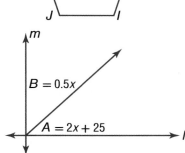

$B = 0.5x$

$A = 2x + 25$

Connect & Extend **For each expression, copy the diagram and shade an area that matches the expression.**

20. $(2x)^2$

21. $2x^2$

22. $x(2x + 1)$

23. $2x + 2$

24. $(2x + 2)^2$

25. $x(2x + 2)$

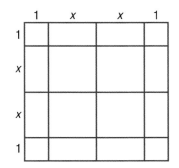

Expand each expression.

26. $x(a - b + c)$

27. $\frac{k}{7}(21a - 0.7)$

28. $\frac{x}{3}\left(\frac{a}{2} - \frac{b}{3} + \frac{c}{4}\right)$

29. $4w(4w - 2x - 1)$

30. This diagram shows a drawer and the surrounding cabinet. The drawer is 2 inches wider than it is tall. There is a 1-inch gap between the drawer and the outside of the cabinet on all sides. The length of both the drawer and the cabinet is *y*.

Write two equivalent expressions for the volume of the drawer, one in factored form and one in expanded form.

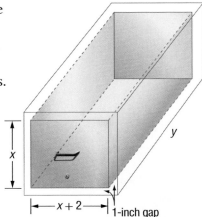

x

$x + 2$

1-inch gap

y

31. Ben and Lucita are discussing algebraic expressions.

a. After the one-cm strip is removed, there are two ways to fold the remaining piece of paper. For each possibility, write an expression for the area of the final piece, after the paper is folded and then cut in half. Are the expressions equivalent? Explain.

b. Ben posed a new situation. *Imagine a square with side length x. Fold the square in half, cut it, and throw away one half. Now remove a one-cm strip from the remaining half. Write an expression for the area of the remaining piece.*

Lucita said there are two ways to interpret Ben's instructions. Find both ways. Write an expression for each. Are the two expressions equivalent? Explain.

32. A construction worker made a stack of bricks 16 layers high. Each layer consists of three bricks arranged in the pattern shown below. Each brick is twice as long as it is wide, and each has a thickness $1\frac{1}{2}$ inches less than its width.

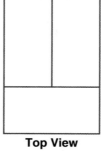

Top View   **Side View**

Layer 4
Layer 3
Layer 2
Layer 1

a. How many bricks are in the 16-layer stack?

b. Write an expression for volume of the stack, using *w* for the width of each brick.

33. Garrett and his dad needed to roughly measure the length and width of their land before they could determine a possible selling price. Garrett's dad had a long piece of rope that they used as a measuring tool. His dad was not certain of the length of the rope. They paced off the land and found the following.

   a. The length of the land was 11 times the length of the rope.

   b. The width of the land was seven times the length of the rope plus six more feet.

   c. They folded the rope into fourths. It measured ten feet long.

   Find the length and width of the land. Use expressions, equations, and sketches to show your thinking.

Mixed Review

Find the value of *c* in each equation.

34. $\sqrt[c]{8} = 2$

35. $\sqrt[c]{81} = 3$

36. $\sqrt[4]{1} = c$

37. Science Civil engineers are designing a holding tank for the water system of a small community. The tank will be in the shape of a cylinder and will have a height of 15 meters. They are trying to determine the best size for the tank's radius.

15 m

   a. Fill in a table like the one below for the volume of the holding tank for each given radius.

Radius (m)	4	5	6	7	8	9
Volume (m³)						

   b. Plot the points from your table on a grid like this one.

   c. Use your graph to estimate what the radius of the tank should be if the tank is to hold 3,500 m³ of water.

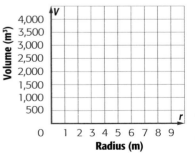

Tank Specifications

Find an equation of the line passing through the given points.

38. $(9, -0.5)$ and $(-1, 4.5)$

39. $(1, -2)$ and $(-4, -3)$

Monomials, Binomials, and Trinomials

In Lesson 5.1, you examined rectangle diagrams consisting of a square with a rectangular strip added to or taken away from one side. What if you added or removed *two* rectangular strips?

Vocabulary

binomial

monomial

trinomial

Think & Discuss

Start with a square with side length m cm. Add a three-centimeter strip to one side.

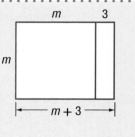

Now add a one-centimeter strip to an adjacent side of the new rectangle.

What is the area of the final large rectangle? Describe how you found it. Are there other ways to find this area?

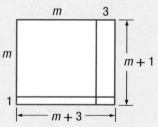

In Lesson 5.1, you used the distributive property to multiply expressions in the forms $m(m + a)$ and $m(m - a)$. That is, you found the product of a number or variable and a binomial. A **binomial** is the sum or difference of two unlike terms.

The expression $x + 5$ is a binomial because it is the sum of two unlike terms. Similarly, $x^2 - 7$ is a binomial. It is the difference of two terms that cannot be combined into one term.

The expression $x + 2x$ is not a binomial. Its terms are like terms. The expression is equivalent to $3x$, a **monomial**. Expressions such as $x^2 + x - 1$ have three unlike terms, it is called a **trinomial**.

The area of the final large rectangle in Think & Discuss is the product of *two* binomials, $m + 3$ and $m + 1$. In this lesson, you will learn how to multiply two binomials.

Investigation 1 Multiply and Divide by Monomials

In Lesson 5.1, you worked with rectangle diagrams to understand and show equivalent expressions. For example, you determined that $h(h + 1) = h^2 + h$.

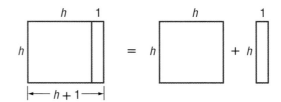

This helped you to see that to multiply h times $h + 1$, you had to *distribute* the multiplication over both terms in $h + 1$.

✅ Develop & Understand: A

1. Sketch an area model to help you find $3p(2p + 4)$. What are the dimensions of the rectangle? What is the area of the rectangle?

2. What do you notice about multiplying a monomial with the variable p times another term with the variable p? What type of result occurs with an expression like $2n(5n)$?

3. Determine the product of the following multiplication expressions. How is the distributive property being used in these exercises?

    a. $a(3a + 5)$

    b. $2w(2w)$

    c. $4m(m - 2)$

    d. $3(2k + 3)$

    e. $3x(7 - x)$

4. What multiplication expression, including a monomial and binomial, could produce the product $8d^2 + 12d$?

Fact families help you think about the relationships between inverse operations. In the past, you have used fact families to solve expressions such as the following.

$\dfrac{15}{n} = 3$ is rewritten as

$15 = 3n$ so that you are now thinking "3 times what number equals 15?"

$15 = 3 \cdot 5$

Therefore, $n = 5$

You may recall this kind of fact family from elementary school.

$3 \cdot 5 = 15$ $5 \cdot 3 = 15$ $15 \div 5 = 3$ $15 \div 3 = 5$

Use the idea of fact families to think about division of a binomial by a monomial.

✅ Develop & Understand: B

5. Consider $\dfrac{4p^2 + 2p}{2p}$. Using the idea of fact families, how would you rewrite this as a multiplication expression with a missing fact? What does this expression tell you?

6. What is one binomial that makes this related fact family true?

Fill in the boxes in the equations below.

7. $2p \cdot \boxed{} = 4p^2$

8. $2p \cdot \boxed{} = 2p$

9. $2p \cdot \boxed{} = 4p^2 + 2p$

10. Look at Exercises 7–9. What pattern do you notice in the expressions on the right side of the equation? What do you notice about the answers you found for the same exercises?

Think & Discuss

Using Exercises 7–9 as an example, discuss two other situations where you are using related facts to divide a binomial by a monomial. Do you always see the same pattern?

✅ Develop & Understand: C

In the division chart, the expression in each cell is the quotient of the first expressions in that row and column. For example, the quotient of $8t^2 - 14t$ and 2 is $4t^2 - 7t$.

11. Copy and complete this chart by finding the missing expressions.

÷	$8t^2 - 14t$	$2t^2 + 16t$
2	$4t^2 - 7t$	
8		
4t	$2t - 3.5$	

12. Use the chart to find the missing expressions in the algebraic fact family.

a. $\dfrac{(8t^2 - 14t)}{4t}$ _____ $\cdot \, 4t = 8t^2 - 14t$

b. $\dfrac{2t^2 + 16t}{8}$ _____ $\cdot \, 8 = 2t^2 + 16t$

Share & Summarize

1. Use the distributive property to expand $bx(cx + d)$.

2. Describe how you would divide $\dfrac{ax^2 + bx}{cx}$.

Investigation 2   Use Geometric Models to Multiply Binomials

The geometric model you used to think about multiplying a term and a binomial can be adapted for multiplying two binomials.

✅ Develop & Understand: A

1. Look at this rectangle.

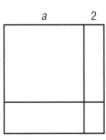

a. Write two expressions, one for the length of the large rectangle and one for the width.

b. Use your expressions to write an expression for the area of the large rectangle.

c. Use the diagram to expand your expression for the area of the large rectangle. That is, write the area of the large rectangle without using parentheses.

2. Arturo wanted to expand $(x + 4)(x + 3)$. To do this, he drew a rectangle $x + 4$ units wide and $x + 3$ units high.

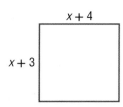

He then drew lines to break the rectangle into four parts.

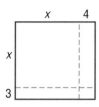

Arturo wrote, "Area of large rectangle $= (x + 4)(x + 3)$."

He then found the area of each of the four smaller parts and used them to write another expression for the area of the large rectangle. Finally, he simplified his expression by combining like terms. What was his final expression?

3. For Parts a–c, do the following.
- Draw a rectangle diagram to model the product.
- Use your diagram to help expand the expression.

  a. $(m + 7)(m + 2)$ **b.** $(w + 2)^2$ **c.** $(2n + 3)(n + 1)$

4. A certain rectangle has area $y^2 + 6y + 3y + 18$.

  a. Draw a rectangle diagram that models this expression.

  b. Use your diagram to help you rewrite the area expression as a product of two binomials.

5. Challenge Another rectangle has an area of $y^2 + 5y + 6$.

  a. Draw a rectangle diagram that models this expression.

  b. Use your diagram to help you rewrite the area expression as a product of two binomials.

Real-World Link

A geometric model of carbon-60 is composed of 60 interlinked carbon atoms arranged in 12 pentagons and 20 hexagons. Because of its structural similarity to the geodesic dome, designed by U.S. architect R. Buckminster Fuller, it was named *buckminsterfullerene.*

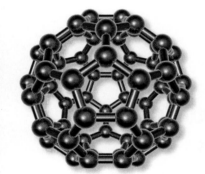

Share & Summarize

Ben thinks that $(n + 3)(n + 5) = n^2 + 15$. Show him why he is incorrect. Include a rectangle diagram that models the correct expansion of $(n + 3)(n + 5)$.

Rectangle models can help you understand how the distributive property works. For example, the expression $m(m + 3)$ is expanded below with the distributive property and a rectangle diagram.

$$m(m + 3) = m \cdot m + m \cdot 3$$
$$= m^2 + 3m$$

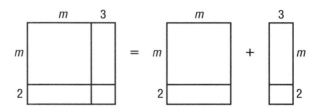

You can also use the distributive property and rectangle diagrams to expand such expressions as $(m + 2)(m + 3)$. Just think of $m + 2$ in the same way you thought about the first variable m in the expression $m(m + 3)$. That is, multiply $m + 2$ by each term in $m + 3$.

$$(m + 2)(m + 3) = (m + 2) \cdot m + (m + 2) \cdot 3$$
$$= m(m + 2) + 3(m + 2)$$

The distributive property can then be used to simplify each term. Start by simplifying the first term, $m(m + 2)$.

$$m(m + 2) = m \cdot m + m \cdot 2$$
$$= m^2 + 2m$$

Then simplify the second term, $3(m + 2)$.

$$3(m + 2) = 3 \cdot m + 3 \cdot 2$$
$$= 3m + 6$$

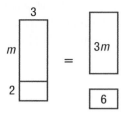

Finally, put everything together and combine like terms.

$$(m + 2)(m + 3) = m(m + 2) + 3(m + 2)$$
$$= m^2 + 2m + 3m + 6$$
$$= m^2 + 5m + 6$$

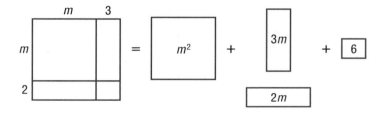

✓ Develop & Understand: A

Expand each expression using the distributive property. Draw a rectangle diagram to model each expression. Check your expansion.

1. $(x + 3)(x + 4)$

2. $(k + 5)^2$

3. $(x + a)(x + b)$

4. $(2x + 1)(3x + 2)$

4. Complete the table by substituting values for x into this equation.

$$y = x(x + 1) - 3(x + 1)(x + 2) + 3(x + 2)(x + 3) - (x + 3)(x + 4)$$

x	0	1	2	3	4	5
y						

5. Make a conjecture about the value of y for other values of x. Test your conjecture by trying more numbers, including some decimals and negative numbers. Revise your conjecture if necessary.

6. Prove It! Use your knowledge of expanding binomial products to show that your conjecture is true.

Share & Summarize

Use the distributive property to expand $(a + b)(c + d)$. Draw a rectangle diagram. Explain how it shows your expansion is correct.

Investigation 4 — Multiply Binomials That Involve Subtraction

You have used rectangle models to think about multiplying binomials involving addition, expressions of the form $(a + b)(c + d)$. Now you will learn to expand products of binomials that involve subtraction.

Example

Here is one way to create a rectangle diagram to represent $(d - 1)(d + 3)$.

First, draw a square with side length d.

Then subtract a 1-cm strip from one side. Add a 3-cm strip to the adjacent side.

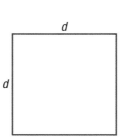

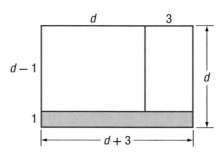

The unshaded rectangle that remains has an area of $(d - 1)(d + 3)$.

> ### Think & Discuss
>
> Expand the expression $(d - 1)(d + 3)$ using either the distributive property or the diagram in the page 223 example. Describe how you found the answer.

✓ Develop & Understand: A

Expand the expression using the distributive property. Draw a rectangle diagram that represents the expression. Use shading to indicate areas that are being removed. Use your diagram to check that your expansion is correct.

1. $(b - 2)(b + 3)$

2. $(a + 1)(a - 4)$

3. $(2 + e)(3 - e)$

4. A certain rectangle has area $y^2 + 5y - 2y - 10$.

    a. Draw a rectangle diagram to represent this expression.

    b. Write the area of this rectangle as a product of two binomials.

You will now use the distributive property to expand products of two binomials that both involve subtraction.

✓ Develop & Understand: B

Expand each expression using the distributive property. Then combine like terms. Your final answer should have no parentheses and no like terms.

5. $(x - 4)(x - 5)$

6. $(R - 2)^2$

7. $(2 - f)(3 - f)$

8. $(a - 2b)(3a - b)$

You will now apply what you have learned about expanding products of binomials to analyze some number tricks.

✅ Develop & Understand: C

Olivia thinks she has found some number tricks.

Prove It! In Exercises 9 and 10, determine whether Olivia's trick really works. If it does, prove it. If not, give a counterexample.

Math Link
A *counterexample* is an example for which a conjecture does not work.

9. Olivia said, "Take any four consecutive integers. Multiply the least number and the greatest number. Then multiply the remaining two numbers. If you subtract the first product from the second, you will always get 2.

   For example, for the integers 3, 4, 5, and 6, the product of the least and greatest numbers, 3 and 6, is 18. The product of the remaining two numbers, 4 and 5, is 20. The difference between these products is 2." (Hint: If the least integer is x, what are the other three?)

10. Olivia said, "Take any three consecutive integers and multiply them. Their product is divisible by 4. For example, the product of 4, 5, and 6 is 120, which is divisible by 4."

11. Here is another number trick Olivia proposed. "Take any two consecutive even integers, multiply them, and add 1. The result is always a perfect square. For example, 4 and 6 multiply to give 24. Add 1 to get 25, which is a perfect square."

    a. Since the numbers are both even, they have 2 as a factor. Suppose the lesser number is $2x$. What is the greater number?

    b. Using $2x$ and the expression you wrote for Part a, find the resulting expression, which Olivia claims is always a perfect square.

    c. Assume the result in Part b is a perfect square. Try to draw a rectangle diagram showing the binomial that can be squared to get that product.

    d. What binomial is being squared in Part c?

    e. Have you proven that Olivia's trick always works? Explain.

Share & Summarize

Use the distributive property to expand $(a - b)(c + d)$. Then draw a rectangle diagram to show why your expansion is correct.

Investigation (5) Shortcuts to Multiply Binomials

You have been using the distributive property and rectangle diagrams to think about how to multiply two binomials. Have you noticed any patterns in your computations? In this investigation, you will look at patterns that can help you multiply binomials quickly and efficiently.

Here is how Mikayla multiplied $(2x + 7)(x - 3)$.

 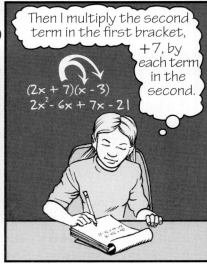 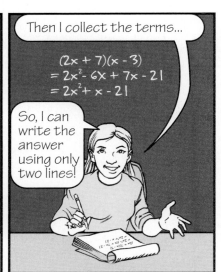

✓ Develop & Understand: A

Use Mikayla's method to multiply each pair of binomials.

1. $(y + 6)(y - 3)$

2. $(p + 4)(p + 3)$

3. $(t - 11)(t - 3)$

4. $(2x + 1)(3x + 2)$

5. $(2n + 3)(2n - 3)$

Think & Discuss

Why does Mikayla's method work?

Here is how Tamika approaches these equations.

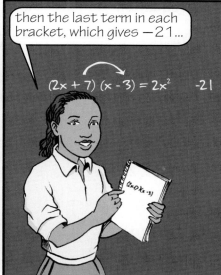

✅ Develop & Understand: B

Use Tamika's method to multiply each pair of binomials.

6. $(y + 7)(y - 4)$

7. $(p + 1)(p - 5)$

8. $(t - 4)(t - 4)$

9. $(2x - 1)(x - 2)$

10. $(3n + 2)(2n - 3)$

Think & Discuss

Why does Tamika's method work?

In the next exercise set, you will apply what you have learned about expanding binomials.

✅ *Develop & Understand: C*

For each equation, do Parts a and b.

   a. Decide whether the equation is true for all values of x, for some but not all values of x, or for no values of x.

   b. Explain how you know your answer is correct. If the equation is true for some but not all values of x, indicate which values make it true.

11. $(x - 2)(x - 3) = 0$

12. $(x + 3)(x + 2) = x^2 + 5x + 6$

13. $(x - b)^2 = x^2 - 2xb + b^2$

14. $(x + 3)(x - 1) = x^2 + 2x + 3$

15. $(x - 3)(x + 3) = x^2 - 6x - 9$

16. Brian thinks he has found some number patterns on a calendar. He says his patterns work for any 2-by-2 square on a calendar. Decide whether each of his patterns works. Justify your answers.

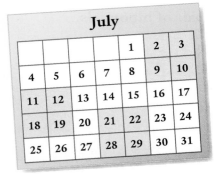

   a. Find the product of each diagonal. The positive difference is always 7. For example, for the square containing 2, 3, 9, and 10, the products of the diagonals are $2 \cdot 10 = 20$ and $3 \cdot 9 = 27$. The difference is $27 - 20 = 7$.

   b. Find the product of each column. The positive difference is always 12. For example, for the square containing 2, 3, 9, and 10, the products of the columns are $2 \cdot 9 = 18$ and $3 \cdot 10 = 30$. The difference is $30 - 18 = 12$.

   c. Find the product of each row. The difference is always even. For example, for the square containing 2, 3, 9, and 10, the products of the rows are $2 \cdot 3 = 6$ and $9 \cdot 10 = 90$. The difference is $90 - 6 = 84$, which is even.

Share & Summarize

You have seen several methods for expanding the product of two binomials.

1. Choose the method you like best. Explain how to use it to expand $(2x + 3)(x - 1)$.

2. Why do you like your chosen method?

Practice & **Apply**

For Exercises 1–3, write the multiplication equation for each rectangle diagram. Multiply to find the product, which is the area of the rectangle.

1.

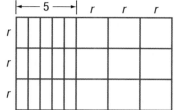

2.

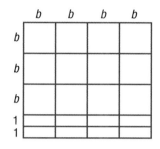

3.

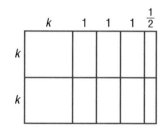

4. In Exercises 1–3, you wrote a multiplication equation with one factor that was a monomial and one factor that was a binomial and the product. Using fact families, write a division expression with a quotient related to each multiplication exercise above, with the monomial as the divisor.

5. Consider the expression $6x^2 + 15x$.

a. Draw a rectangle diagram to represent the expression. Find the dimensions of the rectangle.

b. Write the multiplication expression and related division expression that this rectangle diagram represents.

6. This diagram shows a rectangle with area $(a + 10)(a + 2)$.

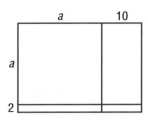

    a. Write an expression for the area of each of the four regions.

    b. Use your answer from Part a to expand the expression for the area of the large rectangle. That is, express the area without using parentheses. Simplify your answer by combining like terms.

7. This diagram shows a rectangle with area $(y + 9)(y + 8)$.

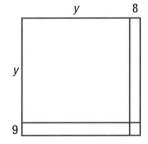

    a. Write an expression for the area of each of the four regions.

    b. Use your answer from Part a to expand the expression for the area of the large rectangle. That is, express the area without using parentheses. Simplify your answer by combining like terms.

For Exercises 8 and 9, draw a rectangle diagram to model each product. Use your diagram to expand the product. Simplify your answer by combining like terms.

8. $(3 + 2k)(4 + 3k)$

9. $(1 + 4x)(x + 2)$

10. A certain rectangle has area $y^2 + 5y + 2y + 10$.

    a. Draw a rectangle diagram that models this expression.

    b. Use your diagram to help you rewrite the area expression as a product of two binomials.

11. A certain rectangle has area $2y^2 + 6y + y + 3$.

    a. Draw a rectangle diagram that models this expression.

    b. Use your diagram to help you rewrite the area expression as a product of two binomials.

12. Consider the expression $(p + 3)(p + 5)$.

    a. Use the distributive property to expand the expression.

    b. Draw a rectangle diagram to model the expression. Check your expansion.

Use the distributive property to expand each expression.

13. $(1 + 3a)(5 + 10a)$

14. $3(2x + \frac{1}{3})(x + 2) - (x + 3)(1 + x)$

Draw a rectangle diagram to model each product. Then expand the product using your diagram. Simplify your answer by combining like terms.

15. $(x + 3)(x - 3)$ **16.** $(p - 4)(3p + 2)$

17. Consider the expression $(h - 2)(h + 2)$.

   a. Use the distributive property to expand the expression.

   b. Draw a rectangle diagram to represent the expression. Shade areas that are being removed. Use your diagram to check that your expansion is correct.

Expand each expression using the distributive property.

18. $(x - 7)(x - 2)$ **19.** $(3 - g)(4 - g)$

20. $(4 - 2p)(4 - p)$ **21.** $(2w + 1)(w - 6)$

22. $(1 - 5q)(2 + 2q)$ **23.** $(3v - 5)(v + 1)$

24. A certain rectangle has area $y^2 - 4y + 8y - 32$.

   a. Draw a rectangle diagram to represent this expression.

   b. Use your diagram to help you rewrite the area expression as a product of two binomials.

25. Challenge A certain rectangle has area $2y^2 + 4y - 3y - 6$.

   a. Draw a rectangle diagram to represent this expression.

   b. Use your diagram to help you rewrite the area expression as a product of two binomials.

Prove It! In Exercises 26 and 27, determine whether each number trick works. If it does, prove it. If not, give a counterexample.

26. Take any three consecutive integers. Multiply the least and the greatest. That product is equal to the square of the middle integer minus 1.

27. Think of any two consecutive odd integers. Square both integers, and subtract the lesser result from the greater. The result is always evenly divisible by 6.

Expand each expression. Simplify your expansion if possible.

28. $(4x + 1)(4x - 1)$

29. $(r - 12)(r - 12)$

30. $(2x + 2)(x - 2)$

31. $(4x + 1)^2$

32. $(5M + 5)^2$

33. $(n + 1)^2 + (n - 1)^2$

Decide whether each equation is true for all values of x, for some but not all values of x, or for no values of x.

34. $(x + 3)(x - 4) = x^2 + 7x - 12$

35. $(2x + 1)(x - 1) = 2x^2 - x - 1$

36. $(3x + 1)(3x - 1)x = 9x^3 - x - 1$

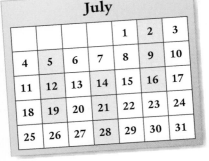

July

37. After working with the calendar situations in Exercise 16 in Develop & Understand C, Chapa wrote one. "Choose a block of three dates in a column so they are all the same day of the week. Square the middle date, and then subtract the product of the first and the last dates. The result is always 49."

Does Chapa's number trick always work? If so, show why. If not, give a counterexample.

38. Consider the graph at the right.

a. Write an expression for the area of the unshaded region.

b. Write two expressions, one with and one without parentheses, for the area of the shaded region.

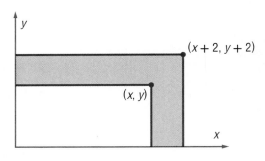

39. Write an expression for the area of the shaded region in the graph below.

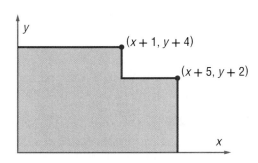

Expand each expression. Simplify your results by combining like terms.

40. $(1 + x^2)(x + y) + (1 + y)(x + y) - x^2(x + y) - xy$

41. $2(x + y) + x(3 + y)(x + 2)$

Expand the following products of a trinomial and a binomial. You may find it helpful to use a rectangle diagram.

42. $(x + y + 1)(x + 1)$

43. $(a + b + 1)(a + 2)$

44. $(x + y + 2)(x + 1)$

45. Consider the product $(2x + y)(x + 2y)$.

   a. Expand the expression using the distributive property.

   b. Draw a rectangle diagram that represents the expression. Use your diagram to check that your expansion is correct.

46. A block of wood has length y and a square base with sides of length x. A woodworker will cut the block of wood twice, taking off two strips from adjacent sides. Each cut removes $1\frac{1}{8}$ inches.

   a. Write an expression for the volume of the wood before any cuts are made.

   b. Write an expression without parentheses for the volume of the wood after the cuts are made.

47. Consider the product $(2a + 3)(3a - 4)$.

   a. Expand and then simplify the expression.

   b. Draw a rectangle diagram to represent the expression. Use your diagram to check that your expansion is correct.

Expand the following products of a trinomial and a binomial.

48. $(x + y + 1)(x - 1)$

49. $(x - y - 1)(x + 1)$

50. In Your Own Words Explain why $(x + a)(x + b) \neq x^2 + ab$ by applying the distributive property and by using a rectangle diagram.

51. **Preview** Expand and simplify the expressions in Parts a–d.

   a. $(x - 3)^2$

   b. $(x - 4)^2$

   c. $(x - 5)^2$

   d. $(x - a)^2$

   e. Look for a pattern relating the factored and expanded forms of the expressions in Parts a–c. Describe a shortcut for expanding the square of a binomial difference.

52. Consider the expression $y^2 - 9$.

   a. Write the expression as the product of two binomials.

   b. Create a diagram to illustrate $y^2 - 9$ as a rectangular area.

Expand and then simplify each expression.

53. $(n + 2)^2 - n(n + 4)$ 54. $(n + p)^2 - n(n + 2p)$

55. Fill in the boxes in the equations below.

   a. $4p \cdot \boxed{} = 20p^2$

   $4p \cdot \boxed{} = 8p$

   $4p \cdot \boxed{} = 20p^2 + 8p$

   b. $2m \cdot \boxed{} = 5m^2$

   $2m \cdot \boxed{} = 12m$

   $2m \cdot \boxed{} = 5m^2 + 12m$

Examine the expression $12x^2 + 20x$. Sometimes in algebra, you will see the direction to "factor out" a factor that is shared by terms in an expression. For example, to "factor out" $2x$ from the expression $12x^2 + 20x$, you would first mentally divide $12x^2 + 20x$ by $2x$ and then write the two factors as a multiplication expression. The result would be $2x(6x + 10)$.

56. For the expression $12x^2 + 20x$, "factor out" $4x$. Write the multiplication expression.

57. Consider the expression $8h^2 + 24h$.

   a. Factor out $2h$. Write the multiplication expression.

   b. What other factors do $8h^2$ and $24h$ have in common that could be factored out?

Mixed Review

Write an equation of a line that is parallel to the given line.

58. $y = 7x - 6$

59. $2x = 4y$

60. $x = -2$

61. Match each equation to a line.

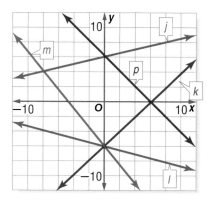

   a. $y = -\dfrac{5}{4}x - 6$

   b. $y = 0.25x + 6$

   c. $y = -0.25x - 6$

   d. $y = -x + 6$

   e. $y = x - 6$

62. Refer to the graph in Exercise 61. On which line or lines does each point lie?

   a. $(0, -6)$

   b. $(6, 0)$

   c. $(-4, -1)$

   d. $(-2, 8)$

   e. $(-8, -4)$

   f. $(-2, -2)$

Evaluate without using a calculator.

63. $\sqrt{(-18)^2}$

64. $-\sqrt{7^2}$

65. $-\left(\sqrt{64}\right)^2$

66. $\left(-\sqrt{49}\right)^2$

Special Products

You have learned several methods for expanding products of binomials. Some binomials have products with identifiable patterns. Recognizing these patterns will make your work easier.

Explore

Expand and simplify each product.

$(x + 1)^2 = (x + 1)(x + 1) =$

$(x + 2)^2 =$

$(x + 3)^2 =$

$(x + 4)^2 =$

Describe the pattern that you see in your work. Use the pattern to predict the expansion of $(x + 10)^2$.

Check your prediction by expanding and simplifying $(x + 10)^2$.

Investigation 1 Perfect Square Trinomials

In this investigation, you will learn some shortcuts for expanding squares of binomials into *trinomials*.

✓ Develop & Understand: A

1. Apply the pattern that you discovered in Explore to expand these squares of binomials.

   a. $(m + 9)^2$ **b.** $(m + 20)^2$ **c.** $(m + 0.1)^2$

2. Apply the pattern to predict the expansion of $(x + a)^2$. Check your answer by using the distributive property.

3. Use the pattern that you discovered to explain, without calculating, why $100^2 \neq 93^2 + 7^2$. Is 100^2 greater than or less than $93^2 + 7^2$?

4. In Exercise 3, you saw that $(93 + 7)^2 \neq 93^2 + 7^2$. Are there any values of x or a for which $(x + a)^2$ *does* equal $x^2 + a^2$? If so, what are they? How do you know?

You have just studied expressions of the form $(a + b)^2$. Now you will look at expressions of the form $(a - b)^2$.

✅ Develop & Understand: B

5. Recall that $(x - 1)^2$ can be thought of as $(x + (-1))^2$.

   a. Use this fact, along with your findings in Exercises 1–4, to expand $(x - 1)^2$.

   b. Check your answer to Part a by using the distributive property to expand $(x - 1)^2$.

6. Expand each expression using any method you like.

   a. $(m - 9)^2$

   b. $(m - 20)^2$

   c. $(m - 0.1)^2$

7. What is the expansion of $(x - a)^2$?

8. Use a rectangle diagram to help explain your answer to Exercise 7.

By now, you have seen that these two statements are true.

$$(a + b)^2 = a^2 + 2ab + b^2 \qquad (a - b)^2 = a^2 - 2ab + b^2$$

The variables a and b can represent any expressions.

$$(2x + 3y)^2 = (2x)^2 + 2 \cdot 2x \cdot 3y + (3y)^2 = 4x^2 + 12xy + 9y^2$$

Expand each expression.

9. $(3m + 2)^2$

10. $(2x - y)^2$

11. $(2m - 4n)^2$

12. $(g^2 - a^4)^2$

13. Imagine a square garden surrounded by a border of square tiles that are one unit by one unit. A variety of sizes are possible.

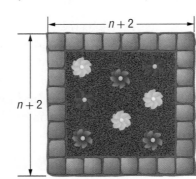

a. The garden has sides of length *n*, not including the tiles, where *n* is divisible by the length of a tile. Write an expression for the area of the ground covered by both the garden and the tiles.

b. Write an expression for the number of tiles needed for the garden in Part a.

14. Prove It! Ellis noticed that when he found the difference of 3^2 and 4^2, or $16 - 9 = 7$, the result was odd. This is also true for the difference of 6^2 and 7^2, which is $49 - 36 = 13$. He conjectured that the difference of squares of consecutive numbers is always odd.

Prove Ellis' conjecture, if possible. If not, give a counterexample.

> ### *Share & Summarize*
>
> Sabrina is confused about squaring binomials. She thinks that for any numbers *a* and *b*,
>
> $$(a + b)^2 = a^2 + b^2 \text{ and } (a - b)^2 = a^2 - b^2.$$
>
> Write her a letter explaining why she is incorrect. Include the correct expansions of $(a + b)^2$ and $(a - b)^2$.

Investigation (2) Differences of Squares

In Investigation 1, you used a shortcut to expand squares of binomials such as $(a + b)^2$ and $(a - b)^2$. In this investigation, you will find a shorter method for a different kind of product of binomials.

✅ Develop & Understand: A

1. Expand each expression.

   a. $(x + 10)(x - 10)$ **b.** $(k + 3)(k - 3)$

   c. $(S + 1)(S - 1)$ **d.** $(x + 5)(x - 5)$

   e. $(2t + 5)(2t - 5)$ **f.** $(3y - 7)(3y + 7)$

2. How are the factors of the original products in Exercise 1 similar?

3. How are the expansions of the products in Exercise 1 similar?

4. **Prove It!** Expand $(x + a)(x - a)$. Show that the pattern you noticed in Exercise 3 will always be true for this kind of product.

5. If the expression fits the pattern of the products in Exercise 1, expand it using the pattern you described in Exercise 3. If it does not fit the pattern, state it in your answer.

   a. $(x + 20)(x - 20)$

   b. $(b + 1)(b - 1)$

   c. $(n - 2.5)(n - 2.5)$

   d. $\left(2m - \frac{1}{2}\right)\left(m + \frac{1}{2}\right)$

   e. $(J + 0.2)(J - 0.2)$

   f. $(z + 25)(z - 100)$

   g. $\left(2n - \frac{1}{3}\right)\left(2n + \frac{1}{3}\right)$

   h. $(3 - p)(3 + p)$

6. Find two binomials with a product of $x^2 - 49$.

7. Some people call the expanded expressions that you wrote in Exercises 1 and 5 *differences of squares*. Explain why this name makes sense.

Example

The shortcut that you found in Exercises 1–7 can help you efficiently solve some difficult-looking computations.

Develop & Understand: B

8. Show how to use Lydia's method to calculate 99 · 101 without using a calculator.

Use a difference of squares to calculate each product. Check the first few until that you are confident that you are doing it correctly.

9. 49 · 51

10. 28 · 32

11. 43 · 37

12. 35 · 25

13. 4.1 · 3.9

14. −14 · 16

For which products in Exercises 15–18 do you think using a difference of squares would be a reasonable method of calculation? If it seems reasonable, find the product.

15. 41 · 38

16. 99 · −101

17. $10\frac{1}{4} \cdot 9\frac{3}{4}$

18. 1.2 · −0.7

19. Think about the kinds of products for which it is helpful to use differences of squares.

a. Create a set of three multiplication expressions for which you might want to use a difference of squares. Do them yourself. Record the answers.

b. Give your exercises to a partner to solve while you do your partner's set. Check that you agree on the answers and that the method works well.

20. If you combine the difference-of-squares method of fast calculation with some other mathematical tricks, you can do even more computations in your head.

  a. Consider the product $32 \cdot 29$.

    i. Explain why $32 \cdot 29 = 31 \cdot 29 + 29$.

    ii. Now use the difference-of-squares pattern to help compute the value of $31 \cdot 29$.

    iii. Finally, use the product of 31 and 29 to compute $32 \cdot 29$.

  b. Compute $21 \cdot 18$ in your head.

  c. How did you find the answer to Part b?

Share & Summarize

1. Suppose you want to expand an expression containing two binomials multiplied together.

  a. How do you know whether you can apply the shortcut you used to expand some of the products in Exercise 5? Write two unexpanded products to help show what you mean.

  b. Describe the shortcut you use to write the expansion.

2. You may have noticed that when you multiply two binomials, you sometimes end up with another binomial and other times with an expression with three unlike terms, or a *trinomial*.

  a. Create a product of two binomials that results in a binomial.

  b. Create a product of two binomials that results in a trinomial.

Practice & Apply

Expand and simplify each expression.

1. $(a + 5)^2$

2. $(m + 11)^2$

3. $(x + 2.5)^2$

4. $(t - 11)^2$

5. $(p - 2.5)^2$

6. $(2.5 - k)^2$

7. $\left(q - \frac{1}{4}\right)^2$

8. $(g^2 - 1)^2$

9. $(s^2 - y^2)^2$

10. $(3f + 2)^2$

11. $(3x + y)^2$

12. $(3m - 2n)^2$

13. Imagine a rectangular swimming pool in which the bottom is made of large square tiles. The pool is 25 tiles longer than it is wide. Around the edge of the pool, at the top, is a border made of the same square tiles. The border is one tile wide.

  a. If n represents the number of tiles in the width of the bottom of the pool, write an expression for the total number of tiles in the bottom of the pool.

  b. Write an expression for the number of tiles in both the bottom of the pool and the border.

Expand and simplify each expression.

14. $(10 - k)(10 + k)$

15. $(3h - 5)(3h + 5)$

16. $(0.4 - 2x)(0.4 + 2x)$

17. $\left(\frac{1}{5} + k\right)\left(\frac{1}{5} - k\right)$

Write each expression as the product of binomials.

18. $4x^2 - 1$

19. $16 - 25x^2$

20. $x^2 - y^2$

Write each product as a difference of squares. Use this form to calculate the product in Exercises 21–24.

21. $35 \cdot 45$

22. $27 \cdot 33$

23. $207 \cdot 193$

24. $111 \cdot 89$

Look back at your work in Exercise 20 on page 249. Combine the difference-of-squares method with addition to calculate each product in Exercises 25 and 26.

25. $12 \cdot 9$

26. $37 \cdot 25$

Connect & Extend **Expand and simplify each expression.**

27. $\left(\frac{x}{2} + \frac{y}{2}\right)^2$

28. $(3 - xy)^2$

29. $(xy - x)^2$

30. $(2xy - 1)^2 + (2xy - y)^2$

31. Write an expression without parentheses for the area of the shaded triangle.

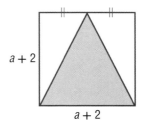

$a + 2$

$a + 2$

32. Challenge For what values of x and a is $(x + a)^2 > x^2 + a^2$? Justify your conclusion. (Hint: Expand the expression on the left side of the inequality.)

Expand and simplify each expression.

33. $(x^2 - y^2)(x^2 + y^2)$

34. $(1 - y^3)(1 + y^3)$

35. $(xy - x)(xy + x)$

Find the values of a and b that make each equation true.

36. $2x^2 + 7x + 3 = (a + bx)(3 + x)$

37. $20 - x - x^2 = (a + x)(b - x)$

38. $21 - 23x + 6x^2 = (3 - ax)(7 - bx)$

39. Physical Science An object is thrown straight upward from a height of four feet above the ground. The object's initial velocity is 30 feet per second. The equation relating the object's height in feet to the time t in seconds since it was released is $h = 30t - 16t^2 + 4$.

a. Find a and b to make this equation true.

$$30t - 16t^2 + 4 = (2 - t)(a + bt)$$

b. The statement $xy = 0$ is true whenever one of the factors, x or y, is equal to 0. For example, the solutions to $(k - 1)(k + 2) = 0$ are 1 and -2. These values make the factors $k - 1$ and $k + 2$ equal to 0. Use this fact and your result from Part a to find two solutions to the equation $30t - 16t^2 + 4 = 0$.

c. One of your solutions tells you at what time the object hit the ground. Which solution is it? How do you know?

40. This rectangle is divided into three triangular regions. Find an expression without parentheses for each of the three areas.

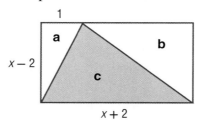

41. Challenge Katie and Gilberto used paper and scissors to convince themselves that $(a + b)(a - b)$ really is equal to $a^2 - b^2$.

a. Gilberto started with this paper rectangle. He wrote labels on the rectangle to represent lengths.

Gilberto then cut the rectangle and rearranged it.

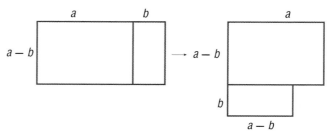

Explain why this shows that $(a + b)(a - b)$ is equal to $a^2 - b^2$.

b. Katie started with this paper square.

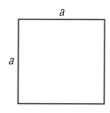

Then she cut the square and rearranged it.

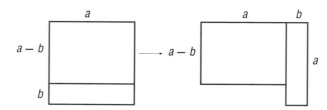

Explain why this shows that $(a + b)(a - b)$ is equal to $a^2 - b^2$. (Hint: The areas of the two diagrams above must be the same, so expressions representing those areas must be equal.)

42. **Challenge** Prove that for a quadratic equation, $y = ax^2 + bx + c$, the second differences in a table with consecutive inputs must be constant. (Hint: Suppose the inputs for the table are x, $x + 1$, $x + 2$, $x + 3$, and so on.) What are the corresponding outputs?

43. **In Your Own Words** Write two expressions involving multiplication that look difficult but that you can solve easily in your head using differences of squares. Explain how using differences of squares can help you find the products.

Mixed Review

Write an equation of a line that is parallel to the given line.

44. $2(y - 3) = -7x + 1$

45. $x = -2 - y$

In Exercises 46–49, tell which of the following descriptions fits the relationship.

- a direct variation
- linear but not a direct variation
- nonlinear

46. $r = 25v + 32$

47. $a = -\dfrac{5}{6}j$

48. $k = \dfrac{3}{n}$

49.

x	2	4	6	8	10	12
y	25	36	47	58	69	80

50. $x + 12x + 17 = x^2 + 12x + 36 - \underline{\hspace{1cm}} = (x + 6)^2 - \underline{\hspace{1cm}}$

51. $k^2 - 14k + 70$

52. $b^2 + 5b - \dfrac{3}{4}$

Review & Self-Assessment

Chapter Summary

This chapter focused on expanding and simplifying algebraic expressions. The main themes were using the distributive property and working with monomials, binomials, and trinomials.

You simplified expressions that involved multiplying binomials, requiring you to use the distributive property to expand the product and then to simplify by combining like terms.

Strategies and Applications

The questions in this section will help you review and apply the important ideas and strategies developed in this chapter.

Using geometric models to expand expressions

1. Draw a rectangle diagram that models the expression $(x + 3)(x + 6)$. Use it to expand the product of these binomials.

2. Draw a rectangle diagram that models the expression $(3t - 1)(t + 1)$. Use it to expand the product of these binomials.

3. Consider this graph.

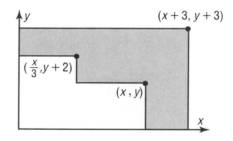

a. Write an expression for the area of the unshaded region.

b. Write an expression, with and without parentheses, for the area of the shaded region.

Using the distributive property to expand expressions

4. Describe the steps required to simplify $x(1 - x) + (1 - x)(3 - x)$. Give the simplified expression.

5. Simplify the expression $(x^2 - 1)(y^2 - 1) - (1 + xy)(1 + xy)$.

6. *Molding* is a strip of wood placed along the base of a wall to give a room a "finished" look. The diagram shows a floor plan of a room with one-inch-thick molding along the edges. Measurements are in inches.

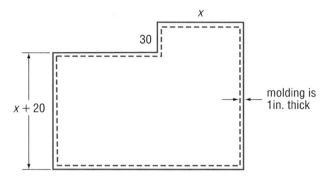

a. Write an expression for the floor area before the molding was installed. Simplify your expression as much as possible.

b. Write an expression for the remaining floor area after the molding was installed. Simplify your expression as much as possible.

Expanding expressions of the forms $(ax + b)^2$, $(ax - b)^2$, and $(ax + b)(ax - b)$

7. Consider expressions of the form $(ax + b)^2$.

a. Describe a shortcut for expanding such expressions. Explain why it works.

b. Use your shortcut to expand $(x + 13)^2$.

8. Consider expressions of the form $(ax - b)^2$.

a. Describe a shortcut for expanding such expressions. Explain why it works.

b. Use your shortcut to expand $\left(\frac{x}{2} - 1.5\right)^2$.

9. Consider expressions of the form $(ax + b)(ax - b)$.

a. Describe a shortcut for expanding such expressions. Explain why it works.

b. Use your shortcut to expand $(xy + 3)(xy - 3)$

10. Rewrite as the product of two binomials. Then use the difference of squares pattern to compute the product.

Demonstrating Skills

Rewrite each expression without parentheses.

11. $-6(x + 1) + 2(5 - x) - 9(1 - x)$

12. $-a(1 - 3a) - (2a^2 - 5a)$

13. $(2b + 1)(4 + b)$

14. $(x + 1)(5 - x)$

15. $(2c - 8)(c - 2)$

16. $(2x + y)(y - xy)$

17. $(L - 8)^2$

18. $(x - xy)(x + xy)$

Simplify each expression.

19. $(5xy - 2)(1 - 2x)$

20. $(d + 2)(1 - d) + (1 - 2d)(3 - d)$

21. Using the number 6, give three different expressions that result in 25.

22. Using the variable n to represent 6, write an expression for each of the mathematical phrases you wrote for Exercise 1.

23. Use the value of x to determine the measure of each angle.

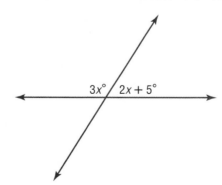

Multiply or divide the following expressions.

24. $\dfrac{8t^2 + 12t}{2t}$

25. $3x(4x + 9)$

26. $\dfrac{48x^2 + 12x}{6x}$

Test-Taking Practice

SHORT RESPONSE

1 $10x^3(x - 3)$ can be written as a product of $5(x - 3)$ and what term?

Show your work.

Answer _____

MULTIPLE CHOICE

2 Which expression is the quotient of $3n^4 + 9n^2 - 12n$ divided by $3n$?

A $n^3 + 6n + 5$

B $n^3 + 3n - 4$

C $n^5 + 3n^3 - 4n^2$

D $3n^3 + 3n + 4$

3 If a backyard has the dimensions shown below, what expression represents its total area?

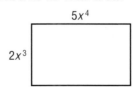

$5x^4$

$2x^3$

F $14x^7$ square feet

G $10x$ square feet

H $10x^7$ square feet

J $7x^{12}$ square feet

4 Which solution is correct?

$7(3t^2 - 4t) - 5(t^2 + 4)$

A $21t^2 - 33t - 20$

B $16t^2 + 28t + 20$

C $16t^2 - 4t + 4$

D $16t^2 - 28t - 20$

5 What is the area of a square piece of fabric that has a side measure of $(s - 4)$ inches?

F $s^2 - 8s + 16$ square inches

G $s^2 + 8s - 16$ square inches

H $s^2 + 16$ square inches

J $2s - 8$ square inches

Transformational Geometry

Real-Life Math

A History of Symmetry Symmetry is a type of balance created by repeating a basic shape or design in a regular pattern. Many cultures have used symmetry in their clothing, pottery, and artwork. Perhaps one of the greatest uses of symmetry in art can be found in the work of the Dutch artist M.C. Escher. Symmetry can also be seen in many architectural structures, such as the Taj Mahal, shown below.

Think About It Can you find some basic designs that are repeated in the structure shown?

Math Online

Take the **Chapter Readiness Quiz** at glencoe.com.

Dear Family,

The next chapter is about *transformational geometry*. We will learn about four basic transformations that can be applied to two-dimensional objects. They are reflections (flips), rotations, translations (slides from one place to another), and dilations (enlargements and size reductions).

Key Concepts—Transformations

Using these transformations, we can move an object, reposition it, or place it on top of another object of the same shape. We will learn to recognize these types of transformations, describe them, and create symmetric designs using them.

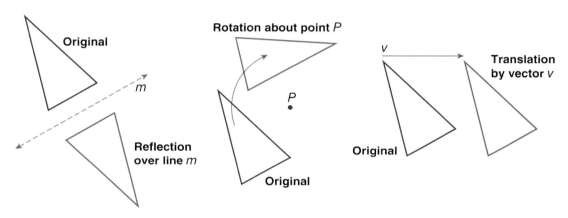

Some of the most interesting border patterns, wallpaper designs, and quilt patterns are created by combinations of transformations.

Chapter Vocabulary

dilation	rotation
image	rotation symmetry
line of reflection	scale drawing
line of symmetry	scale factor
line symmetry	transformation
reflection over a line	translation
reflection symmetry	vector

Home Activities

- Create designs by performing transformations.
- Be on the lookout for patterns or figures with symmetry and arrangements of dilated figures on buildings, in wallpaper and quilts, and in other places in your home and neighborhood.

LESSON 6.1

Symmetry and Reflection

In this chapter, you will learn about transformations. To *transform* means to change. A **transformation** is a way to take a figure and create a new figure that is similar or congruent to the original.

One helpful way to begin thinking about transformations is through *symmetry*. **Symmetry** is a form of balance in figures and objects.

When studying different cultures, archaeologists and anthropologists often look at the symmetry of designs found on pottery and other crafts. Chemists and physicists study how symmetry in the arrangements of atoms and molecules is related to their functions. Artists often use symmetry in their creations.

Vocabulary

symmetry

transformation

Materials

• scissors

Explore

Follow these directions to create a symmetric "snowflake" design.

• Start with a square piece of paper. Fold it in half in one direction and then in half in the other direction. This will make a smaller square.

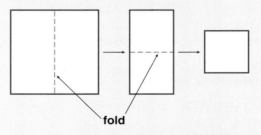

fold

• Along each edge of the folded square, draw a simple shape, something like these.

• Cut out the shapes you drew, and unfold the square.

Describe the ways your snowflake design looks "balanced," or symmetric.

Investigation (1) Lines of Symmetry

Vocabulary

line of symmetry

line symmetry

reflection symmetry

Materials

• GeoMirror

There are different types of symmetry. One type is **reflection symmetry**, also called **line symmetry**. With this type of symmetry, a line can be drawn between two halves of a figure or between two copies of a figure. The line is like a mirror between them, with one half of the image identical to the other half but flipped. The line is called a **line of symmetry**.

The dashed lines below are lines of symmetry.

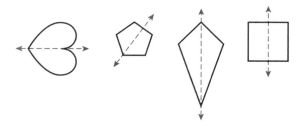

The dashed lines below are *not* lines of symmetry. They do cut the figures in half, but they do not create "mirror image" halves.

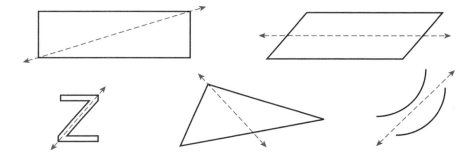

Héctor and Dante are talking about how they find lines of symmetry.

I use my GeoMirror. I move it around until I can see the reflection on top of the other half of the figure. Then my mirror is sitting on a line of symmetry.

I cut the shape out and fold it so one part fits exactly on top of the other.

✓ Develop & Understand: A

For each figure in Exercises 1–6, find all the lines of symmetry you can. Record how many lines of symmetry the figure has. Then sketch the figure and show the lines of symmetry.

1. This is an equilateral triangle. All three sides are the same length.

2. This is a scalene triangle. All three sides are different lengths.

3. These are two overlapping polygons.

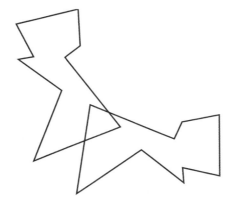

4. This is a rectangle.

5. This is a circle.

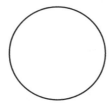

6. This is a parallelogram.

Share & Summarize

Find the lines of symmetry in this square. Using dashed lines, sketch the square to show the lines of symmetry. How many lines of symmetry does a square have?

Investigation ② Reflection

Vocabulary

image

line of reflection

reflection over a line

Materials

- GeoMirror
- protractor
- metric ruler
- graph paper
- tracing paper

You can create figures with reflection symmetry using a transformation, a **reflection over a line**. Suppose you have a figure and a line.

Imagine that the line is a line of symmetry and that the curve is only half of the figure. The other half is a *reflection* of the curve.

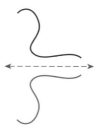

The reflection of a figure is its mirror image. The line acts like a mirror and is called a **line of reflection**. The result of a reflection, or of any transformation, is called an **image**.

Each point in the original figure has an image. In a reflection, the image of a point *P* is the point that matches with point *P* when you look through a GeoMirror or when you fold along the line of reflection.

The image of point *P* is called point *P′*, pronounced "*P* prime." In the drawing below, points *P′* and *Q′* are the images of points *P* and *Q*.

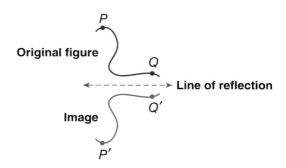

Real-World Link

As with real faces, this mask is not exactly symmetric.

· ·

✓ Develop & Understand: A

Exercises 1 and 2 show a figure and a line of reflection. Carefully copy each picture. Then, using a GeoMirror, reflect the figure over the line to create the image. Mark the image of point A and name it point A'. Make your drawings as accurately as you can.

1.

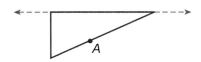

2.

3. When you reflect a figure over a line, is the image you create *congruent* to the original (the same size and shape), *similar* to the original (the same shape but possibly a different size), or *neither*? Explain how you know.

4. For each drawing you made in Exercises 1 and 2, connect points A and A' with a straight line. Measure the angle between each line of reflection and Segment AA' $(\overline{AA'})$. What are the measures?

5. For each drawing you made in Exercises 1 and 2, measure these lengths.

    a. the length of the segment between point A and the line of reflection

    b. the length of the segment between point A' and the line of reflection

6. Make a conjecture about the relationship between the line of reflection and the segment connecting a point to its image.

✅ *Develop & Understand: B*

You already know how to use coordinates to graph relationships. Coordinates are also helpful for describing a position of a geometric figure and for performing transformations.

7. Draw the triangle shown onto graph paper.

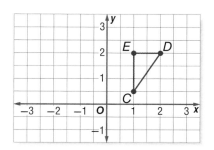

8. Transform each vertex of the triangle using this rule: $(x, y) \rightarrow (-x, y)$. This rule says two things.

- The *x*-coordinate of the image is the opposite of the original *x*-coordinate.

- The *y*-coordinate of the image is the same as the original *y*-coordinate.

On the same set of axes, plot each image point. Connect them in order. Describe how the image is related to the original figure. Is it a *reflection*? Why or why not?

9. Describe the transformation.

Share & Summarize

Describe the relationship between a line of reflection and the segment joining a point to its reflected image.

Investigation 3 | The Perpendicular Bisector Method

Materials

- tracing paper
- ruler
- protractor
- GeoMirror (optional)

Lines called *perpendicular bisectors* are particularly useful in working with reflections. You may recall that a perpendicular bisector of a segment has two important characteristics.

- It meets the segment at its midpoint. It *bisects* the segment.

- It is perpendicular to the segment.

This line is perpendicular to $\overline{AA'}$ but is not a bisector of it.

This line bisects $\overline{AA'}$ but is not perpendicular to it.

This line is a perpendicular bisector of $\overline{AA'}$.

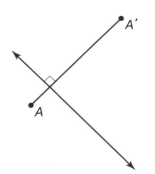

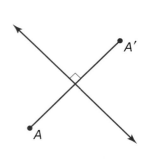

not a perpendicular bisector

not a perpendicular bisector

perpendicular bisector

1. Copy each segment. Draw a perpendicular bisector of it.

a.

b.

2. Explain how you found the perpendicular bisectors in Exercise 1.

3. How many perpendicular bisectors does a segment have?

If you do not have a GeoMirror, you can still reflect a point over a line using the *perpendicular bisector method*. If point A' is the image of point A in a figure with a line of symmetry, the line of symmetry is the perpendicular bisector of the segment connecting the points.

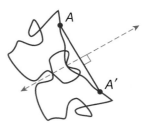

To reflect a point B over a line by the perpendicular bisector method, draw a segment from point B, perpendicular to the line.

Continue the segment past the line until you have doubled its length. At the other endpoint of the segment, mark the image point, point B'.

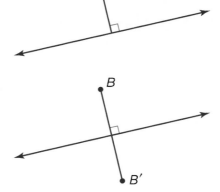

Think & Discuss

Suppose you want to reflect △ABC over line *m*.

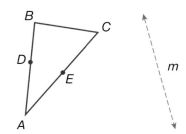

Which points would you reflect? Why?

✓ Develop & Understand: A

For each exercise, copy the figure and the line. Reflect the figure over the line using the perpendicular bisector method. Check your work using a GeoMirror or by folding.

4.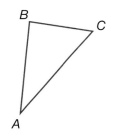

5.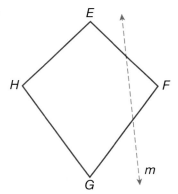

6.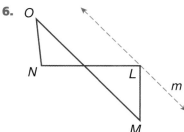

Share & Summarize

Here is a triangle and a line. Explain how you would use the perpendicular bisector method to reflect the triangle over the line.

On Your Own Exercises

Lesson 6.1

Practice & Apply

In Exercises 1–4, copy the figure. Draw all the lines of symmetry you can. Check the lines you drew by folding the paper.

1. This is an isosceles triangle.

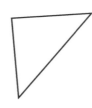

2. This is a trapezoid.

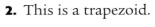

Math Link

An *isosceles trapezoid* is a trapezoid with two opposite, nonparallel sides that are the same length.

3. This is an ellipse.

4. This is an isosceles trapezoid.

Exercises 5 and 6 show an original figure and a line of reflection.

- Copy the picture.
- Reflect the figure over the line. Draw the image.
- Mark the image of point *A* and name it point *A'*.

5.

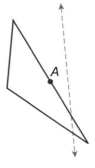

6.

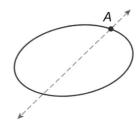

7. This picture shows two lines of reflection.

- Copy the picture.
- Reflect the figure over one line to create an image.
- Now reflect both figures, the original and the image, over the other line.

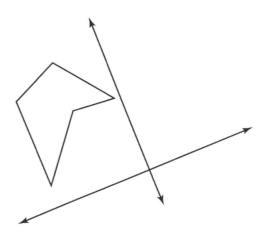

Copy each segment. Draw its perpendicular bisector.

8.

9.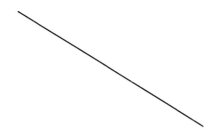

10. Copy this picture.

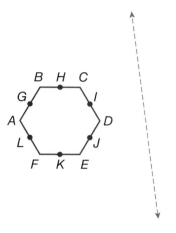

a. Reflect the figure over the line using the perpendicular bisector method.

b. What is the minimum number of points you have to reflect by this method to be able to draw the whole image figure? Explain.

11. Copy this picture.

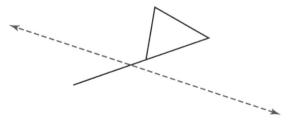

a. Reflect the figure over the line using the perpendicular bisector method.

b. What is the minimum number of points you have to reflect by this method to be able to draw the whole image figure? Explain.

Connect & Extend

12. Create a paper snowflake with four lines of symmetry. Check the lines of symmetry by folding.

13. Show a figure on a coordinate plane and a rule to perform on the coordinates. Do Parts a–d to create an image of the figure.

   a. Explain the rule in words.

   b. Find the coordinates of each vertex. Copy the figure onto graph paper.

   c. Perform the given rule on the coordinates of each vertex to find the image vertices.

   d. On the same set of axes, plot each image point. Connect them in order.

   e. Give the line of reflection. You might want to use your GeoMirror to check.

   Rule: $(x, y) \rightarrow (y, x)$

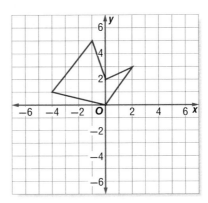

14. Consider a reflection over the line $y = -x$.

   a. Reflect the point $(1, 0)$ over this line. What are the coordinates of the image?

   b. Reflect the point $(0, 1)$ over the line. What are the coordinates of the image?

   c. If you reflect a point (x, y) over the line, what will the coordinates of its image be?

15. **Sports** Bianca enjoys playing billiards. At the end of one close game she played, the only balls on the table were the 8 ball and the cue ball. It was Bianca's turn. To win, all she had to do was hit the 8 ball into a pocket of her choosing.

Bianca hit the cue ball, which collided with the 8 ball. This sent the 8 ball in the direction shown, sinking it into the correct pocket. Bianca won the game.

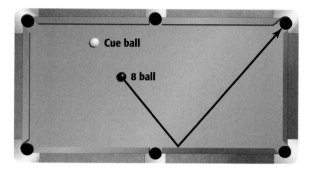

a. Was the path the ball traveled symmetric in some way? If so, find the line of symmetry.

b. Bianca said the angles made by the side of the table and the 8 ball's path, coming and going, are always equal to each other. Is this true in this case? How do you know?

16. Pete said that to reflect this star over the line, he needs to reflect only five points.

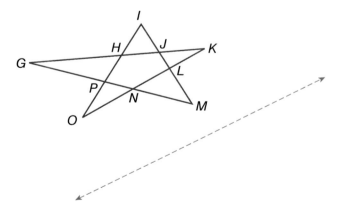

Is Pete correct? If so, explain how he could draw the image. If not, explain why he could not do it.

17. List three natural or manufactured things that have lines of symmetry.

18. **In Your Own Words** Find a picture, design, or object in your home that has at least one line of symmetry. Describe or sketch it, and indicate each line of symmetry.

19. Copy this picture.

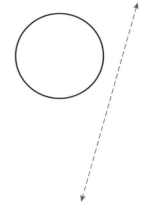

    a. Reflect the circle over the line using the perpendicular bisector method. How many points did you reflect?

    b. Check your work by folding. Do the figures match exactly?

    c. What is the minimum number of points you must reflect by this method in order to reflect the whole figure? Explain.

Mixed Review

Expand each expression.

20. $\frac{4}{7}\left(\frac{1}{2}t + 12\right)$

21. $x(4 - 13x)$

22. $0.2(72v - 3)$

Give the slope and *y*-intercept for the graph of each equation.

23. $y = -3x + 5$ **24.** $2y = 2 - 3.2x$

25. $3y + 1.5 = x$ **26.** $x = y$

27. **Life Science** The data in the table represent how a certain population of bacteria grows over time. Write an equation for the relationship, assuming the growth is exponential.

Hours from Start, *h*	Population, *p*
0	1
1	5
2	25
3	125

Rotation

In Lesson 6.1, you looked closely at reflection symmetry, or line symmetry. These designs show another kind of symmetry.

Think & Discuss

Examine the figures above. In what way does it seem reasonable to say that these figures have symmetry?

Investigation 1 Rotation Symmetry

Vocabulary

rotation symmetry

Materials

- scissors
- tracing paper
- pin
- ruler protractor

At the beginning of Lesson 6.1, you examined a paper snowflake for symmetry. You probably noticed that it had reflection symmetry, but you might not have realized that your creation had other symmetry, as well.

✅ Develop & Understand: A

Follow these directions to create another paper snowflake.

1. Start with a square piece of paper.

   a. Fold it in half three times as shown.

   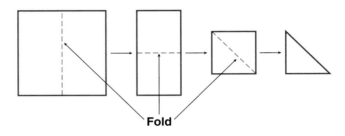

   Fold

   b. On each side of the folded triangle, draw some shapes. For example, you might draw shapes like these.

   c. Cut out along the lines you drew. Then unfold the paper. The design above made this snowflake.

   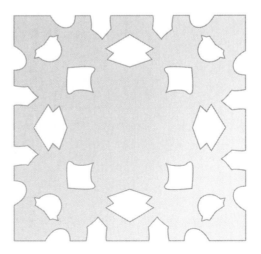

Now place your paper snowflake over a sheet of tracing paper. Copy the design by tracing around the edges, including the holes.

Math Link
A full turn has 360°.

2. Pin the centers of your snowflake and your tracing together. *Rotate,* or turn, the snowflake about its center until the design on the tracing paper coincides with the design on the snowflake. Did you need to turn the snowflake all the way around?

3. Rotate the snowflake again, until the designs match once more. Have you returned the paper to the position it was in before Exercise 2? If not, rotate it again, until the designs match. How many times do you have to rotate the snowflake before you have turned it all the way around?

4. How many *degrees* must you rotate the snowflake each time to get the designs to match?

A figure has **rotation symmetry** if you can copy the figure, rotate the copy about a centerpoint *without turning it all the way around,* and find a place where the copy exactly matches the original. The *angle of rotation,* the smallest angle you need to turn the copy for it to match with the original, must be less than 360°.

✅ Develop & Understand: B

Each of these figures has rotation symmetry. Find the angle of rotation for each figure.

5.

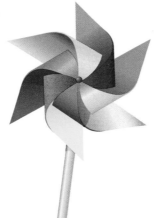

6.

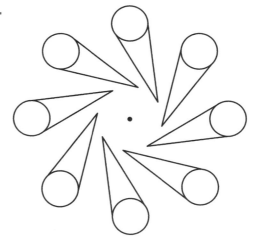

7.

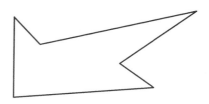

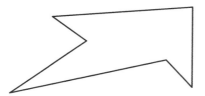

8.

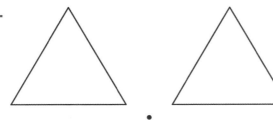

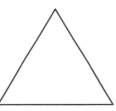

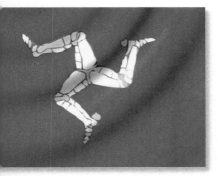

Share & Summarize

Look again at the figures in Exercises 5–8. Make a conjecture about the relationship between the angle of rotation and the number of identical elements in the figure.

Investigation Rotation as a Transformation

Vocabulary

rotation

Materials

- tracing paper
- protractor
- pin

To create your own design with rotation symmetry, you need to have three things: a figure called a *basic design element*, a center of rotation, and an angle of rotation.

By convention, angles of rotation assume a figure is rotated *counterclockwise*. To indicate a clockwise rotation, use a negative sign.

90° rotation about point *P*   −90° rotation about point *P*

Math Link

To measure an angle, place the hole on the straight edge of the protractor over the vertex of the angle. Line up the zero on the straight edge of the protractor with one of the angle sides. Find the point where the second angle side intersects the curved edge of the protractor. The number at the intersection is the angle measure in degrees.

✅ Develop & Understand: A

You will now create a design by rotating the figure below. The figure is your basic design element, and the point is the center of rotation. For this design, you will use an angle of rotation of 60°.

1. Copy this picture, including point *P* and the reference line.

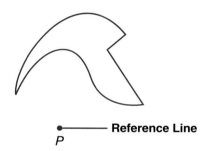

2. Draw a new segment with point *P* as one endpoint and forming a 60° angle with the reference line. Label this segment 1. Then draw and label four more segments from point *P*, each forming a 60° angle with the previous segment, as shown here.

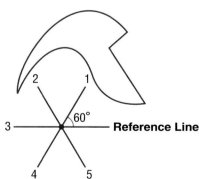

3. Place a sheet of tracing paper over your figure. Pin the papers together through the center of rotation. Trace the figure, including the reference line, but *do not* trace segments 1–5.

4. Now rotate your tracing until the reference line on the tracing is directly over segment 1. Trace the original figure again. Your tracing should now look like this.

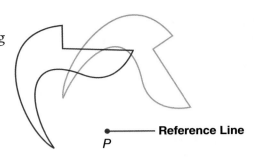

Reference Line

P

5. Rotate the tracing until the reference line on the tracing is directly over segment 2. Trace the original figure again.

6. Repeat the process, rotating to place the reference line over the next segment and tracing the figure. Do this until the reference line on the tracing is back on the original reference line.

In Lesson 6.1, you learned about the reflection transformation. You have now used the second transformation, **rotation**. In a rotation, a figure is turned abut a point. Each time you turned the basic design element and traced it, you performed a rotation.

✅ Develop & Understand: B

Use the given basic design element, center of rotation, and angle of rotation to create a design with rotation symmetry.

7. angle of rotation: 120°

8. angle of rotation: 90°

9. Do your completed designs have lines of symmetry? If so, where?

> ### Share & Summarize
>
> 1. When you reflect a figure over a line once, you create a design with reflection symmetry. If you rotate a figure about a point once, does that always create a design with rotation symmetry?
>
> 2. Look at the designs you created in Exercises 7 and 8. Is there a pattern between the number of identical elements in the finished designs and the angle of rotation? If so, what is it?

Investigation ③ The Angle of Rotation

Materials

- tracing paper
- ruler
- protractor
- colored pencils (optional)

In Investigation 2, you used the *rotation* transformation to create figures with rotation symmetry. Now you will look more closely at an important part of these transformations, the *angle of rotation*.

✅ Develop & Understand: A

△*A′C′D′* is the image of △*ACD* when it is rotated a certain angle about point *P*. Carefully make a copy of the entire picture.

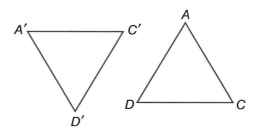

1. Consider point *A* and its image, point *A′*.

   a. Draw $\overline{AP}$ and $\overline{A'P}$. You may want to use a different color for these segments than you used to make your drawing.

   b. Measure the two segments you just drew. What do you find?

   c. Measure ∠*APA′*.

2. Now consider point *C* and its image, point *C′*.

   a. Add $\overline{CP}$ and $\overline{C'P}$ to your picture. Again, you may want to use a different color.

   b. Measure these two segments. What do you find?

   c. Measure ∠*CPC′*.

3. Finally, consider point *D* and its image, point *D′*.

   a. Add $\overline{DP}$ and $\overline{D'P}$ to your picture.

   b. Measure these two segments. What do you find?

   c. Measure ∠*DPD′*.

4. Compare the three angles you measured. What do you notice?

Real-World Link

The recycle logo, with its ever-revolving arrows, has rotation symmetry. Can you determine the angle of rotation?

✅ *Develop & Understand: B*

5. Sketch this picture. Use what you discovered in Exercises 1–4 to rotate the figure about point *P*, using 140° as the angle of rotation.

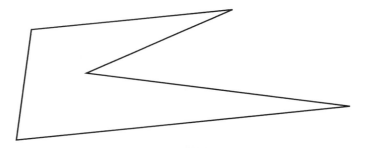

• *P*

6. Sketch this picture. Draw a copy of the figure rotated about point *Q*, using −80° as the angle of rotation.

Math Link

A negative angle of rotation means to rotate clockwise.

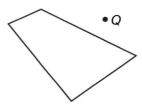

7. When you rotate a figure, is the image *congruent* to the original, *similar* to the original, or *neither*? Explain how you know.

Develop & Understand: C

8. Use this figure and the rule to perform on the coordinates to create an image of the figure in Parts a–d.

Rule: $(x, y) \rightarrow (-y, x)$

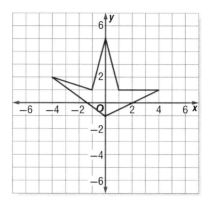

a. Explain the rule in words.

b. Find the coordinates of each vertex, and copy the figure onto graph paper.

c. Perform the given rule on the coordinates of each vertex to find the image vertices.

d. On the same set of axes, plot each image point. Connect them in order.

e. Give the center and the angle of rotation. You might want to use tracing paper and a protractor to check.

Share & Summarize

Suppose you have a protractor, a ruler, and a pencil, but no other paper. You are asked to perform a rotation of the segment at right, using the point as the center of rotation.

1. What additional information do you need?

2. Suppose you are given the information you need. How would you perform the rotation of the segment?

On Your Own Exercises
Lesson 6.2

1. Copy this figure onto tracing paper. Then rotate the tracing until the copy matches with the original.

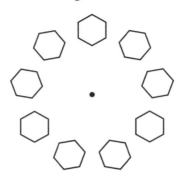

   a. How many times do you have to rotate the tracing, matching the tracing to the original, to return it to its starting position?

   b. What is the angle of rotation?

   c. Describe how the angle of rotation is related to the number of identical elements.

2. Use this as a basic design element.

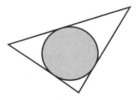

   a. Choose a center of rotation. Make a design with a 72° angle of rotation. Use tracing paper if you need it.

   b. Make another design using this basic design element and a 72° angle of rotation, using a different center of rotation.

3. Copy △ABC, and rotate it −60° about point Q. (Remember: A negative angle of rotation means to rotate clockwise.)

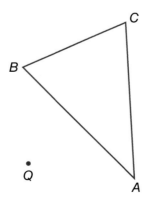

Do your work for Exercises 4 and 5 without using tracing paper.

4. Draw a quadrilateral *ABCD*. Choose a point to be the center of rotation and rotate your quadrilateral 80° about that point. Use prime notation to label the image vertices. For example, the image of vertex *A* will be vertex A′.

5. Draw a pentagon *EFGHI*. Choose a point to be the center of rotation, and rotate your pentagon −130° about the center. Use prime notation to label the image vertices. For example, the image of vertex *E* will be vertex E′.

Connect & Extend

6. Three-dimensional objects can have rotation symmetry. A three-dimensional object with rotation symmetry has an *axis of rotation* instead of a center of rotation.

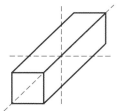

While a two-dimensional figure can have only one center of rotation, a three-dimensional object can have more than one axis of rotation. This rectangular prism has three axes of rotation.

> **Math Link**
> A prism is named for the shape of its bases.

a. The prism below has bases that are equilateral triangles. How many axes of rotation does this triangular prism have? Explain where they are. You may use a diagram to explain, if necessary.

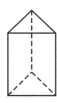

b. For each axis of rotation you found, what is the angle of rotation?

7. Create your own design with rotation symmetry.

a. What is the angle of rotation?

b. How many identical elements are there in your design?

c. Does your design have reflection symmetry? If so, how many lines of reflection does it have?

8. List at least four natural or manufactured objects or figures that have rotation symmetry. Do your examples also have line symmetry?

9. In Your Own Words Describe how rotation and reflection are similar and how they are different. Give examples.

Math Link

The greatest common factor is the greatest number that will exactly divide into two or more integers.

10. Choose a point A and a center of rotation. Rotate point A 135°. Rotate the image 135°. Keep rotating the images until you return to the original point. When you perform the rotations, you will pass the original point because you have made one full turn.

  a. How many full circles did you make?

  b. How many copies of the point do you have in your drawing?

  c. There is an angle of rotation smaller than 135° that you could have used to create this same design. What is its measure?

  d. Now find the *greatest common factor* (GCF) of 135 and 360.

  e. Divide 135 and 360 by your answer to Part d.

  f. Compare your answers for Parts a–c to your answers for Parts d and e. What do you notice?

  g. Suppose you created a figure by rotating a basic design element 80° each time. What angle of rotation will the final design have? Test your answer by rotating a single point.

11. Triangle ABC has vertices A (2, 5), B (−2, 4), and C (0, 3). A certain transformation of these vertices gives A' (−5, 2), B' (−4, −2), and C' (−3, 0).

  a. Draw the two triangles on a grid. Is this a reflection or a rotation? If it is a reflection, give the line of reflection. If it is a rotation, give the center and angle of rotation.

  b. What is the image of (x, y) under this transformation?

Mixed Review

12. Examine how this pattern grows from one stage to the next.

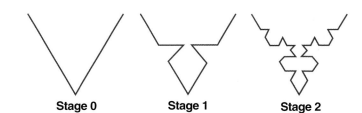

Stage 0   Stage 1   Stage 2

  a. Copy and complete the table.

Stage, n	0	1	2	3
Line Segments, s	2	8		

  b. What kind of relationship is this?

  c. Write an equation to describe the relationship between the number of line segments in a stage s and the stage number n.

Evaluate each expression for $a = 2$ and $b = 3$.

13. $\left(\dfrac{1}{9}\right)^{a}$ **14.** $\left(9 \cdot \dfrac{1}{b}\right)^{b}$ **15.** $\left(\dfrac{b}{a} \cdot \dfrac{a}{b}\right)^{a}$

LESSON 6.3

Translations, Dilations, and Combined Transformations

You will now explore two more transformations, a *translation* and a dilation. You can think of translating a figure as moving it a specific distance in a specific direction. Unlike rotations and reflections, the image of a translation and the original figure have the same *orientation,* the top is still the top and the bottom is still the bottom. Some people call a translation a *slide* or a *glide.*

Dilation is refered to as *scaling,* which allows you to create similar figures.

Materials

• ruler

Think & Discuss

Look at this series of figures.

If the original figure is the one on the left, how far was it translated at each stage and in what direction?

If the original figure is the one on the right, how far was it translated at each stage and in what direction?

Investigation 1 Translation

Vocabulary

translation

vector

Materials

• tracing paper
• metric ruler
• graph paper

To describe a **translation**, you need to give both a distance and a direction. *Vectors* can be used to describe translations. A **vector** is a line segment with an arrowhead. The length of the segment tells how far to translate, and the arrowhead gives the direction.

The two vectors below indicate translations of the same length but different directions.

Translate this hexagon using the given vector.

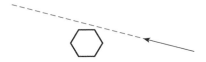

First, trace the hexagon and the vector. Extend the vector with a dotted line. This line will help you keep the figure's orientation.

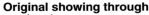

Now slide the *original* in the direction indicated by the vector until the *head* of the traced vector touches the *tail* of the original vector. Keeping the original vector under the dotted line on the tracing, trace the original hexagon and vector again.

Original showing through

You can repeat the process. When you slide the original hexagon and vector, position the original vector so that its tail touches the head of the second vector.

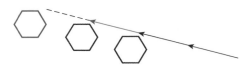

✅ Develop & Understand: A

Translate the figure using the given vector to create a design with four copies of the figure.

1.

2.

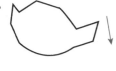

3. Now draw your own figure and vector of translation. Use tracing paper to create a design with at least three copies of your figure.

4. Are a figure and its translated images *congruent, similar,* or *neither*? Explain.

Real-World Link

If you could continue this pattern forever, in both directions, the result would have *translation symmetry*.

✅ Develop & Understand: B

Triangle *ABC* has been translated by the given vector to create an image, △*A′B′C′*.

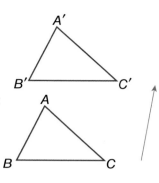

5. Trace the picture. On your copy, connect point *A* to its image, point *A′*. Also connect points *B* and *C* to their images.

6. Measure the lengths of $\overline{AA'}$, $\overline{BB'}$, and $\overline{CC'}$. Compare these lengths to the length of the vector.

7. Imagine extending $\overline{AA'}$, $\overline{BB'}$, and $\overline{CC'}$ to form lines. Suppose you also extended the vector to form a line. What would be true about these four lines?

8. Suppose you have a single point and a vector. How can you translate the point by the vector without using tracing paper?

✅ Develop & Understand: C

You saw that you can perform a rule on the coordinates of a figure to rotate or reflect that figure in the plane. In a similar way, some rules will produce a translation of a figure.

Exercises 9 and 10 each show a figure on a coordinate plane and a rule to perform on the coordinates. For each exercise, follow these steps to create an image of the figure.

- Find the coordinates of each vertex, and copy the figure onto graph paper.
- Perform the given rule on the coordinates of each vertex to find the image vertices.
- On the same set of axes, plot each image point. Connect them in order.

9. Rule: $(x, y) \rightarrow (x + 2, y)$. That is, to get the image point, add 2 to the *x*-coordinate and leave the *y*-coordinate the same.

10. Rule: $(x, y) \rightarrow (x, y + 3)$

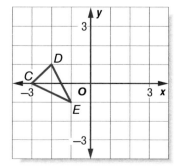

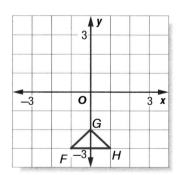

Lesson 6.3 Translations, Dilations, and Combined Transformations **287**

11. All the rules you have seen for translations have the same form. The numbers added might be positive, negative, or 0. What kind of numbers would you add to each coordinate of a point to move the point:

  a. straight down (not to the left or right)?

  b. up and to the left?

You will now practice writing rules to give a desired translation.

12. Here is a quadrilateral on a coordinate plane.

  a. Copy the quadrilateral onto graph paper. Then draw an image of the quadrilateral that has been translated 2 units to the right and 1 unit down.

  b. What rule performed on the coordinates would create this translation?

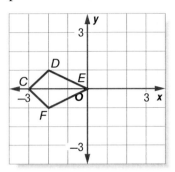

Share & Summarize

Explain what a *translation* is in your own words. Be sure to include what information is needed to perform one.

Investigation 2 Combined Transformations

Materials

- tracing paper
- GeoMirror (optional)
- protractor
- ruler

What happens when you combine two transformations? For example, suppose you reflect a figure and then reflect its image.

✓ Develop & Understand: A

Consider what happens when you reflect over two lines that intersect. Copy each picture. Reflect the figure over line *l* and then reflect the image over line *m*.

1.

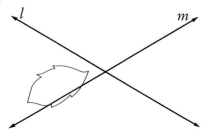

2.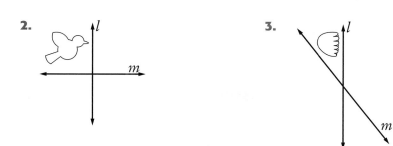

3.

4. The final images in Exercises 1–3 could have been created using only one transformation of the original figure. Describe that single transformation.

You have looked at combinations of two reflections. There are many other ways to combine transformations. Now you will explore combining a reflection and a translation.

✓ Develop & Understand: B

5. Vector *c* is parallel to the line of reflection, line *m*. Trace △*FGH*, vector *c*, and line *m*.

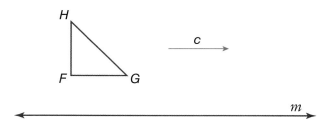

    a. Reflect △*FGH* over line *m*.

    b. Translate the image in Part a by vector *c* to get a second image.

6. Now try reversing the order of the transformations.

    a. Translate △*FGH* from Exercise 5 by vector *c*.

    b. Reflect the image you created in Part a over line *m*.

7. What do you notice about the relationship between the final images in Exercises 5 and 6?

The combination of a reflection over a line and a translation by a vector parallel to that line is called a *glide reflection*. Patterns on wallpaper borders can show glide reflection.

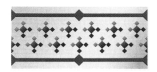

Share & Summarize

    1. What is the relationship between an original figure and the final image when you reflect the figure over two intersecting lines?

    2. Describe and give an example of a glide reflection.

Inquiry

Investigation ③ Making Tessellations

Materials

- stiff paper squares
- scissors
- tape
- poster board
- markers or crayons

Some artists translate, rotate, or reflect a figure or a collection of figures to create fascinating images.

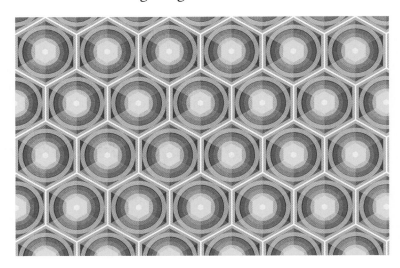

Real-World Link

Many works of art are tessellations made up of irregular shapes. These are known as non-regular tessellations.

A design using one or more shapes to cover the plane without any gaps or overlaps is called a *tessellation*. Many shapes will *tessellate*. One of the easiest shapes to tessellate is a square.

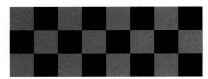

You can use a square and the techniques of reflection, rotation, and translation to create your own tessellation artwork.

Try It Out

1. Start with a paper square. On one side of the square, draw a shape. The shape should be a single piece that starts and ends on the same side. Here are some examples.

Carefully cut out the shape you drew, keeping it in a single piece.

2. You now have some options for moving your shape.

- You could translate it to the other side of the square.

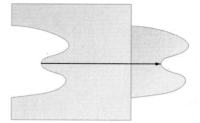

- You could translate it to the other side of the square, as above, and then reflect it over a line through the center of the square and parallel to the direction you translated.

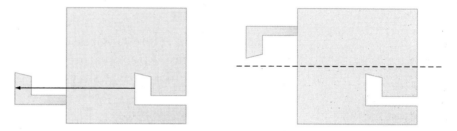

- You could rotate it 90° about one of the vertices adjacent to your shape. The shape below has been translated about the upper side of the square.

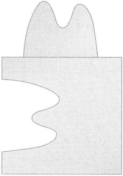

Choose one of these three options. Tape the shape to the square at the place where you moved it.

3. Now you can create a tessellation. Place your figure on a large sheet of paper or posterboard and trace around it. Then decide how to move your figure (by translation, rotation, or some combination of transformations) so that the cutout on the tracing fits together with the attached piece on your figure. Trace the figure again.

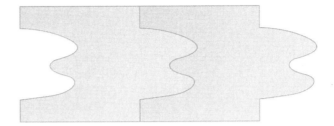

4. Repeat the process several times, filling your paper or posterboard with your shape. You have a tessellation.

Try It Again

To create more interesting tessellations, you can cut shapes from two sides of a square. After creating the tessellation, you can color each figure to look like birds, fish, people, or just about anything else you can imagine.

5. Start with a new square. On one side of the square, draw a shape. Cut out the shape and translate, translate and reflect, or rotate it as described in Step 2. Reattach the shape to the square.

6. You have now used two of the four sides of your square. Choose one of the other sides and draw a new shape. Cut out the shape and translate, translate and reflect, or rotate it. You will have to choose your rule based on which side of the square is left to attach the shape.

For example, here is a translation of one shape and a translation and reflection of another shape.

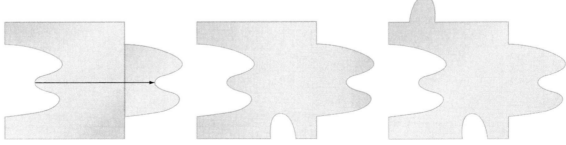

7. Trace your figure onto a large sheet of paper or posterboard.

    a. Decide how to move the shape so that the *first* cutout on the tracing fits correctly with the attached piece. Trace the figure again.

    b. Now move the shape so that the *second* cutout fits correctly. Trace it.

    c. Continue the process until you have created a tessellation. Decorate the figures however you like.

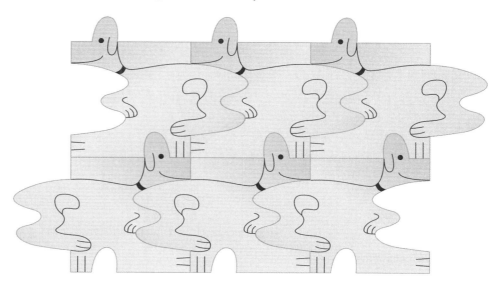

What Have You Learned?

8. Suppose you *translate* a cutout to create a figure. How would you move the figure to make a tessellation?

9. Suppose you *translate and then reflect* a cutout to create a figure. How would you move the figure to make a tessellation?

10. Suppose you *rotate* a cutout to create a figure. How would you move the figure to make a tessellation?

Investigation 4 Dilation

Vocabulary

dilation

scale drawing

scale factor

Materials

- graph paper
- ruler
- protractor
- tracing paper

You have studied three transformations that create congruent figures. The fourth transformation, **dilation**, which is sometimes called *scaling*, creates figures that are similar but not necessarily congruent to the original.

You might recall that matching, or *corresponding*, sides of similar polygons like the triangles below are in proportion. That is, the side lengths share a common ratio. Corresponding angles of similar figures are congruent.

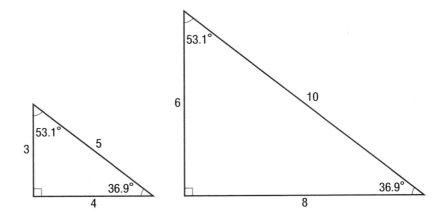

In this investigation, you will learn two ways to create similar figures by dilating.

Math Link

Congruent angles have the same measure.

✓ Develop & Understand: A

Copy the outline of a cat's head onto a coordinate grid.

1. Find the coordinates of each point.

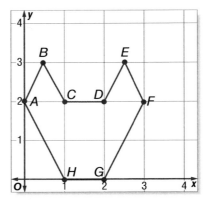

2. Multiply the coordinates of each point by 2, creating points A' through H'. In other words, if the coordinates of point A are (x, y), the coordinates of point A' are $(2x, 2y)$.

3. Plot these new points on the same grid. Connect them in the same order.

4. You have just *scaled* or *dilated* the original cat's head. Is the new, dilated figure similar to the original? How do you know?

When you dilated the cat's head, you were using the *coordinate method* to create a similar figure. You can use this method, which involves multiplying coordinates of vertices by a number, to make *scale drawings* of figures on a coordinate grid.

A **scale drawing** is a drawing that is similar to some original figure. The **scale factor** is the ratio between corresponding side lengths of the similar figures.

Every pair of similar figures of different sizes has *two* scale factors associated with it. One describes the dilation from the small figure to the large figure. The other describes the dilation from the large figure to the small figure.

For example, the lengths in your scale drawing of the cat's head are all twice the corresponding lengths in the original figure, so the scale factor from the small figure to the large figure is 2. And since the lengths in the small figure are $\frac{1}{2}$ the lengths in the large figure, you can also say that the scale factor from large to small is $\frac{1}{2}$.

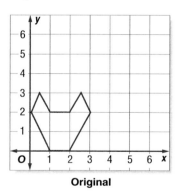

Original   Scale Drawing

5. Dilate the figure below by a scale factor of $\frac{1}{3}$.

6. Dilate the figure below by a scale factor of 3.

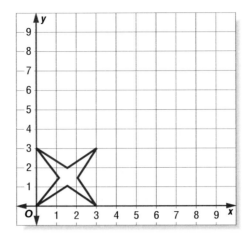

Some computer drawing and animation programs use a technique similar to the coordinate method to create scale drawings. The software treats the screen as a coordinate plane and calculates the placement of points. Slide projectors also make similar figures, but they work differently.

Slide projectors use a bright bulb and a series of lenses to project enlarged images onto a screen. The most important purpose of the lenses is to focus the light so that it comes from a single point. The focused light passes through the slide and onto the screen, spreading out to create a larger image.

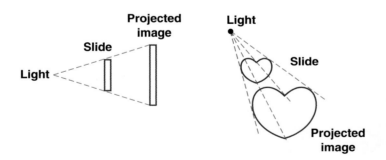

You can make scale drawings using this concept by applying a technique called the *projection method*.

Example

Here is how to use the projection method to make a figure similar to polygon *ABCDE*. This method will result in sides that are half as long. It will make a figure that is dilated by $\frac{1}{2}$.

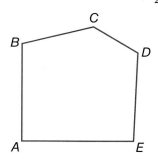

- First, choose any point. Although the point can be on the polygon, you may find it easier to work with a point that is inside or outside of the figure, like point *F*.

  After you have chosen the point, draw segments from it to every vertex of the polygon.

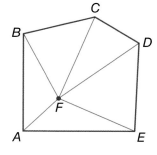

Math Link

The *midpoint* of a segment is the point that lies halfway between the endpoints of the segment.

- To *halve* the polygon, find the *midpoints* of segments *FA* through *FE*.

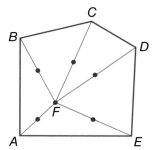

- Finally, connect the midpoints, in order, to form the new polygon.

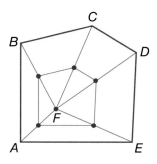

A point that helps you to make a similar figure, such as point *F*, is called a *projection point*.

✅ Develop & Understand: C

7. Below is a figure of a house.

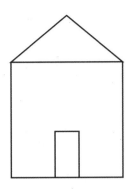

    a. Use the projection method to create a similar figure smaller than this original figure.

    b. Now modify the projection method to enlarge the original figure by a scale factor of 2.

Share & Summarize

You will now dilate this figure using both methods you have learned.

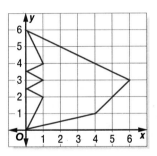

1. Dilate the figure by $\frac{1}{2}$ using the coordinate method.

2. Dilate the figure by $\frac{1}{2}$ using the projection method. Use the origin as the projection point.

Practice & **Apply**

In Exercises 1–3, translate the figure by the vector to create a design with three elements.

1.

2.

3.

4. Santiago drew a picture using line segments on a coordinate grid. He then multiplied the coordinates of all the endpoints by 1.5, plotted the resulting points on a new grid, and connected them to form a new picture.

    a. One segment in Santiago's original drawing was 2 in. long. How long was the corresponding segment in the new drawing?

    b. One segment in the new drawing was 2 in. long. How long was the corresponding segment in Santiago's original drawing?

5. Carefully copy this picture, including lines *l* and *m*, which are perpendicular.

    a. Reflect the figure over line *l* and its image over line *m*.

    b. Describe the single transformation that would give you the same final image. Give as much detail as you can.

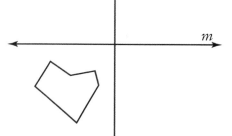

6. Dilate this figure by a factor of 2 using the coordinate method.

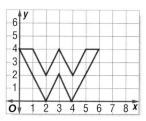

7. Use the projection method to scale this figure.

    a. Dilate the figure by a factor of $\frac{1}{4}$.

    b. Dilate the figure you created in Part a by a factor of 4.

    c. How does the figure you created in Part b compare to the original?

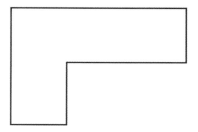

8. Carefully copy this picture.

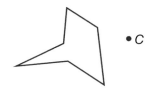

a. Rotate the figure about point *C* through an angle of 30°. Then rotate the image about point *C* through an angle of 40°.

b. What single transformation of the original figure would give the same final image as in Part a?

c. Rotate the original figure about point *C* through an angle of 30°. Then rotate the image about point *C* through an angle of −40°.

d. What single transformation of the original figure would give the same final image as in Part c?

9. Carefully copy this picture, including the line of reflection and the vector. Follow the directions to perform a glide reflection.

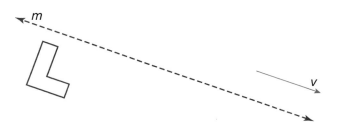

a. Reflect the figure over line *m*.

b. To complete the glide reflection, translate the image by vector *v*.

c. Take the final image from Part b and reflect it over line *m*. Then translate that figure by vector *v*.

d. Your image in Part c can be created by a single transformation of the original. Give as much information as possible about that transformation.

Real-World Link

Glide reflections are the only type of symmetry that involves more than one step.

Connect & Extend

10. Draw a figure of your own using line segments on a coordinate grid. Enlarge your figure using the coordinate method. Record the scale factor you used. Verify the scale factor from the original figure to the enlargement by checking at least two corresponding pairs of line segments.

11. Tonya and Aaron were talking about this figure.

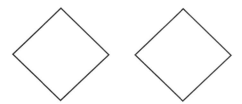

Tonya said it was created using a reflection, but Aaron thought it was created using a translation.

a. Could Tonya be correct? Try to find a line of reflection that would work.

b. Could Aaron be correct? Try to find a vector of translation that would work.

c. Here is a pair of figures that could have been created by either translation or reflection. Find the line of reflection and a vector of translation that would work.

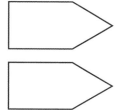

d. Create a design that could be created by translation or reflection.

e. A figure must have a particular kind of symmetry to look the same reflected or translated. Look back over the basic elements in this exercise. What kind of symmetry is necessary?

12. In Parts a–c, an equation is given.
The three equations are graphed at
right. Draw what each graph would
look like translated by the given vector.

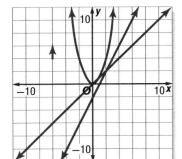

a. $y = x$

b. $y = x^2$

c. $y = 2x - 2$

d. For Parts a–c, write the equation
of the new graph you created.

13. Isabel reflected this figure over line p and reflected its image over
line r.

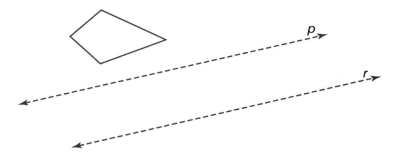

What single transformation, a *reflection*, a *rotation*, or a *translation*,
would give Isabel the same final image?

14. Maggie wants to impress her friends with her skill at billiards.
She clears the table of all the balls except the cue ball.

When a ball *banks*, or bounces, off a side of the table, the angle it
makes as it travels away from the side is the same as the angle it
made when it approached the side.

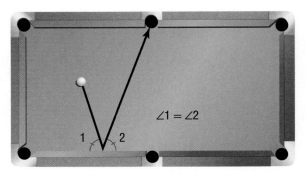

a. Maggie challenged her friend Marlon to sink the cue ball into a side pocket after banking it off the sides of the table exactly twice. Marlon made the following shot, which failed.

Carefully copy the billiard table below. Find the path that the ball traveled. Marlon tried to put the ball in the marked side pocket.

Target Pocket

b. Maggie used her understanding of reflections to estimate a shot that would work. Complete these steps to find the shot Marlon should have used.

- On another sheet of paper, copy the table and ball from Part a.
- Imagine that the inner edge of the right side of the table is a line of reflection. Find the image of the target side pocket when you reflect over that line.
- Now imagine that the inner edge of the bottom side of the table is a line of reflection. Reflect the image from the previous step over that line.
- Draw a line connecting this final image to the cue ball. This line shows the direction in which the shot should be made.
- Verify the shot by finding the path the ball would follow.

15. In Your Own Words Suppose you have two congruent triangles on a piece of paper. Do you think you can always transform one into the other using some combination of translations, rotations, and reflections? Explain your thinking.

16. Mile Square Park in Fountain Valley, California, is exactly one mile on each side. The city officials want to create a park map, showing visitors where playgrounds, drinking fountains, restrooms, and paths are located. They want the map to fit on a standard, $8\frac{1}{2}$ in. by 11 in. sheet of paper.

Because the officials want the map to be easy to read, it should be as large as possible. That is, the scale factor from the map to the park should be as small as possible.

If you consider only whole numbers for scale factors, what is the smallest possible scale factor that they could use?

17. Challenge Carefully copy this picture.

a. Rotate the flag 90° about point *P*. Then rotate the image −90° about point *Q*. What single type of transformation of the original flag would give the same final image?

b. Make a new copy of the picture. Rotate the flag 90° about point *P*, and then rotate the image 90° about point *Q*. What single transformation of the original flag would give the same final image?

c. Now rotate the original flag 50° about point *P*, and then rotate the image −30° about point *Q*. Extend the flagpoles of the original flag and the final image, until the two segments intersect. What is the measure of the angle between them?

d. Add the two angles of rotation in Part a. Then do the same for the pairs of angles in Parts b and c.

e. How are these sums connected to your answers to Parts a, b, and c?

18. **Technology** Many photocopy machines reduce and enlarge figures automatically. However, copy machines often have only a limited number of scale factors. Suppose you are using a photocopy machine that has three settings for reducing or enlarging. They are 50%, 150%, and 200%.

    a. Will wants to reduce a picture using a scale factor of $\frac{1}{4}$. How would he do it?

    b. Neela wants to enlarge a picture using a scale factor of 3. How would she do it?

    c. How can you reduce a picture to 75% its original size?

Mixed Review

19. Find an equation for the line through the points (3, 2) and (8, −5).

Simplify each expression.

20. $\dfrac{m-3}{m(2m-6)}$

21. $\dfrac{7}{k-2} - \dfrac{5}{2(k-2)}$

Each table in Exercises 22–24 describes a linear relationship, an exponential relationship, or an inverse variation. Write an equation describing each relationship.

22.

x	y
−2	500
−1	200
0	80
1	32
2	12.8

23.

x	y
−6	500
−3	200
−2	80
2	12
3	8

24.

x	y
−4	10
−3	3
−1	−11
2	−32
3	−39

Review & Self-Assessment

Chapter Summary

In this chapter, you studied transformational geometry. You learned how to perform four transformations. They are *reflection, rotation, translation,* and *dilation.* You also learned how to recognize designs with both reflection and rotation symmetry.

Three of the translations, *reflection, rotation,* and *translation,* produce images that are congruent to the original figures. Dilation produces images that are similar but not necessarily congruent.

Vocabulary

dilation

image

line of reflection

line of symmetry

line symmetry

reflection over a line

reflection symmetry

rotation

rotation symmetry

scale drawing

scale factor

transformation

translation

vector

Strategies and Applications

The questions in this section will help you review and apply the important ideas and strategies developed in this chapter.

Recognizing reflection and rotation symmetry

Each of the figures in Questions 1–3 has reflection symmetry, rotation symmetry, or both. Copy each figure. Then do Parts a–c.

    a. Determine the type or types of symmetry the figure has.

    b. Indicate the line or lines of symmetry, if any, and the center of rotation, if any.

    c. If the figure has rotation symmetry, determine the angle of rotation.

1.   **2.**    **3.**

Performing reflections

Three of the methods for reflecting figures that you learned are folding, using a GeoMirror and using perpendicular bisectors.

4. Choose one of these three methods. Explain in your own words how to reflect a figure using that method.

5. A particular rule for reflecting a figure on a coordinate grid changes an original coordinate (x, y) to the image coordinate $(-y, -x)$. What is the line of reflection for this rule?

Performing rotations

Two of the methods for rotating figures that you learned are using tracing paper and using a protractor and a ruler.

6. Choose one of these two methods. Explain in your own words how to use it to rotate a figure about a point with a given angle of rotation.

7. A particular rule for rotating a figure on a coordinate grid changes an original coordinate (x, y) to the image coordinate $(-y, x)$. What are the center and angle of rotation for this rule?

Performing translations

8. Explain in your own words how to translate a figure by a given vector.

9. A particular rule for translating a figure on a coordinate grid changes an original coordinate (x, y) to the image coordinate $(x + 4, y - 3)$. On a coordinate grid, show the translation vector for this rule.

Performing dilations

In this chapter, you learned two methods for dilating figures. They are the coordinate method and the projection method.

10. Choose one of these two methods. Explain in your own words how to use that method to dilate a figure by a given scale factor.

11. A polygon has a vertex at the point (8, 6). A dilated version of the polygon has a corresponding vertex at the point (12, 9). What scale factor was used to dilate the original polygon?

Combining transformations

12. Figure Z is the image of Figure A.

a. Find a way to transform Figure A into Figure Z. Use *two* transformations.

b. Does it matter which order you perform your transformations?

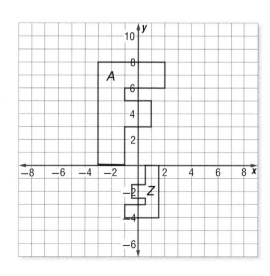

Performing constructions

13. Use a straightedge and a compass to construct a right triangle.

14. Using a compass and a straightedge, construct a line and its perpendicular bisector.

Demonstrating Skills

15. Copy this picture. Reflect the figure across the line using a method other than the one you described in Question 4.

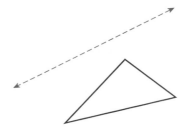

16. Copy this picture. Rotate the figure around point *P* with a −40° angle of rotation. Use the method you did not describe in Question 6.

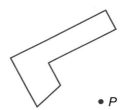

• *P*

Bisect each angle.

17.

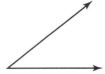

18.

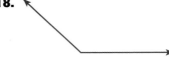

For each of the following angles, use a compass and a straightedge to construct a congruent angle. Then bisect the angle that you drew.

19.

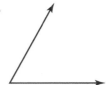

20.

Test-Taking Practice

SHORT RESPONSE

1 Reflect figure *ABC* over the *y*-axis. Then reflect figure *A'B'C'* over the *x*-axis. Write the coordinates of the vertices for each image on the lines below and graph the images.

Figure ABC	After Reflection Over *y*-axis	Figure A'B'C' Reflected Over *x*-axis
$A(-4, 3)$		
$B(-2, 4)$		
$C(-1, 1)$		

Show your work.

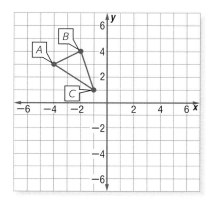

Answer _____

MULTIPLE CHOICE

2 Figure *MNPQ* has vertices $M(-4, -1)$, $N(3, -1)$, $P(3, -4)$, $Q(-4, -4)$. Which will be the coordinates of vertex *N* after the rectangle is rotated 90° clockwise about the origin?

A $(-1, 3)$

B $(-1, -3)$

C $(-3, -1)$

D $(3, 1)$

3 If point $A(4, -6)$ is translated 6 units to the left and 3 units down, which coordinates represent *A'*?

F $(10, -3)$

G $(-2, -9)$

H $(-2, -3)$

J $(2, 0)$

Inequalities and Linear Systems

Real-Life Math

Get with the Program Mathematical programming is a technique used in a wide variety of fields. While much of the mathematics you have studied has been around for centuries, mathematical programming did not exist until the late 1940's, just after World War II.

Mathematical programming problems are also called optimization problems. They involve looking for an optimal solution, one that gives a maximum or minimum value for a variable such as profit, cost, or time. They usually involve several equations and inequalities with many variables.

Petroleum companies use mathematical programming to find the best ways to blend gasoline. Due to limited space on Space Shuttle missions, this technique is also used to determine the minimum amount of food needed to ensure that astronauts get the proper nutrition.

Think About It List some of the variables that a manufacturing company might need to consider in order to maximize its profits.

Math Online
Take the **Chapter Readiness Quiz** at glencoe.com.

Dear Family,

The next chapter in mathematics is about solving equations and inequalities. This is one of the most important and most frequently used mathematical skills, one with many applications in the sciences, social studies, and everyday life.

Students have learned several methods for solving equations in earlier grades. In this chapter, the class will review and extend these methods to solve inequalities and systems of two equations with two variables. They will also learn how to use graphing calculators to find approximate solutions to equations and pairs of equations.

Key Concept—Inequalities

Situations involving inequalities are common in daily life. For example, suppose you go to the store with $5.00. You want to buy markers at $1.95 each and ink pads at $0.59 each. The inequality below tells which combinations of markers and pads you will be able to buy.

$$1.95m + 0.59p \leq 5.00$$

In this example, 1 marker and 5 pads, or 2 markers and 2 pads, will satisfy the inequality.

Students will also learn to solve systems of equations with two variables. In such situations, they will have to find a pair of values that satisfies *both* equations.

> In a game of basketball, Corrine scored 10 times for a total of 23 points. Some of the shots were 2-point shots, and some were 3-point shots. How many of each type did Corinne score?

To answer this question, set up the system of equations below, where x represents the number of 2-point baskets and y stands for the number of 3-pointers.

$$x + y = 10 \quad \text{and} \quad 2x + 3y = 23$$

Chapter Vocabulary

elimination　　　　　　　substitution

inequality　　　　　　　system of equations

Home Activities

• Ask your student to show you some of the equations he or she is solving and the methods used.
• Encourage your student to think about ways these skills can be used outside of school.

Equations

Solving mathematical problems often involves writing and solving equations. You have probably already encountered several strategies for solving equations.

Think & Discuss

Describe some strategies you might use to solve this equation.

$$6\left(\frac{2(2x - 6)}{4}\right) = 12$$

Here are three equation-solving methods you have probably used.

- *Guess-check-and-improve*
  Guess the solution. Check your guess by substituting it into the equation. Use the result to improve your guess if needed.

- *Backtracking*
  Start with the output value. Work backward to find the input value.

- *Doing the same thing to both sides*
  Apply the same mathematical operation to both sides of the equation until the solution is easy to see.

In this lesson, you will review backtracking and doing the same thing to both sides.

Investigation 1 Equation-Solving Methods

Suppose you want to solve an equation that consists of an algebraic expression on one side and a number on the other, such as this equation.

$$2\left(\frac{2n}{6} - 1\right) = 16$$

Backtracking may be a good solution method for this type of equation.

Example

Solve $2\left(\dfrac{2n}{6} - 1\right) = 16$ by backtracking.

Think of n as the input and 16 as the output. Make a flowchart to show the operations needed to get from the input to the output.

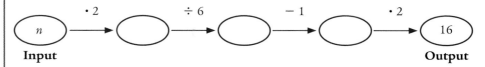

Input   **Output**

This flowchart shows that you multiply the input by 2, divide the result by 6, subtract 1, and then multiply by 2. The output value is 16. To backtrack, start from the output and work backward, *undoing* each operation, until you find the input.

The value in the fourth oval is multiplied by 2 to get 16, so this value must be 8.

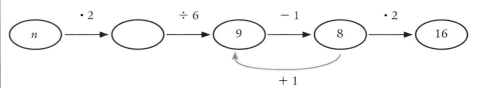

The number 1 is subtracted from the value in the third oval to get 8, so this value must be 9.

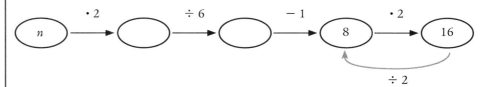

The value in the second oval is divided by 6 to get 9, so this value must be 54.

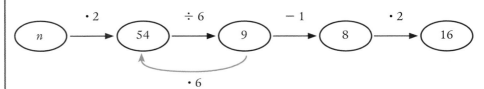

The input is multiplied by 2 to get 54, so the input must be 27. So, the solution of $2\left(\dfrac{2n}{6} - 1\right) = 16$ is 27.

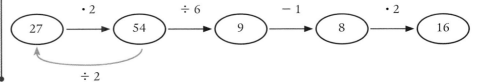

✅ *Develop & Understand: A*

Solve each equation by backtracking. Check each solution by substituting it into the original equation.

1. $3\left(\frac{n}{2} - 1\right) = 15$

2. $\frac{2(n + 1) - 3}{4} = 5$

3. $2\left(\frac{n}{4} - 3\right) + 1 = 5$

For fairly simple equations, you may be able to backtrack in your head. Here is how you might think about solving $\frac{n}{3} + 1 = 4$.

> *This equation means "divide by 3 and then add 1," which gives 4. To backtrack, "subtract 1" from 4 to get 3 and then "multiply by 3" to get 9.*

Solve each equation mentally if you can. If you have trouble, use pencil and paper. Check each solution by substituting it into the equation.

4. $6(3m - 4) = 12$

5. $\frac{3m}{4} + 4 = 12$

6. $\frac{3(n + 4)}{6} = 12$

Some equations cannot be solved directly by backtracking.

Think **&** Discuss

Alfonso tried to solve $3a + 4 = 2a + 7$ by backtracking, but he could not figure out how to make a flowchart. Why do you think he had trouble?

How could you solve $3a + 4 = 2a + 7$?

Equations that cannot easily be solved by backtracking can often be solved by doing the same thing to both sides.

Example

Solve $3a + 4 = 2a + 7$ by doing the same thing to both sides.

$$3a + 4 = 2a + 7$$
$$a + 4 = 7 \qquad \text{after subtracting } 2a \text{ from both sides}$$
$$a = 3 \qquad \text{after subtracting 4 from both sides}$$

✅ Develop & Understand: B

Solve each equation by doing the same thing to both sides. Check your solutions.

7. $3a - 4 = 2a + 3$

8. $11b - 6 = 9 + 6b$

9. $7 - 5x = 12 - 3x$

10. $3y + 7 = 7 - 2y$

✅ Develop & Understand: C

Sometimes you need to simplify an equation before you can solve it.

11. Since one side of this equation is a number, Taci thought she could solve it by backtracking.

$$4(3x + 2 - 2x + 2) = 20$$

a. Taci drew a flowchart for this equation but could not figure out how to backtrack to solve it. Why do you think she had trouble?

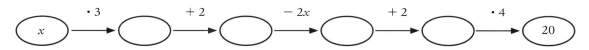

b. Taci thought the equation might be easier to solve if she simplified it, so x appeared only once. Simplify the equation. Then solve it by backtracking.

Simplify each equation. Then solve it using any method you choose. Check your solutions.

12. $5(2a - 3) + 2a + 3 = 0$

13. $-6(8 - 3n) + 2n = -8$

14. $2k + 2\left(\dfrac{k}{4} - 3\right) - k + 1 = 25$

Real-World Link

Flowcharts are often used by builders to organize the steps involved in a construction project.

✅ *Develop & Understand: D*

Now try applying equation-solving techniques to solve situations that are described in words.

15. Jessica said, "I'm thinking of a number. When I subtract 3 from my number, multiply the result by 8, and then divide this result by 3, I get 32. What is my number?"

16. Neva and Owen collect action figures. Owen has five times as many action figures as Neva. Owen also has 60 more than Neva. Write and solve an equation to find the number of action figures Neva and Owen each have.

17. The sum of four consecutive whole numbers is 150.

    a. Let x represent the first number. Write expressions to represent the other three numbers.

    b. Write and solve an equation to find the four numbers.

18. The sum of three consecutive *even* numbers is 78.

    a. Let n represent the first even number. Write expressions for the next two even numbers.

    b. Write and solve an equation to find the three numbers.

19. A distributor has 800 copies of a new DVD to distribute among four stores. She plans to give the first two stores the same number of copies. The third store will receive three times as many DVDs as either of the first two. The fourth store will receive 20 more than either of the first two.

    a. How many copies will each store receive?

    b. The distributor has 700 copies of another DVD that she wants to distribute using the same rules. How many copies do you think she should give to each store? Explain your answer.

> ## *Share & Summarize*
>
> 1. Could you solve all the backtracking exercises on page 314 by doing the same thing to both sides?
>
> 2. Write an equation, similar to those on page 314, that can be solved by backtracking. Solve your equation first by backtracking and then by doing the same thing to both sides. Record each step in your solutions. What connection do you notice between the two methods?

Investigation 2 Equation Applications

Emma's cousin lives in England, where they measure temperature using the Celsius system. Emma is used to the Fahrenheit system of measuring temperature. She wants to know how to convert Celsius temperatures into Fahrenheit so she can understand what her cousin tells her about the weather in England.

Her cousin says that in Celsius, water freezes at 0° and boils at 100°. Emma knows that in Fahrenheit, water freezes at 32° and boils at 212°.

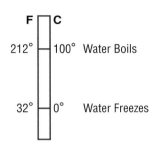

Real-World Link

Most countries, apart from the United States, use the Celsius system to measure temperature. Biologists, chemists, and physicists, even those in the U.S., usually use Celsius or Kelvin. Some U.S. engineers use the *Rankine system*. But otherwise, it is not often used.

✅ *Develop & Understand: A*

1. In Fahrenheit, how many degrees are between the freezing and boiling points of water? In Celsius?

2. Which is greater, a Fahrenheit degree or a Celsius degree?

3. Use a ratio to show how big a Fahrenheit degree is, compared to a Celsius degree.

4. Is the relationship between Fahrenheit degrees and Celsius degrees linear? Why or why not?

5. Write an equation that shows how to find the temperature in Fahrenheit (F) if you know the temperature in Celsius (C).

The strategies you used to solve equations with one variable are also useful for working with equations with two variables. You can *do the same thing to both sides of an equation* or *backtrack* to rearrange an equation into a more convenient form.

✓ Develop & Understand: B

Emma's cousin wanted to know how to convert Fahrenheit temperatures into Celsius.

6. Start with your original Fahrenheit-to-Celsius equation. Rearrange it into a Celsius-to-Fahrenheit conversion equation by doing the same thing to each side of the equation.

7. Use your two conversion equations to fill in this temperature conversion table.

°F		41	0		11		140
°C	30			15		−12	

8. Graph the two conversion equations on separate coordinate planes.

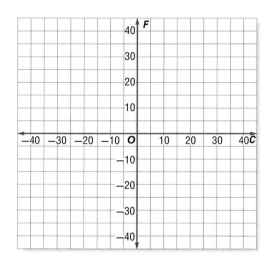

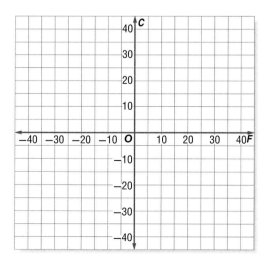

✅ Develop & Understand: C

You are probably familiar with these units of mass and volume.

Mass:

milligrams	1000
grams	1
kilograms	$\frac{1}{1000}$

ounces	16
pounds	1

Volume:

cups	16
pints	8
quarts	4
gallons	1

milliliters	1000
liters	1

9. Write an equation to perform each of the following conversions.

   a. Convert pints to gallons.

   b. Convert milliliters to liters.

   c. Convert kilograms to milligrams.

   d. Convert pounds to ounces.

10. Perform the following conversions.

    a. 4 pints to gallons

    b. 2 quarts to cups

    c. 10 grams to milligrams

    d. 100 grams to kilograms

11. Which two units could the conversion equation $y = \frac{x}{1000}$ represent?

✅ Develop & Understand: D

12. The Peterson Tablecloth Company buys dye from a manufacturer and then dilutes it at its factory. The manufacturer sends bottles of dye measured in milliliters. At the factory, two liters of water are added to each batch of dye.

    Write an equation that allows the factory manager to calculate the final volume of a batch of diluted dye, in liters, depending on what the initial volume was.

13. The dye manufacturer started shipping pre-diluted dye to the Peterson factory. For every 250 mL of dye, the manufacturer adds 500 mL of water. The factory still measures the dye in liters.

   Write an equation to calculate the final amount of dye, in liters, depending on the original amount of dye before the manufacturer added water, in milliliters.

14. The company went back to the old system of diluting the dye once it gets to the factory. However, the factory manager decided to change the system for diluting dye so that the amount of water added to each batch of dye is *proportional* to the amount of dye. For every L of dye, the factory will add 1 L of water.

   Write an equation to calculate the final volume, in liters, depending on the initial volume of dye, in mL.

Think & Discuss

Consider the equations you created in the last set of exercises. What would the graphs of the equations look like?

Are there pairs of negative numbers that fit the equations? If so, does it make sense to talk about negative numbers when you are converting measurement units?

Develop & Understand: E

Simon, Kai, and Zoe want to figure out what the meal will cost. They get a 10% discount on the meal, sales tax is 8%, and they plan to leave an 18% tip. But they are not sure how the discount works.

15. How do you think Kai and Simon came up with their equations?

16. Why does Zoe think the two equations are the same?

17. Write an equation to calculate the total cost of a meal if x is the price of the food, sales tax is 8%, and you leave an 18% tip. Write the equation using Kai's method and using Simon's method.

18. Use a flowchart to show how to find the total cost of the meal if you know the price of the food. Does your flowchart follow the steps of Kai's version of the equation or Simon's?

19. Write an equation to calculate the cost of the meal if the price of the food is x, sales tax is 8%, and there is a 10% discount.

✅ *Develop & Understand: F*

Zoe thinks the discount applies to the cost of the meal, including sales tax, but not including tip. Simon thinks the discount applies only to the price of the meal, not tax or tip. Kai thinks the discount applies to the cost of the meal, including tax and tip.

20. Make a flowchart to show how to calculate the cost of the meal in the three situations.

  a. Zoe's interpretation

  b. Simon's interpretation

  c. Kai's interpretation

21. Whose interpretation results in the largest total cost of the meal? Explain why.

22. Write an equation corresponding to one of the friends' interpretations. Then simplify it as much as possible.

Share & Summarize

Explain how to create an equation that converts from one measurement unit to another. Someone should be able to pick any two units that measure the same quantity, for example, temperature or mass. Write a conversion equation.

Practice & Apply

Solve each equation using the method of your choice.

1. $5(n - 2) = 45$

2. $2(n - 5) = 7$

3. $\frac{n}{2} + 5 = 7$

4. $3(4m - 6) = 12$

5. $\frac{3m - 3}{6} = 12$

6. $3(\frac{m}{6} - 4) = 12$

7. $2x + 3 = x + 5$

8. $7y - 4 = 4y - 13$

Simplify and solve each equation.

9. $4 - 2(-5a - 10) = 30$

10. $\frac{b - 2}{4} = \frac{6}{5}$

11. You work at a clothing store. A customer wants to use two discount coupons on the same order. One gives $10 off, and the other gives 15% off. You are not sure which discount to apply first.

  a. Write an equation to calculate the final price of the order, if you apply the $10 discount and then the 15% discount.

  b. Write an equation to calculate the final price of the order, if you apply the 15% discount first and then the $10 discount.

  c. Which order gives the customer a better deal? Does it depend on the original cost of the customer's order?

In Exercises 12–14, write and solve an equation to find the number of coins each friend has.

12. Ken has three more coins than twice the number Javier has. Khalid has five fewer coins than Javier. They have 50 coins altogether.

13. Da-Chun has five times as many coins as Austin. Da-Chun also has 16 more than Austin.

14. Kavel has two less than double the number of coins Ty has. Kavel also has 23 more than Ty.

15. Lindsey said, "I'm thinking of a number. If I multiply it by 5, subtract 4, and then multiply the result by 2, I get 62. What is my number?"

Connect & Extend

16. Write two equations that cannot be solved directly by backtracking. Explain why backtracking will not work.

17. Tory made this toothpick pattern. She described the pattern with the equation $t = 5n - 3$, where t is the number of toothpicks in stage n.

| Stage 1 | Stage 2 | Stage 3 |

    a. Explain how each part of the equation is related to the toothpick pattern.

    b. How many toothpicks would Tory need for stage 10? For stage 100?

    c. Tory used 122 toothpicks to make one stage of her pattern. Write and solve an equation to find the stage number.

    d. Is any stage of the pattern composed of 137 toothpicks? Why or why not?

    e. Is any stage of the pattern composed of 163 toothpicks? Why or why not?

    f. Tory has 250 toothpicks and wants to make the largest stage of the pattern she can. What is the largest stage she can make? Explain your answer.

18. Recall that the absolute value of a number is its distance from 0 on the number line. You can solve equations involving absolute values. For example, the solutions of the equation $|x| = 8$ are the two numbers that are a distance of 8 from 0 on the number line, 8 and -8.

    Solve each equation.

    a. $|a| = 2.5$ **b.** $|2b + 3| = 8$

    c. $|9 - 3c| = 6$ **d.** $\dfrac{|5d|}{25} = 1$

    e. $|-3e| = 15$ **f.** $20 + |2.5f| = 80$

19. Kenya and Lora were making hair ribbons to sell at a crafts fair. Kenya cut seven segments from one length of ribbon and had two feet left. Lora said, "I'm cutting segments twice as long as yours. If your length of ribbon had been just one foot longer, I could have cut four segments from it."

From their conversation, determine how long Kenya's and Lora's segments were.

20. **In Your Own Words** How do you decide whether to solve an equation by backtracking or by doing the same thing to both sides?

21. Nine square tiles are used to cover a floor that covers an area of 36 square feet.

   a. Write and solve an equation to find the dimensions of the tiles.

   b. Draw and label a picture to show how these nine tiles could be used to cover a floor that measures 3 feet by 12 feet. Assume that you are able to cut tiles in half if needed.

Mixed Review

22. **Physical Sciences** Scientists often measure temperature using the Kelvin scale. Kelvin degrees are the same size as Celsius degrees, but 0 degrees Kelvin is absolute zero, the coldest possible temperature. Atoms stop moving at absolute zero. 0 degrees Kelvin, written 0 K, equals $-273.15°C$, or $-459.67°F$.

   a. Write an equation to convert temperature in Celsius into Kelvin.

   b. Write an equation to convert temperature in Kelvin into Celsius.

   c. Using the conversion equations you developed during Investigation 2, write an equation to convert Fahrenheit degrees into Kelvin. Show or explain how you got your answer.

   d. The Renkine temperature system puts 0 degrees at absolute zero but uses degrees the size of Fahrenheit degrees. So, $0°F$ equals $459.67°R$. Write an equation converting Fahreheit temperature to Rankine.

   e. Suppose someone invented a temperature system in which 0 degrees was equal to $0°C$ (the freezing point of water) but with degrees the same size as Fahrenheit degrees. How would the equation that converted Fahrenheit into this new system look similar to the other conversion equations? How would it look different?

Solve.

23. 32.2 is 92% of what number?

24. What percent of 125 is 90?

25. What is 81% of 36?

Find the value of n in each equation.

26. $3.582 \times 10^n = 3,582,000$ 27. $n \times 10^7 = 34,001$

28. $82.882 \times 10^3 = n$ 29. $28.1 \times \dfrac{1}{10^3} = n$

Inequalities

The equals sign, =, expresses a comparison between two quantities. It states that two quantities are *equal*. The table lists other mathematical symbols you have used to express comparisons.

Symbol	What It Means	Examples
$<$	is less than	$5 < 7 \quad 4 + 8 < 15 \quad -17 < -4 \cdot 3$
$>$	is greater than	$7 > 5 \quad\quad 80 > 20 + 9 \quad\quad 3^4 > 2^4$
$\leq$	is less than or equal to	$5 \leq 7 \quad\quad 6 \cdot -5 \leq -30 \quad\quad \frac{1}{3} \leq \frac{1}{2}$
$\geq$	is greater than or equal to	$7 \geq 5 \quad\quad 10^8 \geq 10^7 \quad\quad \frac{1}{3} \geq \frac{3}{9}$

Vocabulary

inequality

An **inequality** is a mathematical statement that uses these symbols to compare quantities. The inequalities above compare only numbers. You have also used inequalities to compare expressions involving variables.

- The statement $x > 7$ means that the value of the variable x is greater than 7. Some possible values for x are 7.5, 12, 47, and 1,000.

- The statement $n - 3 \leq 12$ means that the value of "n minus 3" is less than or equal to 12. Possible values for n are 15, 1.2, 0, and -5.

Think & Discuss

How is the meaning of $7 < x$ similar to or different from that of $x > 7$?

It is sometimes convenient to combine two inequalities.

$$7 < x < 10$$

What do you think $7 < x < 10$ means?

How could you write $7 < x < 10$ as two inequalities joined by the word *and*?

List three values of x that satisfy the inequality $7 < x < 10$.

Thermostats are used to control the temperature in a room or a building. How do you think a thermostat might use inequalities to do this?

Investigation 1 Describe Compound Inequalities

The exercises in this investigation will give you more practice working with inequalities.

✅ Develop & Understand: A

1. List six whole numbers that satisfy the inequality $n + 1 < 11$.

2. Consider this statement: k is less than or equal to 14 and greater than 9.

   a. Write this statement in symbols.

   b. List all the whole numbers for which the inequality in Part a is true.

3. Consider the inequality $15 \leq m \leq 18$.

   a. Write the inequality in words.

   b. List all the whole numbers for which the inequality is true.

   c. Are there other numbers, not necessarily whole numbers, for which the inequality is true? If so, give some examples. How many are there?

4. If you consider only whole-number values for n, four out of the five statements below represent the same values. Which inequality below is *not* equivalent to the others?

$$10 < n < 20 \qquad 11 \leq n \leq 19 \qquad 11 \leq n < 20$$

$$11 < n < 19 \qquad 10 < n \leq 19$$

List six values, not necessarily whole numbers, that satisfy each inequality or pair of inequalities.

5. $10.2 < p < 14.7$

6. $q \geq 12$ and $q > 15$
   The values of q must make *both* inequalities true.

7. $20 \geq |r| \geq 17.75$

8. $3s > 12$

9. $t > 0$ and $t^2 \leq 16$

Math Link

The absolute value of a number is its distance from 0 on the number line.

$$|3| = 3 \qquad |-3| = 3$$

Now you will write inequalities to represent situations that are described in words.

✅ *Develop & Understand: B*

Write an inequality or a pair of inequalities to describe each situation. Make your answers as specific and complete as possible.

10. A box of matches contains at least 48 but fewer than 55 matches. Write an inequality for the number of matches n in the box.

11. Tamika expects from 100 to 120 people to buy tickets to the talent show. Each ticket costs $5. Write an inequality to represent the total amount of money m she expects from ticket sales.

12. Four friends, Sandy, Mateo, Felisa, and Destiny, are comparing their heights. Sandy, who is 155 cm tall, is the shortest. Mateo is 165 cm tall. Felisa is not as tall as Mateo. Destiny is at least as tall as Mateo. Write inequalities to represent Felisa's and Destiny's heights, F and D.

13. The Completely Floored store sells square tiles in a variety of sizes with side lengths ranging from 2 cm to 20 cm. Write an inequality to represent the range of possible areas a, in square centimeters, for the tiles.

14. Rondell's walk to school takes 15 minutes, give or take 2 minutes. Write an inequality for the time t it takes Rondell to walk to school.

15. The bed of José's trailer is 120 cm above the ground. Each layer of cartons he loads onto the trailer adds an additional 40 cm to the height. His loaded trailer must be able to pass under a footbridge with an underside 4 meters high.

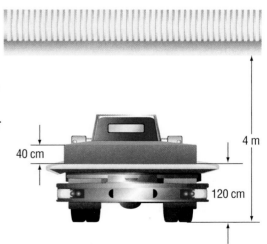

    a. Write an expression for the height of the trailer when it holds n layers of cartons.

    b. Use your expression to write an inequality relating the height of the loaded trailer to the height of the footbridge.

16. The families of Kelsey and her four friends are throwing a party for the girls' sixteenth birthdays. They estimate they will spend $7 per person for food, $3 per person for beverages, and $200 to hire a disc jockey. The total amount budgeted for the party is $500.

a. Let p represent the number of people they can invite. Write an expression for the costs of the party for p people.

b. Use your expression to write an inequality relating the costs, the number of people, and the budget.

17. An architect estimated that the maximum floor area for a square elevator in a particular building is 4 m². Building regulations require that the minimum area be 2.25 m². Write an inequality for the possible side lengths s for the elevator floor.

Share & Summarize

Create a situation that can be represented by an inequality. Express your inequality in symbols.

Investigation 2 Solve Compound Inequalities

Most of the equations you have solved have had one or two solutions. Inequalities, however, can have many solutions. They often have an infinite number.

In Exercise 8 in Investigation 1, you found some of the solutions of the inequality $3s > 12$. For example, the values 4.5, 7, and 10 all satisfy $3s > 12$. In fact, any value greater than 4 will satisfy this inequality. But you certainly cannot list them all.

For this reason, the solutions of an inequality are usually given as another inequality. For example, you can express all the solutions of $3s > 12$ by writing $s > 4$.

✅ Develop & Understand: A

1. Solve each equation or inequality. That is, find the value or values of a that make the equation or inequality true.

a. $3a - 10 = 35$ **b.** $3a - 10 > 35$ **c.** $3a - 10 < 35$

2. This is how Tamika thought about Exercise 1.

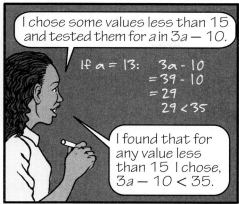

a. Test at least four other values for a. Does Tamika's conclusion seem correct?

b. Use this graph of $y = 3a - 10$ to explain why Tamika's conclusion is correct.

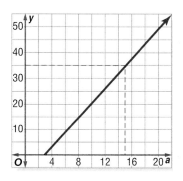

Use Tamika's method to solve each inequality.

3. $5a + 7 < 42$

4. $\dfrac{b}{7} + 1 \geq 6$

5. $4c - 3 > 93$

6. $-6(d + 1) \geq 24$

Think & Discuss

To solve a linear inequality of the form $mx + b < c$ or $mx + b > c$ for x, where a, b, and c are constants, you can solve the related equation $mx + b = c$. Then test just *one* value greater than or less than the solution. Explain why this method works.

Ben solved the inequality $3a - 10 < 35$ by doing the same thing to both sides.

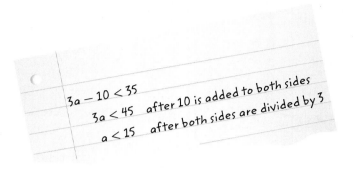

$3a - 10 < 35$
$3a < 45$ after 10 is added to both sides
$a < 15$ after both sides are divided by 3

✅ Develop & Understand: B

Use Ben's method, doing the same thing to both sides, to solve each inequality. Then solve the inequality by using Tamika's method, solving the related equation and testing one value greater than or less than the solution. If you get different solutions, tell which solution is correct.

7. $7x + 2 \geq 100$ **8.** $-7x + 2 \leq 100$

9. Consider these inequalities.

$$5 < 6 \qquad -4 > -10 \qquad 16 \geq 2 \qquad -7 < 4$$

   a. Multiply both sides of each inequality by a negative number. Is the resulting inequality true? If not, how can you change the inequality symbol to make it true?

   b. Now *divide* both sides of each inequality above by a negative number. Is the resulting inequality true? If not, how can you change the inequality symbol to make it true?

   c. Use what you discovered in Parts a and b to explain why Ben's and Tamika's methods sometimes give different results.

   d. How can you alter Ben's method to always give the right solution?

Solve each inequality.

10. $2b > 15$ **11.** $-2c > 15$ **12.** $12 + 6d \leq 54$

13. $12 - 6e \leq 54$ **14.** $-3(f - 12) < -93$ **15.** $\dfrac{5g}{-7} + 1 \geq -\dfrac{23}{7}$

16. Solve the inequality $0 < 1 - u < 1$. Explain your solution method.

Share & Summarize

Suppose one of your classmates has been absent during this investigation. Explain to him or her, in writing, how to solve a linear inequality. Describe some common mistakes to avoid.

Investigation ③ Graph Compound Inequalities on a Number Line

You can use number line graphs to show values that satisfy inequalities.

Example

This number line shows the solutions of the inequality $-3 \leq n < 1$. The filled-in circle indicates that -3 is included in the solution. The open circle indicates that 1 is *not* included. The heavy line shows that all numbers between -3 and 1 are included.

Below is the graph of $x \geq 1$. The arrow indicates that the solution extends to the right, beyond the part of the number line shown.

This is a graph of $|z| \leq 2$. It shows that values of z less than or equal to 2 and greater than or equal to -2 are solutions of the inequality.

Real-World Link

Size regulations for taking certain fish species from public waters can often be stated as an inequality. During 1999 in Florida, a black drum fish could only be kept if it was not shorter than 14 in. and not longer than 24 in.

✓ Develop & Understand: A

1. Consider the inequality $0 < b < 4.7$.

   a. List all the *integer* values that satisfy the inequality.

   b. Use your answer to Part a to help you graph *all* the values that satisfy $0 < b < 4.7$.

2. In Parts a–d, you will consider the inequalities $x < 10$ and $x > 5$. Draw a separate number-line graph for each part.

   a. Graph all x values for which $x < 10$.

   b. Graph all x values for which $x > 5$.

   c. Graph all x values for which $x < 10$ *and* $x > 5$.

   d. Graph all x values for which $x < 10$ *or* $x > 5$.

   e. Explain how the words *and* and *or* affect the graphs in Parts c and d.

3. In Parts a–c, you will consider the inequalities $c \geq 10$ and $c \leq 5$. Draw a separate graph for each part.

   a. Graph all c values for which $c \geq 10$ and $c \leq 5$.

   b. Graph all c values for which $c \geq 10$ or $c \leq 5$.

   c. Explain how the words *and* and *or* affect the graphs in Parts a and b.

4. Consider inequalities that involve absolute values.

   a. Graph all m values that satisfy the inequality $|m| \geq 2.5$.

   b. Graph all t values that satisfy the inequality $|t| < 3$.

5. Now consider the inequality $x^2 < 16$.

   a. List the *integer* values that satisfy the inequality.

   b. Use your answer to Part a to help you graph *all* x values that satisfy the inequality $x^2 \leq 16$.

   c. Express the solution of $x^2 \leq 16$ as an inequality.

✅ Develop & Understand: B

6. Graph all m values for which $|m| - 3 \geq 1.75$.

7. Express the solution to Exercise 6 as an inequality.

8. Graph all r values for which $|r - 3| \geq 3$.

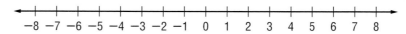

9. Express the solution to Exercise 8 as an inequality.

Recall what multiplying and dividing an equality by a negative number does. Use that knowledge in the next exercise set.

✓ Develop & Understand: C

Solve each inequality. Graph the solution on a number line.

10. $-5x + 10 > 20$

11. $\frac{n}{3} < 9$

12. $\frac{t}{8} - 4 < -5$

13. $3(n + 2) > 15$

14. $-2(k - 5) \leq 10$

15. Eva said it would take her at least an hour and a half, but no more than two hours, to finish her homework. Write an inequality to express the number of hours h that Eva thinks it will take to do her homework.

16. Wednesday nights are special at the video arcade. Customers pay $3.50 to enter the arcade and then only $0.25 to play each game. Samuel brought $7.50 to the arcade and still had some money when he left. Write an inequality for this situation, using n to represent the number of games Samuel played.

Share & Summarize

Consider these statements.

$$x \geq a \text{ and } x \leq b \qquad x \geq a \text{ or } x \leq b$$

1. Describe how the number-line graph of $x \geq a$ and $x \leq b$ is different from the number-line graph of $x \geq a$ or $x \leq b$.

2. Give values of a and b for which the number-line graph of $x \geq a$ or $x \leq b$ includes all numbers.

3. Compare the solutions to $|x| < 4$ and $x^2 < 16$. How are they related?

Investigation 4 — Graph Inequalities on a Coordinate Plane

Materials

- graph paper

You already know how to show solutions of a two-variable equation such as $y = 2x + 3$ by making a graph of the equation on a coordinate plane. Any point (x, y) on the graph satisfies the equation. You will now learn how to graph inequalities with two variables.

✅ Develop & Understand: A

1. Graph the equation $y = x$.

2. List the coordinates of three points that are above the line $y = x$ and three points that are below the line.

3. Which inequality, $y > x$ or $y < x$, describes the coordinates of points that are above the line $y = x$?

4. Which inequality, $y > x$ or $y < x$, describes the coordinates of points that are below the line $y = x$?

5. Based on your answers to Exercises 3 and 4, predict whether each given point will be above, below, or on the line $y = x$. Test each prediction by plotting the point.

   a. $(5, 11)$ **b.** $(7, 3)$ **c.** $(0, -6)$ **d.** $(1, 1)$

6. List three (x, y) pairs that make the inequality $y > 3x$ true.

7. Graph the equation $y = 3x$. Then plot the points you listed in Exercise 6. Where do they appear on your graph?

You can use what you learned in Exercises 1–7 to graph inequalities.

Example

Graph the inequality $y > 3x$.

First, graph the *equation* $y = 3x$. Use a dashed line because the points on the line do not make the inequality true.

Then shade the area containing the points that make the *inequality* $y > 3x$ true. You saw in Exercises 6 and 7 that these are the points above the line.

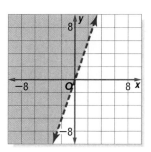

✅ *Develop & Understand: B*

Graph each inequality.

8. $y < 2x + 3$

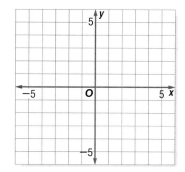

9. $y > x + 2$

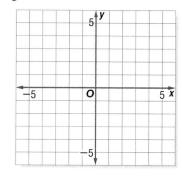

10. $y \leq -4x$

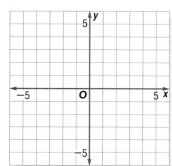

Convert each inequality into linear form and graph.

11. $4y + x < -3x - 12$

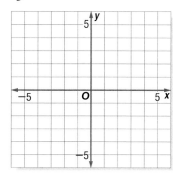

12. $x \geq 2(y - 2)$

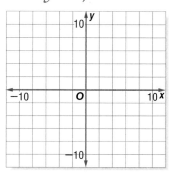

13. $-4x + 81 - 9y > 14x$

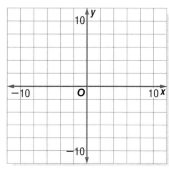

Write a linear inequality that corresponds to each graph.

14.

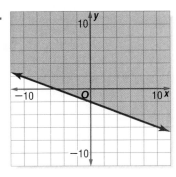

15.

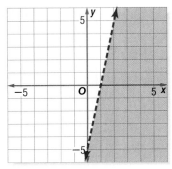

16.

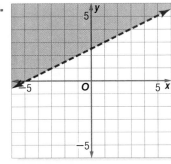

17. Challenge Consider the relationship "*y* is more than *x* and less than three times *x*."

a. What would the graph of this situation look like?

b. Where are the points that satisfy the inequality?

c. What about the relationship "*y* is more than *x* *or* less than three times *x*"? Now which points satisfy the inequality?

> ### Share & Summarize
>
> How is graphing an inequality on the coordinate plane similar to graphing an equation on the coordinate plane? How is it different?
>
> How is graphing an inequality on the coordinate plane similar to graphing it on a number line? How is it different?

Practice & Apply

1. Consider the inequality $-4 < x \leq 0$.

    a. List all the integers that satisfy the inequality.

    b. List three non-integers that satisfy the inequality.

List five values that satisfy each inequality. Include negative and positive values, if possible.

    2. $-2n \geq 6$ **3.** $p^2 < 4$ **4.** $6 < x < 7$

    5. $y > 5$ and $y > 12$ **6.** $m - 3 > 9$ **7.** $1 \leq -x \leq 5$

    8. $|q| < 5$ **9.** $0 < |b| \leq 6$ **10.** $|s| - 5 \geq 6$

11. Sareeta earns \$0.12 for each pamphlet she delivers. She thinks she can deliver 500 to 750 pamphlets on Sunday. Write an inequality for the number of dollars d Sareeta expects to earn on Sunday.

12. Dan's Delivery Service charges \$9 to ship any package weighing five pounds or less. Mr. Valenza wants to send a box containing tins of cookies to his daughter in college. Each tin weighs 0.75 pound, and the packing materials weigh about one pound.

    a. Write an expression, using t to represent the number of tins in the box, to represent the weight of the box.

    b. Use your expression to write an inequality representing the number of tins Mr. Valenza can send for \$9.

13. Solve each equation or inequality.

    a. $1 - d = 1$ **b.** $1 - d < 1$ **c.** $1 - d > 1$

Solve each inequality.

    14. $5(e - 2) > 10$ **15.** $\dfrac{f}{2} + 5 > 10$

    16. $\dfrac{g + 2}{5} > 10$ **17.** $-3 \leq h - 2 \leq 1$

18. Consider the inequality $x^3 \leq 27$.

    a. Express the solution of $x^3 \leq 27$ as an inequality.

    b. Graph the solution on a number line.

Graph the solution of each inequality on a number line.

19. $-\dfrac{3p}{4} < 6$

20. $12 - 5q \geq 32$

Graph each inequality.

21. $y < 3x + 7$ **22.** $y \leq -3x - 7$ **23.** $y > -2x + 4$

Write the inequality that corresponds to each of these graphs.

24.
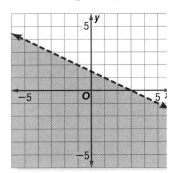

25.
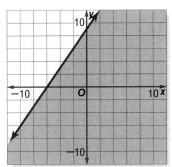

Graph each inequality. Rewrite the inequality first, if necessary.

26. $2y + 18 > 4x$ **27.** $y - 3x + 4 \leq 10$

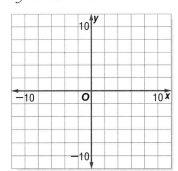
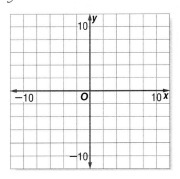

28. Show how you would change the graph of $y = \dfrac{2}{3}x - 4$ to represent $y > \dfrac{2}{3}x - 4$. Explain the changes that you made.

29. Show how you would change the graph in Exercise 28 to show $-y > \frac{2}{3}x - 4$. Explain the changes that you made.

30. What geometric transformation relates the graph from Exercise 28 and the graph from Exercise 29?

Connect & Extend

31. In U.S. amateur boxing, fighters compete in classes based on their weight. Five of the 12 official weight classes are listed below. Copy and complete the table to express the weight range for each class as an inequality.

Weight Class	Weight Range (pounds)	Inequality
Super heavyweight	over 201	
Heavyweight	179–201	
Welterweight	140–147	$140 \leq w \leq 147$
Featherweight	120–125	
Light flyweight	under 107	

32. Mia is writing a computer game in which a player stands on the balcony of a haunted house and drops water balloons on ghosts below. The player chooses where the balloon will land and then launches it.

Since water splatters, Mia's game gives the player points if a ghost is anywhere within a square centered where the balloon lands. The square extends 15 units beyond the center in all four directions. That is, if both of the ghost's coordinates are 15 units or less from the center, the player has scored a hit.

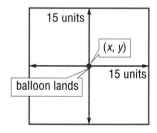

Suppose the balloon lands at (372, 425). The nearest ghost has the coordinates (x, y), and it counts as a hit by the game. Use inequalities to describe the possible values for x and y. (Hint: You will need two inequalities, one for x and one for y. Should you say "and" or "or" between them?)

When you draw a graph, you have to decide the range of values to show on each axis. Each exercise below gives an equation and a range of values for the *x*-axis. Use an inequality to describe the range of values you would show on the *y*-axis. Explain how you decided. It may help to try drawing the graphs.

33. $y = 2x + 7$ for $0 \leq x \leq 10$

34. $y = 2x - 10$ for $-5 \leq x \leq 5$

35. $y = x^2 + 1$ for $-5 \leq x \leq 5$

36. $y = -2x$ for $-5 < x < 0$

37. Physical Science The pitch of a sound depends on the frequency of the sound waves. High-pitched sounds have higher frequencies than low-pitched sounds. Most animals can hear a much greater range of frequencies than they can produce. The table shows the range of frequencies various animals can produce and hear.

Animal	Frequencies of Sounds Produced (hertz)	Frequencies of Sounds Heard (hertz)
Human being	$85 \leq f \leq 1{,}100$	$20 \leq f \leq 20{,}000$
Bat	$10{,}000 \leq f \leq 120{,}000$	$1{,}000 \leq f \leq 120{,}000$
Dog	$452 \leq f \leq 1{,}080$	$15 \leq f \leq 50{,}000$
Grasshopper	$7{,}000 \leq f \leq 100{,}000$	$100 \leq f \leq 15{,}000$

a. Make a number-line graph showing the range of frequencies a grasshopper can produce but cannot hear.

b. Make a number-line graph showing the range of frequencies a dog can hear that a human being cannot produce.

c. Make a number-line graph showing the range of sounds a bat can make that a dog can hear.

d. Make a number-line graph showing the range of frequencies both a dog and a grasshopper can produce.

38. In Your Own Words Describe how graphing an inequality differs from graphing an equation.

39. Make a graph showing the values for which $y < 3x + 2$ and $y > x + 5$.

40. Challenge Use a graph to solve the inequality $-2x \geq x - 3$. Explain how you found your answer.

Mixed Review

41. Copy this figure and vector. Translate the figure using the vector.

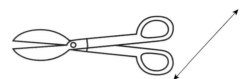

42. Suppose r is a number between -1 and 0. Order these numbers from least to greatest.

$$r \qquad r^{-3} \qquad r^2 \qquad r^3$$

Find the value of m in each equation.

43. $\sqrt[m]{512} = 8$

44. $\sqrt[3]{1{,}331} = m$

45. $\sqrt[3]{m} = 7$

46. $\sqrt[4]{m^4} = 10$

Simplify.

47. $\sqrt{50}$

48. $\sqrt{150}$

49. $\sqrt{162}$

50. $\sqrt{210{,}000}$

Solve Systems of Equations

You can think of an equation as a condition that a variable or variables must satisfy. For example, the equation $x + 2y = 6$ states that the sum of x and $2y$ must equal 6. An infinite number of pairs of values satisfy this condition, such as (0, 3), (2, 2), and (−2.3, 4.15).

When there is a second equation, or condition, that the same variables must satisfy, sometimes only one number pair satisfies *both* equations. For example, the only solution satisfying both $x + 2y = 6$ and $x = y$ is (2, 2).

In the activity that follows, you and your classmates will make "human graphs" to find pairs of values that satisfy two equations.

Explore

Select a team of nine students to make the first graph. The team should follow these rules.

- Line up along the *x*-axis, from −4 to 4.

- Multiply the number on which you are standing by 2. Subtract 1.

- When your teacher says "Go!" walk forward or backward to the *y* value equal to your result from the previous step.

Describe the resulting "graph." What is its equation? Have two students explain why their coordinates are solutions of this equation.

You will next make two graphs on the same grid. Select two teams of nine students. Team 1 members should follow their instructions and then stay on their points while Team 2 members follow their instructions.

Team 1: Graph of $y = 2x$

- Line up along the *x*-axis.

- Multiply the number on which you are standing by 2.

- When your teacher says "Go!" walk forward or backward that number of paces.

Team 2: Graph of $y = -x + 3$

- Line up along the *x*-axis.

- Multiply the number on which you are standing by −1 and add 3.

- When your teacher says "Go!" walk forward or backward that number of paces.

Are some students trying to stand on the same points? If a shared point is on both graphs, it should be a solution of both equations. Check that this is true.

Now choose two new teams. Make new human graphs to find ordered pairs that are solutions of both $y = 3x$ and $y = x + 4$.

Investigation 1 Graph Systems of Equations

Vocabulary

system of equations

Materials

- graph paper
- graphing calculator

In Explore on page 342, you found one (x, y) pair that was the solution of two equations. A group of two or more equations is called a **system of equations**. A solution of a system of equations is a set of values that makes all the equations true.

In Exercises 1–8, you will concentrate on the method of graphing to solve systems of equations.

✅ Develop & Understand: A

1. Consider these five equations.

   i. $2x + 3 = 7$

   ii. $(x - 2)^2 = 0$

   iii. $2x + y = 7$

   iv. $3x - y = 3$

   a. Equations i and ii each involve one variable. How many solutions does each equation have? Find the solutions using any method you choose.

   b. Equations iii and iv each involve two variables. How many solutions does each equation have? Can you list them? Explain why or why not.

   c. Consider equations iii and iv. Could you graph the solutions of $2x + y = 7$? Could you graph the solutions of $3x - y = 3$? Explain.

   d. Draw graphs of $2x + y = 7$ and $3x - y = 3$. At what point or points do they meet?

   e. What values of x and y make both $2x + y = 7$ and $3x - y = 3$ true? That is, what is the solution of this system of equations? How do these values relate to your answer to Part d?

In Exercises 2 and 3, solve the system of linear equations using the following method.

- Rewrite each equation so y is alone on one side.
- Graph both equations on your calculator.
- If the lines meet, estimate the x- and y-coordinates of their intersection. This may not be an exact solution, but it will give you an estimate of the solution.
- Check that these values satisfy *both* equations.

2. $2y - x = 20$ and $2x = 5 - y$

3. $y = 4$ and $x + y = -1$

4. Consider these two equations.

$$x + y = 5$$

$$2x = 12 - 2y$$

a. Graph both equations on one set of axes.

b. Do the graphs appear to intersect?

c. How could you verify your answer to Part b by looking at the equations?

d. Does this system of equations have a solution? How do you know?

5. Create your own system of equations for which you think there will be no solution.

✅ Develop & Understand: B

6. Consider these three equations.

$$4x + 2y = 7$$

$$y = -2x + 5$$

$$3x = 5y - 4$$

a. Select two that will form a system of equations that has a solution.

b. Select two that will form a system with no solution.

7. Consider this system of equations.

$$2x - y = 4$$

$$5x - 2.5y = 10$$

a. The slope of the line for each equation is 2. How many solutions do you think this system will have?

b. Without using a calculator, draw graphs of these equations.

c. What do you notice about the two graphs?

d. How many solutions do you think this system has?

e. Why do you think this system of equations is different from the system in Exercise 4?

8. Booker is twice as old as his younger brother Andre. Four years ago, Booker was four times as old as Andre.

a. Write two equations that relate Andre's and Booker's ages.

b. Find the ages of Booker and Andre using a graph.

It is possible to find solutions of systems of equations that are not linear.

Real-World Link

Manufacturers write and solve systems of equations for many purposes. For example, a company that makes doors might do so to determine how much glass, wood, and aluminum it needs on hand for a particular week's production schedule.

Share & Summarize

1. Describe how to estimate a solution of a system of two equations using a graph.

2. Describe how you could create a system of linear equations with no solution.

Problems Involving Systems of Equations

Materials

• graph paper

Sometimes the solution to a practical problem requires setting up and solving a system of equations. In this investigation, you will use graphs to find solutions to such situations.

Explore

Two bookstores each fill catalog orders. Each charges the standard price for the books, but they have different shipping and handling charges. Gaslight Bookstore charges a shipping fee of $3 per order plus $1 per book. Crimescene Bookstore charges $5 per order and $0.50 per book.

For each bookstore, write an equation that describes how the shipping cost y is related to the number of books x. Then graph both equations on the same axes.

Under what circumstances would you order from Gaslight? When would you order from Crimescene? Are there circumstances when it makes no difference from which store you buy?

What other, perhaps nonmathematical, things might you consider when making this choice?

✓ Develop & Understand: A

At Adrian's Audio Emporium, books on tape sell for $10 and CDs sell for $15.

1. Rita's family bought 16 items. List some possible values for the number of books on tape t and CDs d they bought.

2. Write an equation to describe the total number of items bought by Rita's family.

3. Plot points on a grid to show all the pairs (t, d) that satisfy the equation in Exercise 2. Remember, t and d must be whole numbers.

4. Suppose that instead of telling you how many items were bought, Rita told you her family spent $220. List some possible values for the number of books on tape t and the number of CDs d that have a total price of $220.

5. Write an equation to describe the total number of dollars Rita's family spent on books on tape and CDs.

6. On the same grid, plot all the pairs of whole-number values that satisfy the equation in Exercise 5. You may want to use a different color or symbol for these points.

7. Use your graph to find the possible values for t and d if both statements in Exercises 1 and 4 are true.

✅ Develop & Understand: B

The owner of a hat business has *fixed expenses* of $3,000 each week. She has these expenses regardless of how many hats she makes.

The remaining expenses are *variable costs,* such as materials and labor. These costs are proportional to the number of hats made. For example, in a week in which 500 hats were made and sold, her variable costs were $7,500. In a week in which 1,000 hats were made, they were $15,000.

All the hats were sold to department stores at $20 each.

8. Write an equation relating the income i to the number of hats sold n.

9. Consider the hatmaker's costs.

a. Write an expression for the *variable costs* to produce n hats. That is, if the shop made n hats in a week, what would be the variable costs?

b. Write an equation relating the total cost per week c and the number of hats made and sold that week n. Remember that the total cost includes both the variable costs and the fixed expenses.

10. The difference between cost and income is the owner's profit. If the hatmaker has no sales, she will still have to pay the fixed costs and will have a large loss. If she has many sales, she will easily cover her fixed costs and make a good profit.

a. In one particular week, the hatmaker made and sold 500 hats. Did she make a profit or a suffer a loss? How much?

b. In another week, she made and sold 1,000 hats. Did she make a profit or suffer a loss? How much?

11. Somewhere between no sales and many sales is a number of sales called the *break-even point*. This is the point at which costs and income are equal, so there is no profit and no loss.

a. On one set of axes, graph the equations for income and costs. Label each graph with its equation. You will need to read amounts from your graph, so use an appropriate scale. Be as accurate as you can.

b. How can you use your graph to estimate the break-even point?

c. From your graph, estimate the number of hats the hatmaker needs to make and sell each week to break even. Calculate the costs and income for that number of items to check your estimate. Improve your estimate if necessary.

d. Use your graph to estimate the number of hats she needs to sell each week to make a profit of $1,000 per week. Calculate the costs and income for that number of items to check your estimate. Improve your estimate if necessary.

e. Use the equations you wrote for Exercises 8 and 9 to write an equation that gives the value of n for which costs and income are equal. Solve your equation to find the break-even point.

Share & *Summarize*

1. Describe a situation that would require solving a system of equations to find the answer.

2. What is the connection between a graph describing income and expenses for a particular business and the break-even point for that business?

3. What is the connection between a system of equations describing income and expenses for a particular business and the break-even point for that business?

Investigation ③ Solve Equations by Substitution

Vocabulary

substitution

Materials

• graphing calculator

You can find or estimate the solution of a system of equations by graphing the equations and finding the coordinates of the points where the graphs intersect. You can also solve systems by working with the equations algebraically. To see how this works, consider the situation described below.

Ana's mother took her shopping for new socks.

To find the cost of each type of socks, you can write and solve a system of equations. Let x represent the price of the plain socks. Let y represent the price of the designer socks.

The designer socks cost $1 more than twice the cost of the plain socks.	The price for 3 pairs of plain socks and 1 pair of designer socks is $11.
$y = 2x + 1$	$3x + y = 11$

To determine the cost of the socks, you need to find an (x, y) pair that is a solution of both $y = 2x + 1$ and $3x + y = 11$. That is, you need to solve this system of equations.

$$y = 2x + 1 \qquad 3x + y = 11$$

Think & Discuss

Graphs of $y = 2x + 1$ and $3x + y = 11$ are shown at the right.

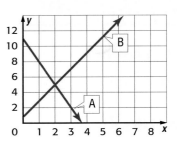

- Which graph is which?
- Use the graph to solve the system of equations. How much does each type of sock cost?

You can also solve this system algebraically. The first equation tells you that y must equal $2x + 1$. Since the value of y must be the same in *both* equations, you can *substitute* $2x + 1$ for y in the second equation. This gives you a linear equation with only one variable, which you already know how to solve.

$$3x + y = 11 \Rightarrow 3x + (2x + 1) = 11$$
$$5x + 1 = 11$$
$$5x = 10$$
$$x = 2$$

So, $x = 2$. Since $y = 2x + 1$, you know that $y = 5$. That means the plain socks cost \$2 and the designer socks cost \$5. Check this solution by substituting it into *both* equations.

This algebraic method for finding solutions is called **substitution** because it involves *substituting* an expression from one equation for a variable in another equation.

✓ Develop & Understand: A

Use substitution to solve each system of equations. Check each solution by substituting it into both original equations.

1. $a = 3 - b$
$4b + a = 15$

2. $x = 2 - y$
$8y + x = 16$

3. Evan suggested solving Exercise 1 by substituting $3 - b$ for a in the *first* equation. Will Evan's method work? Explain why or why not.

As you will discover next, sometimes you must rewrite one of the equations in a system before you can determine what expression to substitute.

✅ *Develop & Understand: B*

4. Consider this system of equations.

$$x + y = 8 \qquad 4x - y = 7$$

 a. Rewrite one of the equations to get y by itself on one side.

 b. Solve the system by substitution. Check your solution by substituting it into both of the original equations.

5. To solve the system in Exercise 4, you could have rewritten the first equation to get x alone on one side. Solve the system this way. Do you get the same answer?

6. Carinne scored 23 points in last night's basketball game. Altogether, she made ten of her shots. Some were 2-point shots, and the others were 3-point shots. Let a stand for the number of 2-point shots. Let b represent the number of 3-point shots.

 a. Write an equation for the total number of shots made by Carinne.

 b. Write an equation for the total number of points Carinne scored.

 c. How many 2-point shots did Carinne make? How many 3-point shots did she make?

7. Consider this system of equations.

$$y - 2x = 3 \qquad 2y + 5x = 27$$

 a. Graph the equations on your calculator. Estimate the point of intersection.

 b. Solve the system of equations by using substitution. Compare your solution with the approximation you found in Part a.

 c. What advantage does solving linear equations by substitution have over solving them by graphing?

Share & Summarize

Describe the steps for solving a system of two equations by substitution.

Materials

• graph paper

Solving systems of equations by *elimination* is sometimes simpler than solving by substitution. The elimination method uses the idea of doing the same thing to both sides of an equation.

Think & Discuss

Consider this system of equations.

$$5x + 4 = 13 + 3y$$
$$2x = 3y$$

To solve the system, Abigail first tried the following.

$$
\begin{array}{r}
5x + 4 = 13 + 3y \\
-\ 2x \qquad\quad -\ 3y \\
\hline
3x + 4 = 13
\end{array}
$$

Abigail said she was doing the same thing to both sides of the first equation. Is this true? Explain.

In a way, what Abigail did above was to subtract the second equation from the first. By doing this, which variable did she *eliminate*?

How do you think eliminating this variable will help her find the solution?

If $3x + 4 = 13$, we know that $x = 3$. Substituting this value into either of the original equations gives $y = 2$. Therefore, the solution of this system is (3, 2).

By subtracting or adding equations, you can sometimes eliminate one variable and solve for the other. This process is called the method of **elimination**.

Sometimes you will need to rewrite one or both equations before you can eliminate one of the variables. The example on the next page demonstrates this technique.

Example

Solve the system consisting of equations A and B.

① The coefficient of y in equation A is -4. Multiply both sides of equation B by 2 so that $-2y$ becomes $-4y$. Rewrite the two equations.

② Now subtract the new equation C from equation A, as Abigail did in Think & Discuss. Since $-4y - (-4y) = 0$, the variable y has been eliminated from the equations.

This leaves just one equation with x as a variable. Solve for x, which gives the equation $x = 20$.

③ Now substitute that value into either of the original equations to find the value of y.

$$7x - 4y = 100 \quad [A]$$
$$3x - 2y = 40 \quad [B]$$

① $\quad 7x - 4y = 100 \quad [A]$
$\quad\;\; 6x - 4y = 80 \quad [C]$

② $\quad 7x - 6x = 100 - 80$
$\qquad\qquad x = 20$

③ Substituting $x = 20$ into $[B]$:
$\qquad 3(20) - 2y = 40$
$\qquad\qquad -2y = -20$
$\qquad\qquad\quad y = 10$

Check in Equation $[A]$ and Equation $[B]$
So, the solution to the system is $(20, 10)$.

Think & Discuss

In the example, x could have been chosen as the variable to eliminate, but it seemed easier to eliminate y. Why might this be so?

✅ Develop & Understand: A

1. Consider this system. You should be able to solve it simply by *adding* the equations to eliminate y. Try it. What is the solution?

$$9x - 2y = 3$$
$$3x + 2y = 9$$

Confirm that the equations in each pair are equivalent. Explain what you must do to both sides of the first equation to obtain the second equation.

2. $3m + 2 = 13.5$
  $12m + 8 = 54$

3. $x + 4y = 2$
  $2x = -8y + 4$

4. $7d + 1 = 4p$
  $21d = 12p - 3$

The first task in using the elimination method is deciding how to eliminate one of the variables. The trick is to write equations, equivalent to those with which you started, where the coefficients of one of the variables are the same or opposites.

5. Consider these four systems of equations.

   i. $x + 2y = 9$ [A]   **ii.** $7x - y = 4$ [C]
        $3x + y = 7$ [B] $2x + 3y = 19$ [D]

  iii. $35x - 6y = 1$ [E]   **iv.** $5x + 3y = 42$ [G]
        $7x + 3y = 10$ [F] $2x + 8y = 78$ [H]

a. Look at system i. You could eliminate x by replacing equation A with an equivalent equation that contains the term $3x$ or $-3x$. What would you do to equation A? What would you get?

b. For system i, you could choose to eliminate y instead of x. How could you do this?

c. Look at system ii. Which variable would you eliminate? Which equation would you rewrite? What would you do then?

d. For system iii, state which variable you would eliminate and how you would do it.

e. For system iv, you will need to write equivalents for both equations. Which variable would you choose to eliminate? How would you do it?

6. Solve systems i, iii, and iv using elimination. Check that each solution fits both equations.

Equivalent equations have the same solutions. You can always change an equation into an equivalent one by doing the same thing to both sides, as long as you do not multiply by 0.

Substitution and elimination work for any system of linear equations. However, some exercises are easier to solve using one method than the other.

✅ Develop & Understand: B

Solve each system of equations two ways, first by *substitution* and then by *elimination*. In each case, compare the two approaches for ease, speed, and likelihood of mistakes. For each system of equations, do you think one method is better? Explain.

7. $3x = y + 7$

$5x = 9y + 41$

8. $5x - 3y = 10$

$15x + 6y = 30$

9. Ryan is puzzled. He has tried to solve this system of equations by substitution and elimination, but neither method seems to work.

$$x + y = 4$$

$$3x + 3y = 11$$

a. Try to solve the equations by making a graph. Explain why Ryan is having trouble.

b. Try to solve the equations by substitution. What happens?

c. Now solve the equations by elimination. What happens?

d. Ryan discovered that she had incorrectly copied the equations. This is the correct system.

$$x + y = 4$$

$$3x + 3y = 12$$

Solve this system of equations by substitution, elimination, or graphing.

Share & Summarize

1. Create a system of linear equations that is easy to solve using substitution. Explain why you think substitution is a good method.

2. Create a system of linear equations that is easy to solve using elimination. Explain why you think elimination is appropriate.

3. Create a system of linear equations that does not have a solution. Explain why there is no solution.

Inquiry

Investigation **5** Use a Spreadsheet

Materials

- computer with spreadsheet software (one per group)

Jeans Universe is having a grand opening sale.

As part of the grand opening celebration, Jeans Universe held a drawing. Lydia won a gift certificate for $150, which she wants to spend entirely on jeans and t-shirts. As she thought about how many of each item to buy, she asked herself several questions.

- How many pairs of jeans and how many t-shirts should I buy to spend as much of my gift certificate as possible? What if I want to buy at least one of each?

- If I want to buy the same number of jeans as t-shirts, how many of each can I buy? What if I want to buy two t-shirts for every pair of jeans? Two pairs of jeans for every t-shirt?

- My mother said, "I really don't think you need more than five new t-shirts." How many pairs of jeans and how many t-shirts can I buy without going over this limit?

Setting Up a Spreadsheet

Lydia had a hard time answering her questions because there are so many possible combinations. She set up the spreadsheet below to help keep track of the possibilities.

When Lydia completes her spreadsheet, it will show the costs of various combinations of jeans and t-shirts. For example, Cell F7 will show the cost for three pairs of jeans and four t-shirts.

	A	B	C	D	E	F	G	H	I
1					jeans	24.95			
2			0	1	2	3	4	5	6
3		0							
4		1							
5		2							
6		3							
7		4							
8	t-shirts	5							
9	11.95	6							
10		7							
11		8							
12		9							
13		10							
14		11							
15		12							

1. To help fill the spreadsheet, Lydia decided she needed an equation showing how the total cost c depends on the number of pairs of jeans j and the number of t-shirts t she buys. What equation should she use?

Lydia then entered formulas in Cells C3–I3.

	A	B	C	D	E	F	G	H	I
1					jeans	24.95			
2			0	1	2	3	4	5	6
3		0	=24.95*0+ 11.95*B3	=24.95*1+ 11.95*B3	=24.95*2+ 11.95*B3	=24.95*3+ 11.95*B3	=24.95*4+ 11.95*B3	=24.95*5+ 11.95*B3	=24.95*6+ 11.95*B3

2. Explain what the formulas in Cells C3–I3 do.

3. Set up your spreadsheet like Lydia's. Use the currency format for the numbers in Rows 3–15 of Columns C–I. What values appear in Cells C3–I3?

Now select the seven cells with formulas, Cells C3–I3. Use the Fill Down command to copy the formulas down to Row 15. You should now have values in all the cells of your table.

4. Look at the formulas in the cells in Column D.

  a. What parts of the formula change as you move down the column from cell to cell? What parts of the formula stay the same?

  b. What do the formulas in Column D do?

Using the Spreadsheet

You can use the values in your spreadsheet to answer Lydia's questions.

5. What combination of t-shirts and jeans would allow Lydia to spend as much of her $150 gift certificate as possible? How much would she spend?

6. What combination of t-shirts and jeans would allow Lydia to spend as much of her $150 gift certificate as possible, if she wants to buy at least one of each item? How much would she spend?

7. What combination should she buy if she wants the same number of t-shirts as jeans? How much would she spend?

8. What combination should she buy if she wants twice as many t-shirts as jeans? How much would she spend?

9. What combination should she buy if she wants twice as many jeans as t-shirts? How much would she spend?

10. If she wants to buy no more than five t-shirts, what combination should she buy? How much would she spend?

Try It Again

What if, instead of jeans and t-shirts, Lydia decides to buy two different items, such as shorts and tank tops or tennis shoes and socks? Instead of creating a separate spreadsheet for each pair of items, Lydia changed her spreadsheet so it would work for any pair of prices.

Notice that, in Lydia's original spreadsheet, the value in Cell F1 is the cost of a pair of jeans and the value in Cell A9 is the cost of a t-shirt. Lydia used this information to type new formulas into Cells C3–I3. She also changed the label "jeans" to "item 1" and the label "t-shirts" to "item 2."

	A	B	C	D	E	F	G	H	I
1					item 1	24.95			
2			0	1	2	3	4	5	6
3		0	=$F$1*0+$A$9*B3	=$F$1*1+$A$9*B3	=$F$1*2+$A$9*B3	=$F$1*3+$A$9*B3	=$F$1*4+$A$9*B3	=$F$1*5+$A$9*B3	=$F$1*6+$A$9*B3
4		1							
5		2							
6		3							
7		4							
8	item 2	5							
9	11.95	6							
10		7							

In place of 24.95 and 11.95, the new formulas use $F$1 and $A$9. The $ symbol tells the spreadsheet to use the value in the indicated cell as a *constant*.

- $F$1 tells the spreadsheet to use the value in Cell F1, 24.95, as a constant.
- $A$9 tells the spreadsheet to use the value in Cell A9, 11.95, as a constant.

When a formula with a constant is copied from one cell to another, the spreadsheet does not "update" the constant to reflect a new row and column.

Set up your spreadsheet in this new way.
- Enter the new formulas shown in Cells C3–I3.
- Select Cells C3–I3, and fill all seven columns down to Row 15.

11. Look at how the formulas change as you move down Column D.

a. What parts of the formula change as you move from cell to cell? What parts of the formula stay the same?

b. What do the formulas in Column D do?

By simply changing the prices in Cells F1 and A9, Lydia can now easily find prices for combinations of any two items.

Suppose shorts cost $19.95 and tank tops cost $13.95. Update your spreadsheet to show prices for combinations of these two items.

- Type "shorts" in Cell E1 and "tank tops" in Cell A8.
- Type the price of the shorts, $19.95, in Cell F1 and the price of the tank tops, $13.95, in Cell A8.

12. Tamika won a $75 gift certificate. She wants to spend as much of it as possible on shorts and tank tops. What should she buy? How much will she spend?

13. Kai wants to buy hats and socks with his $30 gift certificate. Hats cost $6.75, and socks are $2.95 a pair. Alter your spreadsheet to show prices for combinations of hats and socks.

How many of each item should Kai buy to spend as much of his gift certificate as possible? How much will he spend?

What Did You Learn?

14. Imagine you have a part-time job working at a bookstore. You earn $7.25 per hour now but expect a raise soon. Design a spreadsheet you could use to keep track of the hours you work and the money you earn each day of a week. Your spreadsheet should have these characteristics.

- It should allow you to enter the number of hours you work each day, Monday through Sunday.
- It should automatically compute the dollars you earn each day, as well as the total for the week.
- It should use a constant for your hourly rate of pay so you can easily update it each time you get a raise.

Real-World Link

The federal hourly minimum wage was established in 1938 at 25¢ per hour. Seventy years later, the minimum wage had risen about $6, to $6.55 per hour.

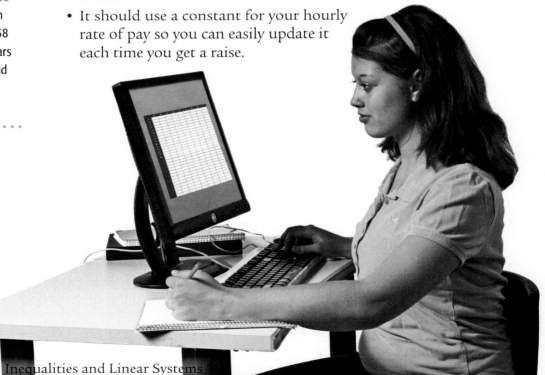

Practice & Apply

1. There are six ways to pair these four equations.

   i. $y = 2x + 4$ **ii.** $y + 2x = -4$

   iii. $x = 4 - \dfrac{y}{2}$ **iv.** $2y - 4x = 10$

   a. Predict which pairs of equations do not have a common solution

   b. Verify your results for Part a by carefully graphing both equations in each pair you selected. Explain how the graphs do or do not verify your prediction.

   c. Predict which pairs, if any, have a common solution.

   d. Verify your results for Part c by graphing both equations in each pair you selected. Explain how the graphs do or do not verify your prediction.

   e. Use your graphs to find a common solution of each pair of equations you listed in Part c.

2. Cheryl and Felipe each have some keys on their key chains. Cheryl said, "I have twice as many as you have." Felipe said, "If you gave me four, we would have the same number."

   a. Write two equations that relate how many keys each has.

   b. How many keys does Cheryl have? How many does Felipe have?

3. A group of friends enters a restaurant. No table is large enough to seat the entire group, so the friends agree to sit at several separate tables. They want to sit in groups of five, but there are not enough tables. Four people would not have a place to sit. Someone suggests they sit in groups of six, which would fill all the tables with two extra seats at one table.

   Answer these questions to find how many people are in the group and how many tables are available.

   a. Write a system of two equations to describe the situation.

   b. Solve your system of equations. How many friends are in the group? How many tables are available? Check your work.

4. **Physical Science** A bicyclist riding from Boston to New York City maintains a steady speed of 18 miles per hour. An hour after the bicyclist started, a car left from the same point, traveling along the same road at an average speed of 50 miles per hour.

Statue of Liberty, New York City

   a. If the car and the bicyclist travel in the same direction, how far from Boston will each be an hour and a half after the bicyclist left town? Who will be in front?

   b. After two hours, who will be in front?

   c. On the same set of axes, draw distance-time graphs for the car and the bicycle. Put distance, in miles from Boston, on the horizontal axis. Put time, in hours from the time the cyclist left Boston, on the vertical axis.

   d. Use your graphs to determine the approximate distance from Boston when the car overtook the bicyclist.

   e. When the car caught up, approximately how long had the cyclist been riding?

   f. Write an equation relating the bicyclist's distance from Boston in miles to the time in hours after the bicyclist left. Using the same variables, write another equation relating the car's distance from Boston to the time after the bicyclist left.

   g. If you were to find the pair of values that fits both of your equations, what would those values represent?

5. Solve one of these systems of equations by drawing a graph. Solve the other by substitution.

   a. $f + g = 20$
      $3f + g = 28$

   b. $y - x = 3$
      $2y + 3x = 16$

6. The sum of two numbers is 31. One of the numbers is 9 more than the other.

   a. Write a system of two equations relating the numbers.

   b. Solve the system to find the numbers.

7. **Economics** Andrew bought 11 books at a used book sale. Some cost 25¢ each, and the others cost 35¢ each. He spent $3.15.

   a. Write a system of two equations to describe this situation. Solve it by substitution.

   b. How many books did Andrew buy for each price?

Solve the systems of equations in Exercises 8–11 by elimination. Check your solutions. Give the following information.

- which variable you eliminated
- whether you added or subtracted equations
- the solution

8. $x + y = 12$
$x - y = 6$

9. $3p + 2q = 13$
$3p - 2q = -5$

10. $5a + 4b = 59$
$5a - 2b = 23$

11. $9s + 2t = 3$
$4s + 2t = 8$

Solve the systems of equations in Exercises 12–15 by elimination. Check your solutions. Give the following information.

- which equation or equations you rewrote
- how you rewrote each equation
- whether you added or subtracted equations
- the solution

12. $3m + n = 7$ $\quad$ [A]
$m + 2n = 9$ $\quad$ [B]

13. $6x + y = -54$ $\quad$ [C]
$2x - 5y = -50$ $\quad$ [D]

14. $2a + 5b = 12$ $\quad$ [E]
$3a + 2b = 7$ $\quad$ [F]

15. $y = \dfrac{3}{4}x - 4$ $\quad$ [G]
$4y = 2x + 3$ $\quad$ [H]

 **16. Economics** A manager of a rock group wants to estimate, based on past experience, how many tickets will be sold in advance of the next concert and how many tickets will be sold at the door on the night of the concert.

At a recent concert, the 1,000-seat hall was full. Tickets bought in advance cost $30, and tickets sold at the door cost $40. Total ticket sales were $38,000.

a. Write a system of two equations to represent this information.

b. On one set of axes, draw graphs for the equations.

c. Use your graphs to estimate the number of advance sales and the number of door sales made that night.

d. Check that your estimates fit the conditions by substituting them into both equations.

17. Three tourists left a hotel one morning and headed in the same direction. Tyrone, the walker, arose early and set off at a steady pace. Manuela, the cyclist, slept in and left two hours later. Kevin caught the bus half an hour after Manuela had left. These graphs show their distances over time.

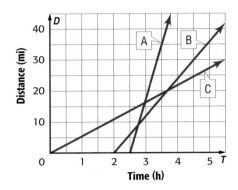

a. Match each graph with the tourist it represents.

b. When and where did Manuela pass Tyrone? When and where did Kevin pass Manuela? When and where did Kevin pass Tyrone?

c. From the graphs, estimate the speed at which each tourist traveled.

d. Write equations for the three lines.

e. Which system of equations would you solve to determine when and where Manuela passed Tyrone? Check your answer by solving the system and comparing the solution with your answer to Part b.

18. Create a situation involving two variables that can be expressed in terms of two linear equations. Write and solve the equations.

19. History Solve this exercise, which was posed by Mahavira, an Indian mathematician who lived around 850 A.D.

The mixed price of 9 citrons and 7 fragrant wood apples is 107; again, the mixed price of 7 citrons and 9 fragrant wood apples is 101. O you Arithmetician, tell me quickly the price of a citron and of a wood apple here, having distinctly separated those prices well.

20. Rachel and Craig are going to the fair. Rachel has $13.50 to spend and calculates that she will spend all her money if she takes four rides and plays three games. Craig has $15.50 and calculates that he will spend all his money if he takes two rides and plays five games.

    a. Choose variables and write equations to describe this situation.

    b. Use substitution to find the costs to go on a ride and to play a game.

21. Economics For a school concert, a small printing business charges $8 for printing 120 tickets and $17 for printing 300 tickets. Both of these total costs include a fixed cost and a unit cost per ticket. That is, the cost for n tickets is $c = A + Bn$, where A is the business' fixed charge per order and B is the charge per ticket.

    a. Using the fact that it costs $8 to have 120 tickets printed, write an equation in which A and B are the only unknown quantities.

    b. Write a second equation using the cost of $17 for 300 tickets.

    c. Use the substitution method to find the business' fixed and variable charges.

22. Economics The Perez and Searle families went to the movies. Admission prices were $9 for adults and $5 for children. The total came to $62 for the ten people who went.

    a. Write two equations to represent this situation.

    b. Solve your system using elimination to determine how many adults and how many children went to the movies.

23. You can use a system of equations to find an equation of a line when you know two points on the line. This exercise will help you find an equation for the line that passes through the points $(1, -2)$ and $(3, 4)$.

    a. An equation of a line can always be written in $y = mx + b$ form. Substitute $(1, -2)$ in $y = mx + b$ to find an equation using m and b.

    b. Substitute the point $(3, 4)$ in $y = mx + b$ to find a second equation.

    c. Find the common solution of the two equations from Parts a and b. Use them to write an equation of the line.

    d. Use the same technique to find an equation of the line passing through the points $(-1, 5)$ and $(3, -3)$. Check your answer by verifying that both points satisfy your equation.

24. Evan and his younger brother Keenan are comparing their collections of CDs. Set up two equations and solve them by elimination to find how many CDs each has.

I have 3 more than you!

That's because you're older. You and I have 11 CD's altogether.

Mom and Dad only have 5, and they're older than both of us.

25. In Your Own Words You have used three methods for solving systems of equations in this lesson. They are graphing, substitution, and elimination. Explain how you decide which solution method to use when solving a particular system of equations.

Mixed Review

26. Copy this picture. Create a design with rotation symmetry by rotating the figure several times about the point using a 30° angle of rotation.

27. A line segment with length 8 and slope $\frac{1}{4}$ is scaled by a factor of 3. What are the length and the slope of the new segment?

28. Five less than four times a number is at least 35. Write an inequality that can be used to find the number.

29. Find the missing measures in the figure below.

    a. $m\angle 1 = ?$

    b. $m\angle 2 = ?$

    c. $m\angle 3 = ?$

    d. $m\angle 4 = ?$

    e. $m\angle 5 = ?$

    f. $m\angle 6 = ?$

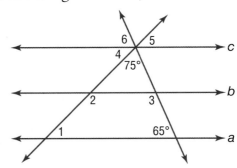

Review & Self-Assessment

Chapter Summary

You began this chapter by reviewing methods for solving equations in preparation for solving inequalities. Graphing was introduced as a way to understand inequalities and to solve equations.

You also discovered that a *solution* of a system of equations is a set of values that satisfies all the equations in the system. You found solutions of systems by graphing and by using the algebraic methods of substitution and elimination.

Vocabulary

elimination

inequality

substitution

system of equations

Strategies and Applications

The questions in this section will help you review and apply the important ideas and strategies developed in this chapter.

Using algebraic methods to solve equations

1. Explain how to solve an equation by backtracking. Illustrate your explanation with an example.

2. Explain how to solve an equation by doing the same thing to both sides. Illustrate your explanation with an example.

3. The sum of three consecutive numbers is 57. Write and solve an equation to find the three numbers.

Understanding and solving inequalities

4. List all the whole numbers that make both of these inequalities true.

$$t < 8 \qquad 2 < t \leq 10$$

5. Write the following inequality in words. Give all the whole-number values for x that make it a true statement.

$$8 < x \leq 11$$

6. A small video store has determined that its monthly profit must be *at least* $1,300 in order to stay in business. Movies rent for $2 each. Business expenses, including rent, for one month are $800.

   Use the formula *total sales − expenses = profit* to write and solve an inequality that shows how many movies m must be rented each month for the store to stay in business.

7. Dinners for large groups can be catered at a banquet hall. The cost to rent the hall for one night is $1,500 plus $40 per person.

Write and solve an inequality that answers this question.

Greg and Jill are planning to have their wedding reception in this banquet hall. How many people can they invite if their total budget for renting the hall is $7,500?

Graphing inequalities

8. Consider these two statements.

$$x > 3 \text{ or } x < 5$$

$$x > 3 \text{ and } x < 5$$

a. Does the set of values that satisfies "$x > 3$ or $x < 5$" include 3 and 5? Draw a number-line graph to illustrate your answer.

b. Does the set of values that satisfies "$x > 3$ and $x < 5$" include 3 and 5? Use a number-line graph to illustrate your answer.

9. Explain the steps involved in graphing an inequality with two variables. Give an example to illustrate your steps.

Solving systems of equations graphically and algebraically

10. Two competing building-supply stores have different prices for structural brick. Store A charges $0.75 per brick and $15 for delivery. Store B charges $1.05 per brick but delivery is free.

a. For each building-supply store, write an equation to represent the total cost for bricks c related to the number of bricks bought, b. Then graph both equations on the same set of axes.

b. Under what circumstances would you order structural bricks from store A? From store B? Are there circumstances when it makes no difference from which store you buy?

11. Solve this system of equations using either substitution or elimination. State which method you used. Explain why you chose it.

$$2x - 2y = 5$$

$$y = -x - 3$$

Demonstrating Skills

12. Solve this equation by backtracking. Check your solution.

$$\frac{1}{3}\left(3 + \frac{x}{2}\right) - 1 = 4$$

13. Solve this equation by doing the same thing to both sides. Check your solution.

$$1 - 2p = 2 + 2p$$

Solve each inequality.

14. $3(j + 1) - 2(2 - j) \leq 9$

15. $6(k - 5) - 2 \leq 10$

16. $-2(b + 1) > -5$

17. On a number line, graph all x values for which $x \leq -1$ or $x > 2$.

18. Graph the solution of this inequality on a number line.

$$\frac{t}{2} + 2 > 3$$

Graph each inequality.

19. $y \geq x - 2$

20. $y < -2x + 1$

Use the following information to answer Exercises 21 through 24.

According to the Kelvin scale, water boils at 373 K and freezes at 273 K.

21. In Kelvin, how many degrees are between the freezing and boiling points of water?

22. Which is bigger, a Celsius degree or a Kelvin degree?

23. Use a proportion to show how big a Celsius degree is, compared to a Kelvin degree.

24. Suppose a scientist states that the temperature of a liquid is 297 K. What is that temperature in Celsius?

25. Use substitution to solve this system of equations. Check your solution.

$$x = y + 8$$
$$3y + 1 = 2x$$

26. Use elimination to solve this system of equations. Check your solution.

$$4x - 10y = 2$$
$$3x + 5y = 9$$

Solve each system of equations. Check your solutions.

27. $-x + 1 = y$

$y - x = 2$

28. $3x = 3y + 1$

$x = 1 - y$

29. $3.5\,y - 1.5x = 6$

$-3t + x = 3$

30. In Parts a–c, you will consider the inequalities $x \leq 4$ and $x \geq 1$. Draw a separate graph for each part.

   a. Graph all x values for which $x \leq 4$ and $x \geq 1$.

   b. Graph all x values for which $x \leq 4$ or $x \geq 1$.

   c. Explain how the words *and* and *or* affect the graphs in Parts a and b.

31. Graph all x values for which $|x - 3| \geq 1$.

32. Ruby's mother said that the trip to the ocean would take them at least eight hours but no more than nine and a half hours. Write an inequality to express the number of hours h that Ruby's mother thinks it will take them to reach the ocean.

33. Explain what type of line you would use for the equation when graphing $y > 3x + 4$ and why.

Convert each inequality into linear form and graph.

34. $2y - 4x \leq 8$

35. $x - y < 3$

36. $6x + 4y \geq 4x$

37. Graph the following inequality: $\frac{1}{2}x < y < 2x$

Test-Taking Practice

SHORT RESPONSE

1 A car rental company charges \$35/day and 6 cents/mile. Marcela rents a car for 3 days, but she does not want to spend more than \$120. What is the greatest number of miles she can drive?

Show your work.

Answer _____

MULTIPLE CHOICE

2 What is the solution of the inequality below?

$$2y + 6 \geq 18$$

A $y \geq 6$

B $y \leq 6$

C $y \geq 12$ç

D $y \leq 12$

3 Five less than four times a number is at least 35. Which inequality can be used to find the number?

F $35n < 20$

G $4n - 5 + 6n$

H $4n - 5 \geq 35$

J $4n - 5 < 35$

4 Which of the following inequalities correctly expresses the values shown on the number line below?

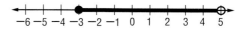

A $x < 2$

B $-3 \leq x < 5$

C $-3 \geq x \geq 5$

D $x \leq -3$ and $x > 5$

5 Which is the solution to the system of equations below?

$$x - 2y = 1$$
$$3y = 16 - 2x$$

F $(5, 2)$

G $(2, 2)$

H $(15, 7)$

J $(0, 2)$

Quadratic and Inverse Relationships

Real-Life Math

Dropping the Ball In the late 1500's, Italian astronomer and physicist Galileo Galilei conjectured that two objects dropped at the same time from the same height would hit the ground at the same time, regardless of their masses. He allegedly proved his theory by dropping one large and one small cannonball from the top of the Leaning Tower of Pisa. His work led to the quadratic equation $d = 16t^2$, which gives the distance an object has fallen in feet t seconds after it has dropped.

Think About It In what situation would it be important for someone to know the relationship described above?

Math Online
Take the **Chapter Readiness Quiz** at glencoe.com.

Dear Family,

In previous chapters, the class has focused on linear and exponential relationships. In this chapter, your student will work with quadratic relationships. The graphs of **quadratic equations** are U-shaped curves called parabolas. Quadratic equations are used to model situations involving projectile motion, like the path of a kicked football.

Key Concept—Quadratic Relationships

In addition to equations, the class will use graphs and tables in its work with quadratic relationships.

Equation: $y = x^2 + 3$

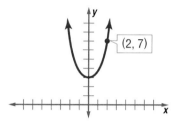

x	y
−2	7
−1	4
0	3
1	4
2	7

The class will also explore **inverse variation**. The number of people working on a task and the amount of time each person works is an example of an inverse relationship. Suppose it takes 10 person-hours to paint a room. The chart below and the equation $x \cdot y = 10$ illustrate this inverse relationship.

Number of People, x	1	2	4	10
Hours Worked per Person, y	10	5	2.5	1

So, the hours of work per person decrease as the number of people increase.

Chapter Vocabulary

conjecture   parabola

hyperbola   vertex

Home Activities
- List examples of sports-related activities that produce projectile motion paths.
- Identify real-world situations that produce inverse relationships.

Use Graphs and Tables to Solve Equations

In earlier chapters, you have worked with linear and exponential relationships. Look at the graphs and tables in Think & Discuss to review these relationships.

Vocabulary

quadratic equation

Think & Discuss

Consider the linear graph and table below. What do you notice?

Linear

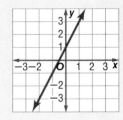

x	y
−1	−1
0	1
1	3
2	5
3	7

Consider the exponential graph and table below. What do you notice?

Exponential

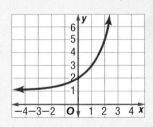

x	y
−1	1.5
0	2
1	3
2	5
3	9

You have probably determined that the linear equation $y = 2x + 1$ is represented by the first pair of tables and graphs. Likewise, the exponential equation $y = 2^x + 1$ is represented by the second pair of tables and graphs.

In this lesson, you will explore *quadratic relationships* and **quadratic equations**. A quadratic equation is one in which one of the variables is squared. Just as with linear and exponential relationships, you will use equations, graphs, and tables to represent quadratics.

Investigation ① Find Values from a Graph

Materials

- graphing calculator

If objects such as tennis balls are thrown straight upward, those with faster starting speeds will go higher. Gravity is the attraction of mass of a celestial body, such as a planet, near or at the body's surface. The pull of gravity will slow each object until it momentarily reaches a maximum height and then the object will begin to fall.

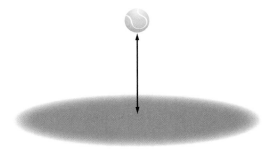

$$\underset{\text{height}}{\longrightarrow} h = \underset{\text{velocity}}{v}\underset{\text{time (in seconds)}}{t} - 16t^2$$

The formula $h = vt - 16t^2$ approximates an object's height above its starting position at a chosen time. The variable h represents the height in feet, t represents the time in seconds, and v represents the starting speed, or initial *velocity*, in feet per second. This formula assumes the starting height is at ground level, and it ignores such complications as air resistance.

Suppose someone throws a tennis ball toward the ground. The ball bounces straight upward with a speed of 30 feet per second (about 20.5 miles per hour) as it leaves the ground. In this case, $v = 30$, so the height as the ball bounces is given by

$$h = 30t - 16t^2$$

where t is the time in seconds after the ball leaves the ground.

You can use the formula to find the height h of the ball for various times t. For example, after one-half second, or at time $t = 0.5$, the formula gives $h = 11$ for height.

$$h = 30(0.5) - 16(0.5)^2$$
$$h = 15 - 4$$
$$h = 11$$

So, the ball is 11 feet above the ground 0.5 second after it leaves the ground.

The graph below shows the relationship between h and t for the equation $h = 30t - 16t^2$. You can see from the graph that the ball is at a height of 11 meters when the time is 0.5 second.

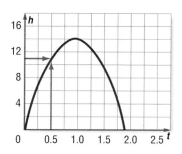

Think & Discuss

After looking at the graph of $h = 30t - 16t^2$, Evan said the ball did not bounce straight upward. Why do you think he said this?

Is he correct? Explain.

✓ Develop & Understand: A

1. You can display the graph of $h = 30t - 16t^2$ on your calculator. When you enter the equation, use x for t and y for h.

   a. Graph the equation $y = 30x - 16x^2$ in the standard viewing window, $-10 \leq x \leq 10$ and $-10 \leq y \leq 10$. Describe the graph.

   b. Find a new viewing window that allows you to identify points along the graph of $y = 30x - 16x^2$. Make a sketch of the graph. Label the minimum and maximum values on each axis.

   c. Estimate the coordinates of the highest point on the graph.

2. You can use the Trace feature on your calculator to estimate when the ball hits the ground.

   a. What is the value of h when the ball is on the ground?

   b. Estimate the values of t that give this value of h.

   c. Write the equation you would need to solve to find the value of t when the ball is on the ground.

   d. Check your estimates from Part b by substituting them into your equation. Are your estimates exact solutions?

3. What does the fact that the equation has two solutions tell you about the flight of the ball?

4. The Zoom feature on your calculator allows you to look more closely at any part of a graph. To see how this works, start with your graph from Exercise 1.

You already used the Trace feature to estimate when the ball hits the ground the second time. Now you can use the Zoom feature to get a better estimate.

- Select Zoom In from the Zoom menu.
- Focus on where the curve crosses the x-axis, or the x-intercept.
- Once again, use the Trace feature on this new graph to try to get a more accurate estimate of a solution of $30t - 16t^2 = 0$.
- Check your estimate by substituting it for t in the equation $30t - 16t^2 = 0$.
- If your estimate is not an exact solution, zoom in once more and refine your estimate again.

5. For the formula $h = 30t - 16t^2$, write an inequality that shows the values t can have. Explain your inequality.

Real-World Link

Because they must be extremely accurate, equations for calculating the height and speed of a rocket account for such factors as air resistance and the decreasing mass of the rocket as it burns fuel.

✅ Develop & Understand: B

6. How could you use this graph of $h = 30t - 16t^2$ to estimate the solution of $30t - 16t^2 = 3$?

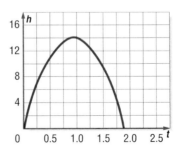

7. Allen said he could tell immediately from the graph that there are exactly two solutions of the equation $30t - 16t^2 = 3$. How did he know?

8. Use your calculator's Zoom and Trace features to find approximate solutions of $30t - 16t^2 = 3$ to the nearest hundredth.

9. What do the solutions of $30t - 16t^2 = 3$ tell you about the ball's flight?

10. Bharati said the equation $30t - 16t^2 = 20$ has no solutions. How could she tell this from the graph?

11. Write another equation of the form $30t - 16t^2 = ?$ that has no solutions.

12. How many solutions does $30t - 16t^2 = 10$ have?

13. Is there a value of c that would make the equation $30t - 16t^2 = c$ have exactly one solution? If so, explain how to find it using the graph, and give the value.

14. For the formula $h = 30t - 16t^2$, write an inequality that shows the values h can have. Explain your inequality.

> ### Share & Summarize
>
> **1.** How can you use a graphing calculator to help estimate solutions of equations? Illustrate your answer by writing an equation and finding solutions to the nearest hundredth.
>
> **2.** How many solutions are possible for an equation of the form $30x - 16x^2 = c$, where c is a constant?

Investigation ② Use Tables to Solve Equations

Materials

• graphing calculator

In Investigation 1, you estimated solutions of equations by making graphs on a calculator. Now you will use your calculator's Table feature to make the guess-check-and-improve process more systematic.

Think & Discuss

Sonia creates and sells large tapestries. A new client wants a tapestry with an area of four square meters. Sonia decides it will have a rectangular shape and a length one meter longer than its width.

- If x represents the width of the tapestry in meters, what equation can Sonia solve to find the length and width of the tapestry?

- Sonia used guess-check-and-improve to try to find the value of x. She first tried a width, x, of 1 meter and found the area would be 2 m^2. Then she found that the area would be 6 m^2 for a width of 2 m. What might Sonia choose as her third guess for x?

- Use guess-check-and-improve to find the value of x to the nearest tenth of a meter. Make a table to keep track of your guesses and the resulting areas.

- Now find the x value to the nearest hundredth of a meter.

- How do you know the value you found for the width gives an area closer to 4 m^2 than any other value to the nearest hundredth?

- How could you find a value of x that gives an even closer approximation to an area of 4?

In Think & Discuss, you chose a first guess for x and then improved your guess one step at a time. Graphing calculators have a Table feature that allows you to examine many guesses at a glance.

Real-World Link
An art form that has been around for centuries, tapestries are colorful, richly woven cloths. Today, they are often used to decorate the walls of homes and offices.

Example

Try Sonia's equation using the Table feature. Follow along with this example using your calculator.

- On the Table Setup screen, set the starting x value to 0 and the increment to 1.

- Enter Sonia's equation in the form $y = x(x + 1)$ or $y = x^2 + x$.

- Look at the Table screen to see which value or values of x give values of y closest to 4.

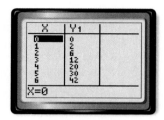

The table below shows part of what may appear on your screen.

From the table, you can see that x must be between 1 and 2 since the corresponding y values go from 2 to 6.

To search for values between 1 and 2, use the Table Setup feature to show x values in smaller increments.

- On the Table Setup screen, set the starting x value to 1 and the increment to 0.1.

When you return to the Table screen, you should see something like the table below.

You can now see that the solution has an x value between 1.5 and 1.6.

In Exercises 1–3, you will use the method shown in the example to find a more accurate estimate for x.

Math Link

The Greek letter *delta*, Δ, is used to represent the increment between consecutive values in a sequence. On many calculators, Δ is used to indicate the amount of change in the variable x from one row of a table to the next.

⊘ Develop & Understand: A

1. In the example, the table shows that the width of Sonia's tapestry must be between 1.5 and 1.6 meters. You can use Table Setup again to improve the accuracy of x to the hundredths place.

    a. What values of x and what increment should you use to get a more exact estimate for x?

    b. Adjust the values of x and the increment in Table Setup and then view the table once more. You will have to scroll down to see all the values you need. Copy and complete the table at the right.

    c. Between which two values of x is the solution? Which of these gives an area closer to 4? Explain.

    d. What are the width and length of the tapestry to the nearest centimeter?

x	y
1.5	3.75
1.51	
1.52	
1.53	
1.54	
1.55	
1.56	
1.57	
1.58	
1.59	
1.6	

2. You can use Table Setup again to improve the accuracy of your estimate for x to the thousandths place, the nearest millimeter.

    a. What should you enter as the starting x value and the increment in Table Setup?

    b. Between which two values of x does the exact answer lie?

    c. Which of the two values gives a better approximation? How do you know?

3. Another client wants a tapestry with an area of 6 square meters and a length 2 meters greater than its width.

    a. Write an equation representing the area of the tapestry, using x to represent the width.

    b. Use the Table and Table Setup features to find the width to the nearest centimeter. For each new table you examine, record the two values of x that give areas closest to 6.

You will now practice using the Table feature to approximate solutions of quadratic equations. You will learn how to find exact solutions of quadratic equations in Chapter 9.

✓ Develop & Understand: B

Each equation below has two solutions. Use your calculator's Table feature to approximate the solutions to the nearest hundredth. Remember, you may have to scroll through the table. Check each answer by substituting it into the equation.

4. $2m(m - 5) = 48$

5. $3t(t - 5) = 48$

6. Since some quadratic equations have two solutions, there may be a second solution of each equation you have explored for the area of Sonia's tapestries.

   a. How could you use a graphing calculator table to determine whether there is another solution of $x(x + 1) = 4$?

   b. Determine whether there is a second solution of $x(x + 1) = 4$. If so, estimate it to the nearest hundredth. Check the solution by substitution.

   c. Determine whether there is a second solution of $x(x + 2) = 6$. If so, estimate it to the nearest hundredth. Check the solution by substitution.

   d. Could Sonia use the solutions you found in Parts b and c to design a tapestry? Why or why not?

Share & Summarize

Think about how you might solve these equations.

   i. $2w - 3 = 3w - 7$ **ii.** $x^3 - 2x^2 = 5$

   iii. $3y(2y + 3) - 15 = 0$ **iv.** $5 = 3z + 1$

1. Which equations would you solve by doing the same thing to both sides or backtracking? Explain.

2. Which equations would you solve using your calculator's Table feature? Explain.

Practice & Apply

1. **Physical Science** A ball is launched straight upward from ground level with an initial velocity of 50 feet per second. Its height h in feet above the ground t seconds after it is thrown is given by the formula $h = 50t - 16t^2$.

    a. Draw a graph of this formula with time on the horizontal axis and height on the vertical axis. Show $0 \le t \le 4$ and $0 \le h \le 40$.

    b. What is the approximate value of t when the ball hits the ground?

    c. About how high does the ball go before it starts falling?

    d. After approximately how many seconds does the ball reach its maximum height?

2. Suppose the height in feet of a ball bouncing vertically into the air is given by the formula $h = 30t - 16t^2$, where t is time in seconds since the ball left the ground. A graph of this equation is shown below.

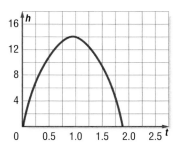

    a. On its way up, the ball reached a particular height at 0.5 second. It reached this height again on its way down. How could you determine at what time the ball reached this height on the way down?

    b. What is the value of this time to the nearest tenth?

3. In Investigation 1, you found that the maximum value of $30t - 16t^2$ is approximately 14. The exact value is 14.0625. Write an equation or an inequality showing the values of h for which $h = 30t - 16t^2$ has each given number of solutions.

    a. two   b. one   c. none

4. Draw a graph of $y = x^2 - 4x + 2$. Use your graph to find a value for d so that the equation $x^2 - 4x + 2 = d$ has each given number of solutions.

    a. two   b. one   c. none

5. Solve the equation $(x + 2)(x - 3) = 14$ by constructing a table of values. Use integer values of x between -6 and 6.

6. Solve the equation $k^2 - k + 3 = 45$ by constructing a table of values. Use integer values of x between -9 and 9.

7. Examine this table.

t	$t(t - 3)$
-2	10
-1	4
0	0
1	-2
2	-2
3	0
4	4
5	10
6	18
7	28

a. Use the table to estimate the solutions of $t(t - 3) = 5$ to the nearest integer.

b. If you were searching for solutions by making a table with a calculator, what would you have to do to find solutions to the nearest tenth?

c. Find two solutions of $t(t - 3) = 5$ to the nearest tenth.

8. The equation $x^3 + 5x^2 + 4 = 5$ has a solution between $x = 0$ and $x = 1$. Find this solution to the nearest tenth.

9. This is a graph of $y = 0.1x^2 + 0.2x + 1$.

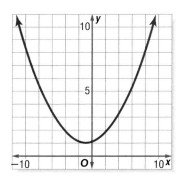

a. Between which two pairs of integer values for x do the solutions of $0.1x^2 + 0.2x + 1 = 4$ lie?

b. Find both solutions of $0.1x^2 + 0.2x + 1 = 4$ to the nearest hundredth.

Connect & Extend

10. Physical Science The formula $h = vt - 16t^2$ approximates a projected object's height above its starting position t seconds after it begins moving straight upward with an initial velocity of v ft/s.

For a thrown ball, the starting position will usually be some distance above the ground, such as 5 feet. So, the height of the ball above the ground will be given by this formula, where s is the initial height at $t = 0$ and v is the initial velocity.

$$h = vt - 16t^2 + s$$

a. Suppose a ball is thrown upward with a starting velocity of 30 feet per second from 5 feet above the ground. Write an equation describing the height h of the ball after t seconds.

b. What equation would you solve to find how long it takes the ball to reach the ground?

c. What equation would you solve to find how long it takes the ball to return to its starting height?

11. Physical Science An object that is dropped, like one thrown upward, will be pulled downward by the force of gravity. However, its initial velocity will be 0. If air resistance is ignored, you can estimate the object's height h at time t with the formula $h = s - 16t^2$, where s is the starting height in feet.

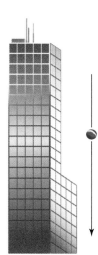

a. Suppose a baseball is dropped from a height of 100 ft. What equation would you solve to determine the number of seconds that would pass before the baseball hits the ground?

b. Solve your equation.

Real-World Link

Air resistance, which depends on the diver's speed and cross-sectional area, must be calculated to accurately determine the height of a falling skydiver.

12. A cartoon cat pushes a piano out of a fourth-story window approximately 50 feet above the sidewalk.

a. An object falls a distance of $s = 16t^2$ feet in t seconds. For how long will the piano fall before it hits the sidewalk?

b. Suppose the cat intended to smash his nemesis, the mouse, on the sidewalk below. But the mouse pulls out a trampoline from his pocket, and the piano bounces back toward the cat at an initial velocity of 100 feet per second.

The height h of the piano t seconds after it hits the trampoline can be found using the formula $h = 100t - 16t^2$. Graph this equation. Use your graph to estimate how long it takes the piano to return to the fourth-story window.

13. A rock climber launched a hook from the base of a cliff. The edge of the cliff is 100 feet above the climber. Use the formula for the height of a thrown object, $h = vt - 16t^2$, to answer these questions. Assume that the hook is thrown straight up and, if it travels high enough, catches the edge of the cliff during its descent.

a. Can the hook reach the top of the cliff if its initial velocity is 70 feet per second, or 70 fps? Explain.

b. Find an initial velocity that would be sufficient to allow the hook to reach the cliff's edge.

c. At an initial velocity of 100 fps, how far above the edge of the cliff will the hook rise?

d. At an initial velocity of 100 fps, how long will it take until the hook catches the top of the cliff on its way back down?

14. Desmond wants to construct a large picture frame using a strip of wood 20 feet long.

  a. Let w represent the width of the frame. Write an equation that gives the height of the frame h in terms of the width w.

  b. Express the area enclosed by the frame in terms of the width.

  c. What dimensions would give Desmond a frame that encloses the greatest possible area?

  d. What dimensions would give Desmond a frame that encloses the least possible area?

  e. If the enclosed area is to be approximately 15 square feet, determine the dimensions of the frame to the nearest tenth of a foot.

15. Physical Science Two objects are projected straight upward from the ground at exactly the same moment. One object is released at an initial velocity of 20 feet per second. The other is released at 30 feet per second. The two equations that relate the heights of the objects to time t are $y_1 = 20t - 16t^2$ and $y_2 = 30t - 16t^2$. Below is a graph of these two equations.

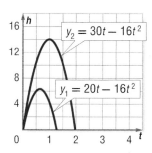

  a. Write two equations to find the times at which the objects hit the ground.

  b. These two objects hit the ground at different times. Use a calculator to find the exact difference between times.

16. Graphs of $y_1 = 4 - (x - 2)^2$ and $y_2 = (x + 2)^2 - 4$ are shown below.

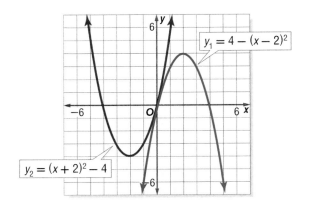

a. The maximum value of y_1 is 4. Explain how you can identify the maximum value of a U-shaped curve using a table of values.

b. The minimum value of y_2 is -4. Explain how you can identify the minimum value of a U-shaped curve using a table of values.

c. For each equation, make a table of x and y values for x values between -3 and 3. Without graphing, determine whether each equation has a maximum or a minimum y value and tell what it is.

i. $y = 21 - 4x - x^2$

ii. $y = (x - 2)^2 - 1$

iii. $y = 8 + 2x - x^2$

17. In Your Own Words Describe how graphing an equation can help you estimate its solutions.

18. Preview Graph the following equations. Be sure to use positive and negative values for x.

How are the graphs similar? How are they different?

$$y = x^2 \qquad\qquad y = -x^2$$

$$y = 0.5x^2 \qquad\qquad y = -0.5x^2$$

$$y = 3x^2 \qquad\qquad y = -3x^2$$

Mixed Review

Find the values of t in each equation.

19. $t^4 = 81$

20. $t^5 = 32$

21. $7t^3 = 7{,}000$

22. $t^6 - 4 = 60$

23. $3^t = 729$

24. $4^t = 1{,}024$

Solve each equation by doing the same thing to both sides.

25. $5m + 19 = 7m - 21$

26. $3(p - 2) = 2p + 18$

27. $\dfrac{3k - 5}{5} + k = 5 + k$

28. $7.5a - 6 = 5a + 4$

Multiply each pair of binomials.

29. $(t + 2)(t + 1)$

30. $(x + 6)(x + 2)$

31. $(b + 5)(b - 4)$

32. $(t + 7)\,(t + 7)$

33. $(4m + 3)(4m - 3)$

Quadratic Relationships

8.2

In Lesson 8.1, you used graphs and tables to explore quadratic relationships. In this lesson, you will deepen your understanding of these relationships.

Explore

Select a team of nine students to make a "human graph." The team should follow these rules.

- Line up along the *x*-axis. One student should stand on −4, another on −3, and so on, up to 4.

- Multiply the number on which you are standing by itself. That is, if your number is *x*, calculate x^2. Remember the result.

- When your teacher says "Go!" walk forward or backward a number of paces equal to your result from the previous step.

- The number you began on is your *x* value. The number of paces you took is your *y* value.

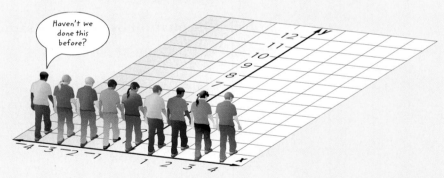

Each of the nine students on the graph should now report his or her coordinates. Record the information in a class table like the one below.

x	−4	−3	−2	−1	0	1	2	3	4
y									

Make a class graph by plotting the points on the board or on a large sheet of paper. Connect the points with a smooth curve.

Describe the graph.

When making the human graph, did any student walk backward? Why or why not?

If someone had started at 1.5 on the x-axis, how far forward or backward would that person have moved? What if someone had started at -1.5?

Could this graph be extended to the left of -4 or to the right of 4? Explain.

Is the graph *symmetrical*? That is, is there a line along which you could fold the graph so that the two halves match up exactly? If so, describe this *line of symmetry*. Tell whether any students were standing on it.

What equation describes the relationship shown in the graph?

Investigation 1 Quadratic Equations and Graphs

Vocabulary

parabola

Materials

• graph paper

The human graph your class created was a graph of $y = x^2$, which is a *quadratic equation*. The simplest quadratic equations can be written in the form $y = ax^2$. This form consists of a constant, represented by the letter a, multiplied by the square of a variable.

The formula for the area of a circle, $A = \pi r^2$, has this form. In the area formula, the constant is π. The equation that gives the distance in meters fallen by an object after t seconds is $d = 4.9t^2$. In this distance equation, the constant is 4.9. In the equation $y = x^2$, the constant is 1.

Examples of quadratic equations are listed below.

$$y = ax^2 \qquad\qquad y = 1x^2$$

$$y = 4.9t^2 \qquad\qquad y = \pi r^2$$

The graph of each of these relationships is a symmetric, U-shaped curve called a **parabola**. In this investigation, you will look at graphs, tables, and equations for more quadratic relationships.

✅ Develop & Understand: A

This box has height 9 cm and a length that is twice its width.

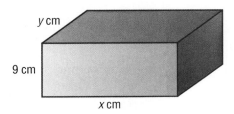

Math Link

The volume of a rectangular prism is the product of its length, width, and height. Volume is measured in cubic units such as cm^3.

1. You can describe the box's volume with either of these quadratic equations.

$$V = 4.5x^2 \qquad\qquad V = 18y^2$$

How can two different equations describe the same volume?

2. On a grid like the one below, plot the equation $V = 4.5x^2$. Start by plotting points (x, V) for x values from 0 to 50. You will look at x values less than 0 a little later. Plot as many points as you need, until you are confident of the shape of the graph. Then draw a smooth curve through the points.

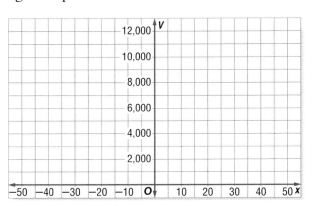

3. Suppose you want to construct a box with a particular volume and a length twice as long as its width. You can use your graph to find possible lengths for the box.

   a. Choose a value for the volume V. Do not use one of the values you plotted in Exercise 2. Use your graph to estimate the length x that corresponds to this volume.

   b. Check your estimate by substituting it for x in the equation $V = 4.5x^2$. How close is this result to the volume you chose in Part a?

4. Of course, boxes do not have negative lengths. But if you think of $V = 4.5x^2$ simply as a rule to generate ordered pairs, you can look at the graph for x values less than or equal to 0. Plot some (x, V) points for $x = 0$ and several negative values of x. Use the points to extend your curve into the second quadrant.

5. What is the line of symmetry for your graph?

Math Link

To guess-check-and-improve, make a reasonable guess of the value, calculate the value of the other variable using your guess, and use the answer to improve your guess.

6. Locate both points on the graph where $V = 9,000$.

   a. Use the graph to estimate the corresponding values of x.

   b. Use a calculator to guess-check-and-improve for a more accurate estimate of the x values for $V = 9,000$.

Share & Summarize

Tell whether each graph could represent a quadratic relationship. Explain how you decided.

1.

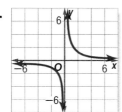

2.

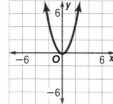

3.

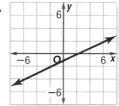

4.

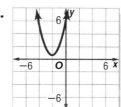

Investigation 2 Quadratic Patterns

Materials

- square tiles (optional)
- graph paper
- graphing calculator

In your study of linear equations, you learned that they can be written in many forms, such as $y = mx + b$ and $Ay + Bx = C$. Quadratic equations can also be written in different forms.

In this investigation, you will explore some geometric patterns that can be represented by quadratic equations in forms other than $y = ax^2$.

✅ Develop & Understand: A

Consider this pattern of square tiles.

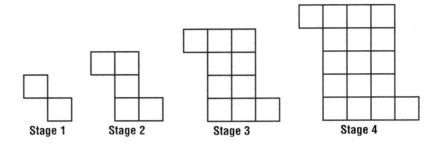

Stage 1   Stage 2   Stage 3   Stage 4

1. Describe the pattern in words.

2. Copy and complete the table to show the number of tiles used in each of the first four stages.

Stage, S	1	2	3	4
Tiles, T				

3. Describe the pattern in the way the number of tiles T increases as the stage number S increases.

4. Use your answer to Exercise 3 to predict the number of tiles in stages 5 through 8. Extend your table to include these values.

5. Check your answers to Exercise 4 by building or drawing the next four stages of the pattern.

6. Gabriel found an equation for the number of tiles T in stage S. He said, "Taking away the two tiles on the corners leaves a rectangle that is $S + 1$ tiles long and $S - 1$ tiles wide. I multiply those numbers to find the number of tiles in the rectangle. Then I add the two corner tiles to find the total number of tiles." What equation do you think Gabriel found?

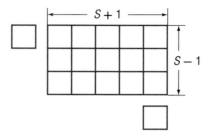

7. Carissa found an equation for T by using her table. "First, I square S. Then I see how S^2 compares to the number of tiles for that value of S."

   Try her idea with your table. Square the numbers in the first row, and see what you could do to the squares to get the numbers in the second row. What equation do you think Carissa found?

8. Jenelle found the same equation as Carissa but in a different way. "There are S tiles across the bottom of stage S. If I remove this row and turn it vertically, I can make a square with S tiles on each side, plus one extra tile." Explain how Jenelle's reasoning leads to the same equation Carissa found.

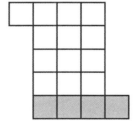 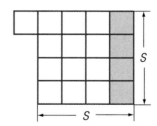

9. Consider the equation Carissa and Jenelle found for the pattern.

   a. Graph the relationship between T and S. Even though it makes no sense to have a negative number of tiles, think of the equation as a rule to generate ordered pairs and consider both positive and negative values of S. Plot enough points so you can draw a smooth curve.

   b. Describe the graph.

   c. Does the graph have a line of symmetry? If so, describe its location.

   d. What is the smallest value of T shown on the graph?

✅ *Develop & Understand: B*

Here is a geometric problem that has a connection to a quadratic equation, although the connection is not obvious at first.

How many diagonals does a polygon with n sides have?

You probably cannot answer this question yet. Start with a few simple polygons. Look for a pattern to help you find the answer.

The drawings show the diagonals for polygons with 3, 4, and 5 sides.

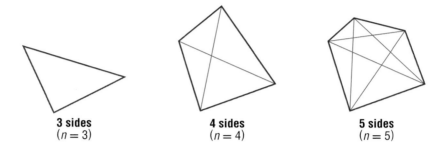

3 sides
($n = 3$)

4 sides
($n = 4$)

5 sides
($n = 5$)

10. Draw a hexagon, $n = 6$, with a complete set of diagonals. Make sure you join each vertex to every other vertex of the hexagon.

11. Copy and complete the table to show the number of diagonals connected to each vertex and the total number of diagonals for each polygon listed.

Sides, n	3	4	5	6
Diagonals Connected to Each Vertex	0	1		
Total Number of Diagonals, d	0	2		

12. Look at the second row of your table.

  a. Describe the pattern in the way the number of diagonals connected to each vertex changes as the number of sides increases.

  b. Is there a linear relationship between the number of sides and the number of diagonals connected to each vertex? Explain.

13. Look at the third row of your table.

  a. Describe the pattern in the way the total number of diagonals changes as the number of sides increases.

  b. Is there a linear relationship between the number of sides and the total number of diagonals? Explain.

Real-World Link

A *polygon* is a closed, two-dimensional figure whose sides are made of line segments.

14. Consider a heptagon, which is a seven-sided polygon.

   a. Use the pattern in your table to predict the number of diagonals connected to each vertex in a heptagon.

   b. Predict the total number of diagonals in a heptagon.

   c. Check your predictions from Parts a and b by drawing a heptagon and carefully drawing and counting its diagonals. Add the data for the heptagon to your table.

15. Without drawing any more polygons, extend your table to include data for 8-, 9-, and 10-sided polygons. Explain how you found your results.

Math Link

A polygon with 20 sides is called an *icosagon.* Mathematicians would also call it a *20-gon*.

16. How many diagonals does a 20-sided polygon have? Find the answer without extending your table, if you can.

17. Write an equation for the total number of diagonals in an *n*-sided polygon. Explain how you found your answer.

18. Use your equation to find the total number of diagonals in a 100-sided polygon.

19. What shape do you think the graph of your equation will have? Use your graphing calculator to check your answer.

Share & Summarize

Consider the quadratic relationships you have explored in this lesson.

1. Write the equations of these quadratic relationships.

2. Look back at the graphs of these equations. What do they all have in common?

3. How can you tell by looking at the equations that their graphs will not be lines?

Practice  **Apply**

1. The area of a circle with radius r is given by the formula $A = \pi r^2$.

   a. Make a table of values for A as r increases from 0 to 10 units.

   b. Plot the values on a graph with r on the horizontal axis and A on the vertical axis. Connect the points with a smooth curve.

   c. The areas of three circles are given. Use your graph to find an approximate value for the radius of each circle.

      i. 25 square units

      ii. 100 square units

      iii. 300 square units

   d. Check your results by substituting each radius into the area formula.

> **Math Link**
>
> If you do not have a calculator with a button for π, you can use the approximation 3.14 in your calculations.

2. Imagine several stacks of blocks like those shown below.

 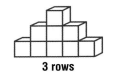

1 row   2 rows   3 rows

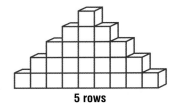

4 rows   5 rows

   a. Complete the table to show the number of blocks in the bottom row as the number of rows increases.

Number of Rows, n	1	2	3	4	5	6	7	8	9	10
Number of Blocks in Bottom Row, b										

   b. How many blocks would be in the bottom row of a stack with 25 rows? With n rows?

c. What is the relationship between the number of rows and the number of blocks in the bottom row?

d. What is the *total* number of blocks *T* needed for a stack with 5 rows?

e. Add a new row to your table for the total number of blocks *T* in each stack. Complete that row.

f. Write an equation that gives the total number of blocks *T* in terms of *n*.

g. Use this diagram to explain your equation from Part f.

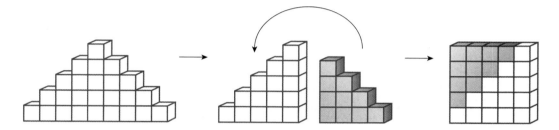

3. Look closely at this sequence of figures. Each square has an area of one square unit.

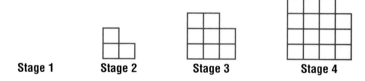

Stage 1   Stage 2   Stage 3   Stage 4

a. Find a formula for the area in square units, *A*, of Stage *n*.

b. Use your equation to explain why there are no squares in Stage 1.

c. Copy and complete the table.

n	1	2	3	4	5	6	7	8	9	10
A	0	3	8	15						

d. Extend your table to include negative integer values of *n*, from −10 to 0. Plot enough points on a graph between −10 and 10 to draw a smooth curve.

e. Describe the graph.

f. Find the line of symmetry for the graph. Through what value of *n* does the line of symmetry pass?

Connect & Extend

4. Consider these three equations.

$$y = x^2 \qquad y = x^2 + 2 \qquad y = x^2 - 2$$

a. Make a table of values for each equation, using positive and negative values of x. Plot all three graphs on the same set of axes.

b. What similarities do you see in the three graphs?

c. What differences do you notice?

5. Consider these three equations.

$$y = -x^2 \qquad y = -x^2 + 2 \qquad y = -x^2 - 2$$

a. Make a table of values for each equation, using positive and negative values of x. Plot all three graphs on the same set of axes.

b. What similarities do you see in the three graphs?

c. What differences do you notice?

6. **Economics** Rosalinda makes birdhouses and sells them at craft fairs. She has found that she can sell more birdhouses when the price is lower. She estimates that for a price of p dollars, she can sell $200 - 2p$ birdhouses during a two-day fair. For example, if she sets the price at $20, she generally sells around 160 birdhouses.

a. Find a formula for Rosalinda's revenue R at a two-day fair if she charges p dollars for each birdhouse.

b. Make a table of values for price p and revenue R. List at least ten prices from $0 to $100.

c. Graph the values in your table with p on the horizontal axis and R on the vertical axis. You may need to find additional points so you can draw a smooth curve.

d. For what price does Rosalinda earn the most revenue? What is that revenue?

Real-World Link

When a person or business sells a product, the money from the sales is called *revenue*. The revenue for a product can be calculated by multiplying the number of items sold by the price.

7. Darlene made this table of ordered pairs.

x	1	2	3	4	5	6	7	8	9	10
y	0	3	8	15	24					

a. Describe the pattern of change in the *y* values as the *x* values increase by 1.

b. Copy and complete Darlene's table.

c. Use your table to plot a graph of these ordered pairs. Connect the points with a smooth curve or line.

d. Describe your graph.

e. Could Darlene's table represent a quadratic relationship? Explain.

8. **Physical Science** The distance it takes for a car to stop depends on the car's speed.

a. The *reaction distance* is the distance the car travels after the driver realizes he needs to stop and before he applies the brakes. It can be represented by the equation $d = 0.25s$, where *s* is speed in kilometers per second and *d* is distance in meters.

The *braking distance* is the distance traveled after the driver applies the brakes until the car comes to a complete stop. It can be represented by $d = 0.006s^2$.

Make a table showing the reaction distance and the braking distance for speeds from 0 km to 80 km.

b. Plot the points for reaction distance and braking distance on one set of axes. Draw a smooth line or curve through each set of points. Label each graph with its equation.

c. To find the total stopping distance, reaction distance and braking distance must be added. Add the values in your table for each speed to find the total stopping distance for that speed.

d. Plot the values in your table for total stopping distance on your graph. Connect them with a smooth line or curve. What is the equation of this relationship? What kind of relationship does this appear to be? (Hint: Think about what you did to generate the points for stopping distance.)

9. Ren made this table of ordered pairs.

x	1	2	3	4	5	6	7	8	9	10
y	7	10	13	16	19					

a. Describe the pattern of change in the *y* values as the *x* values increase by 1.

b. Complete Ren's table.

c. Use your table to plot a graph of these ordered pairs. Connect the points with a smooth curve or line.

d. Describe your graph.

e. Could Ren's table represent a quadratic relationship? Explain.

10. In Your Own Words How do graphs for quadratic relationships differ from those for linear relationships? How do you think the equations differ?

Mixed Review

11. Graph each equation.

a. $y = 2x$

b. $y = 2x + 1$

c. $y = 2x - 2$

d. $y = -2x$

In Exercises 12 and 13, write an equation to represent the situation.

12. Economics The balance *b* in a savings account at the end of any year *t* if $5,000 is deposited initially and the account earns 8% interest per year.

13. Life Science The number of bacteria *b* left in a sample after 24 hours if $\frac{1}{16}$ of the remaining colony of *c* bacteria dies every hour.

Expand and simplify each expression.

14. $-3a(2a - 3)$

15. $(4k - 7)(-2k + 3)$

LESSON 8.3

Families of Quadratics

In Lesson 8.2, you explored several quadratic relationships.

$$y = x^2 \qquad d = \frac{n^2 - 3n}{2}$$

$$V = 4.5x^2 \qquad T = S^2 + 1$$

Any expression that can be written in the form $ax^2 + bx + c$, where a, b, and c are constants and $a \neq 0$, is called a *quadratic expression* in x.

The first term, ax^2, is called the x^2 *term*. The number multiplying x^2, represented by a, is called the *coefficient* of the x^2 term. In the same way, bx is the x *term* and b is its coefficient. The coefficients a and b and the constant c can stand for positive or negative numbers, but a cannot be 0. Can you see why?

A *quadratic equation* can be written in the form $y = ax^2 + bx + c$, where $a \neq 0$. Like those in Lesson 8.2, all quadratic equations have symmetric, U-shaped graphs called *parabolas*.

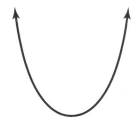

Before you explore the connection among graphs, tables, and equations for quadratic relationships, it will help to review graphs, tables, and equations for linear relationships.

Real-World Link

Mirrors in the shape of parabolas are used in flashlights and headlights to focus light into a narrow beam.

Think & Discuss ·

Equations, graphs, and tables of four linear relationships are shown here. Match each equation with its graph and table. Explain how you made each match.

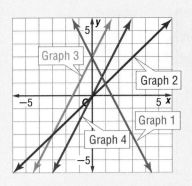

Equation 1: $y = x$

Equation 2: $y = 2x$

Equation 3: $y = 2x + 3$

Equation 4: $y = -2x + 3$

Table 1

x	−2	−1	0	1	2	3
y	7	5	3	1	−1	−3

Table 2

x	−2	−1	0	1	2	3
y	−4	−2	0	2	4	6

Table 3

x	−2	−1	0	1	2	3
y	−2	−1	0	1	2	3

Table 4

x	−2	−1	0	1	2	3
y	−1	1	3	5	7	9

For linear equations of the form $y = mx + b$, how does the value of m affect the graph?

For linear equations of the form $y = mx + b$, how does the value of b affect the graph?

Investigation 1 Quadratic Equations, Tables, and Graphs

Now you will explore the connection among equations, tables, and graphs for quadratic relationships.

✓ Develop & Understand: A

1. The table on the next page has columns for six quadratic equations. Copy and complete the table. You may wish to complete Column A first and then look for connections between its equation, $y = x^2$, and the other equations. For example, if you know the value of x^2 in Column A, how can you easily determine $x^2 + 1$ in Column B?

Math Link

The expression $-n^2$ means $-(n^2)$, not $(-n)^2$. To calculate $-n^2$, square n and take the opposite of the result.

x	**A** $y = x^2$	**B** $y = x^2 + 1$	**C** $y = x^2 - 1$	**D** $y = -x^2$	**E** $y = (x + 1)^2$	**F** $y = (x - 1)^2$
-4	16					
-3.2	10.24					
-2.2						
-1						
-0.5						
0						
0.5						
1						
2.2						
3.2						
4						

2. Compare the values in each of Columns B–F with those in Column A. Explain how each comparison you make is related to the equations.

For the graphs in Exercises 3–8, complete Parts a–c.

3.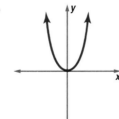

4.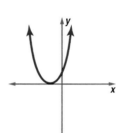

5.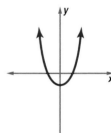

6.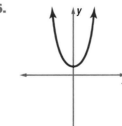

7.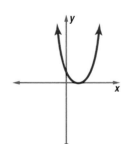

8.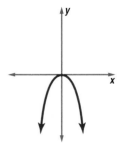

a. Match each graph with one of the quadratic equations in Columns A–F. Explain your reasoning.

b. Describe how the graph differs from the graph of $y = x^2$.

c. Describe the graph's line of symmetry.

✅ *Develop & Understand: B*

For the tables in Exercises 9–16, find a quadratic equation. As you complete Exercises 10–16, you may find it helpful to consider the patterns found in previous exercises.

9.

x	y
−3	9
−2	4
−1	1
0	0
1	1
2	4
3	9

10.

x	y
−3	109
−2	104
−1	101
0	100
1	101
2	104
3	109

11.

x	y
−3	5
−2	0
−1	−3
0	−4
1	−3
2	0
3	5

12.

x	y
−3	13
−2	8
−1	5
0	4
1	5
2	8
3	13

13.

x	y
−3	90
−2	40
−1	10
0	0
1	10
2	40
3	90

14.

x	y
−3	70
−2	20
−1	−10
0	−20
1	−10
2	20
3	70

15.

x	y
−3	36
−2	16
−1	4
0	0
1	4
2	16
3	36

16.

x	y
−3	37
−2	17
−1	5
0	1
1	5
2	17
3	37

Share & Summarize

1. Graph A is the graph of $y = x^2$. Write equations for the other graphs. Explain how you know the equations fit the graphs.

Graph A

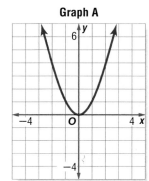

Graph B

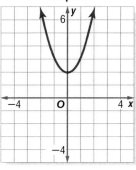

Graph C

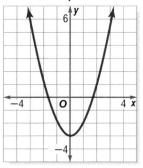

2. Table A represents the relationship $y = x^2$. Write the quadratic equations represented by Tables B and C. Explain how you found your equations.

Table A	
x	**y**
−4	16
−3	9
−2	4
−1	1
0	0
1	1
2	4
3	9
4	16

Table B	
x	**y**
−4	−48
−3	−27
−2	−12
−1	−3
0	0
1	−3
2	−12
3	−27
4	−48

Table C	
x	**y**
−4	36
−3	25
−2	16
−1	9
0	4
1	1
2	0
3	1
4	4

Investigation ② Quadratic Equations and Their Graphs

Vocabulary

vertex

Materials

• graphing calculator

Quadratic equations can be written in the form $y = ax^2 + bx + c$, where a, b, and c are constants and a is not 0. Like the coefficient m and constant b in a linear equation $y = mx + b$, the coefficients and constant in a quadratic equation tell you something about the graph of the equation.

In Exercises 1–5, you will explore how the coefficient a and constant c affect the graphs of quadratic equations. In Exercises 6–8, you will look at the effect of b, which is not as easy to see.

Real-World Link
The cables of a suspension bridge form the shape of a parabola.

✅ Develop & Understand: A

1. Complete Parts a–c for each group of quadratic equations.

Group I	**Group II**	**Group III**	**Group IV**
$y = x^2$	$y = x^2$	$y = x^2$	$y = -x^2$
$y = x^2 + 1$	$y = x^2 - 1$	$y = 2x^2$	$y = -2x^2$
$y = x^2 + 3$	$y = x^2 - 3$	$y = \frac{1}{2}x^2$	$y = -\frac{1}{2}x^2$

   a. Graph the three equations in the same window of your calculator. Choose a window that shows all three graphs clearly. Make a sketch of the graphs. Label the minimum and maximum values on each axis. Also, label each graph with its equation.

   b. For each group of equations, write a sentence or two about how the graphs are similar and how they are different.

   c. For each group of equations, give one more quadratic equation that also belongs in that group.

2. Describe how the graphs in Group I are like the graphs in Group II and how they are different.

3. Describe how the graphs in Group III are like the graphs in Group IV and how they are different.

4. Use what you learned in Exercises 1–3 to predict what the graph of each equation below will look like. Make a quick sketch of the graphs on the same set of axes. Be sure to label the axes. Also, label each graph with its equation. Check your predictions with your calculator.

   a. $y = x^2 + 2$ **b.** $y = 3x^2 + 2$

   c. $y = -3x^2 + 2$ **d.** $y = \frac{1}{2}x^2 - 3$

5. All of the equations you have seen in this exercise set are in the form $y = ax^2 + bx + c$, but the coefficient b is equal to 0. Explain how the values of a and c affect the graph of an equation.

In Exercises 1–5, you probably saw that equations of the form $y = ax^2 + c$ have their highest or lowest point at the point $(0, c)$. The highest or lowest point of a parabola is called its **vertex**.

Not all parabolas have their vertices on the y-axis. In the following exercises, you will look at the properties of an equation that determine where the vertex of its graph will be.

✅ Develop & Understand: B

Consider these four groups of quadratic equations.

Group I	**Group II**	**Group III**	**Group IV**
$y = x^2 + 2$	$y = x^2 + 2$	$y = -2x^2 + 2$	$y = -2x^2 + 2$
$y = x^2 + 3x + 2$	$y = x^2 - 3x + 2$	$y = -2x^2 + 3x + 2$	$y = -2x^2 - 3x + 2$
$y = x^2 + 6x + 2$	$y = x^2 - 6x + 2$	$y = -2x^2 + 6x + 2$	$y = -2x^2 - 6x + 2$

6. In each group, all of the equations have the same values of a and c. Use what you learned about the effects of a and c to make predictions about how the graphs in each group will be alike.

7. For each group of equations, complete Parts a–c.

    a. Graph the three equations in the same window of your calculator. Choose a window that shows all three graphs clearly. Make a sketch of the graphs. Remember to label the axes, and label the graphs with their equations.

    b. Were your predictions in Exercise 6 correct?

    c. For each group of equations, write a sentence or two about how the graphs are similar and how they are different.

8. What patterns can you see in how the locations of the parabolas change as b increases or decreases from 0?

Share & Summarize

1. Imagine moving the graph of $y = -2x^2 + 3$ up 2 units without changing its shape. What would be the equation of the new parabola?

2. Briefly describe or make a rough sketch of the graph of $y = -\frac{1}{2}x^2 - 2$.

3. For each quadratic equation, tell whether the vertex is on the y-axis. Explain how you know.

    a. $y = \frac{1}{2}x^2 - 3$ **b.** $y = x^2 - 3x + 1$

    c. $y = -x^2$ **d.** $y = -3x^2 + x + 13$

    e. $y = -x^2 + 4$ **f.** $y = 7x^2 + 3x$

Investigation 3 Use Quadratic Relationships

Materials

• graphing calculator

Quadratic equations and graphs can help you understand the motion of objects that are thrown or shot into the air. If an object, like a ball, is launched at an angle between 0° and 90° and not straight up or down, its *trajectory*, or the path it follows, approximates the shape of a parabola.

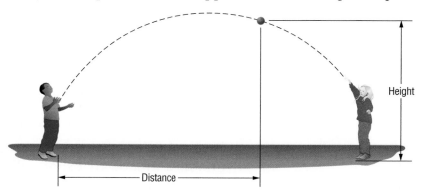

This means that to approximate the relationship between the object's height and the horizontal distance it travels, you can use a quadratic equation. As you will see in Exercises 1–5, there is also a quadratic relationship between the length of time the object is in the air and the object's height.

✓ Develop & Understand: A

A photographer set up a camera to take pictures of an arrow that had been shot into the air. The camera took a picture every half second. The points on the graph show the height of the arrow in meters each half second after the camera started taking photos. A parabola has been drawn through the points.

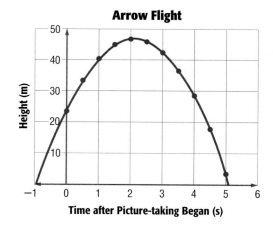

1. Consider the arrow's height when the camera took the first picture.

   a. How high was the arrow at that time? Explain how you know.

   b. How much time elapsed from that point until the arrow returned to that height? Explain how you know.

2. How long after the first photo was taken did the arrow hit the ground? How did you determine this?

3. The arrow was shot from a height of 1.5 m, just above shoulder height. About how long before the time of the first photo was the arrow shot? How did you determine this?

4. Approximately how long was the arrow in flight? Explain how you found your answer.

5. Now think about how high the arrow rose.

   a. What was the arrow's greatest height? Explain how you found your answer.

   b. How many seconds after the arrow was shot did it reach its maximum height? Explain how you found your answer.

✅ Develop & Understand: B

A quarterback threw a football in such a way that the relationship between its height y and the horizontal distance it traveled x could be described by the equation below. Both x and y are measured in yards.

$$y = 2 + 0.8x - 0.02x^2$$

6. Will the graph of this quadratic equation open upward or downward? Explain.

7. Use your calculator to graph the equation. Choose a window that gives a view of the graph for the entire time the ball was in the air. You can change the window settings by adjusting the minimum and maximum x and y values for the axes and the x and y scales. Try an x scale of 10 and a y scale of 2. Sketch the graph, and label the axes.

For Exercises 8–10, use your calculator graph or the equation $y = 2 + 0.8x - 0.02x^2$ to answer the question.

8. From what height was the football thrown? Explain how you found your answer.

9. What was the greatest height reached by the ball? Explain how you know.

10. A receiver caught the ball in the end zone at the same height it was thrown. What distance did the pass cover before it was caught? How did you determine this?

Real-World Link

The path of any projectile, such as a tennis ball hit into the air or a flare launched from a boat, follows the shape of a parabola.

. . . .

Share & Summarize

Luther hit a tennis ball into the air. The graph shows the ball's height *h* in feet over time *t* in seconds. The point *P* shows the ball's height at $t = 1$.

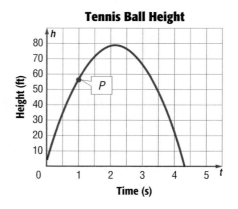

Tennis Ball Height

1. Explain how to use the graph to estimate when the ball reached its highest point.

2. Explain how to use the graph to find another time when the ball had the same height it had at $t = 1$.

Investigation 4 Compare Quadratics and Other Relationships

Vocabulary

cubic equation

Materials

- graphing calculator

The following exercises will help you distinguish between quadratic relationships and other types of relationships.

✅ Develop & Understand: A

One way to distinguish different types of relationships is to examine their equations.

Tell whether each equation is quadratic. If an equation is not quadratic, explain how you know.

1. $y = 3m^2 + 2m + 7$ **2.** $y = (x + 3)^2 + 7$

3. $y = (x - 2)^2$ **4.** $y = 10$

5. $y = \dfrac{2}{x^2}$ **6.** $y = b(b + 1)$

7. $y = 2^x$ **8.** $y = 3x^3 + 2x^2 + 3$

9. $y = -2.5x^2$ **10.** $y = 4n^2 - 7n$

11. $y = 7p + 3$ **12.** $y = n(n^2 - 3)$

Some of the equations in Exercises 1–12 are *cubic* equations. A **cubic equation** can be written in the form $y = ax^3 + bx^2 + cx + d$, where $a \neq 0$. These are all examples of cubic equations.

$$y = x^3 \qquad\qquad y = 2x^3$$

$$y = 0.5x^3 - x + 3 \qquad\qquad y = x^3 + 2x^2$$

The graphs of cubic equations have their own characteristics, different from those of linear and quadratic equations.

Real-World Link

Cubic equations are used by computer design programs to help draw curved lines and surfaces.

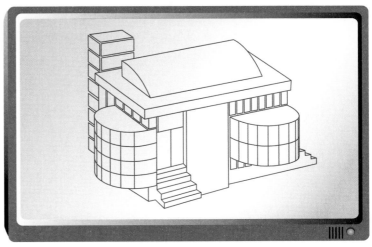

✅ Develop & Understand: B

13. Consider these three equations.

$$y = x \qquad y = x^2 \qquad y = x^3$$

a. Graph the three equations in the same window of your calculator. Choose a window that shows all three graphs clearly. Sketch the graphs. Remember to label the axes and the graphs.

b. Write a sentence or two about how the graphs are the same and how they are different.

c. Give the coordinates of the two points where all three graphs intersect.

Math Link

The words *linear, quadratic,* and *cubic* come from the idea that *lines, squares,* which are also called *quadrangles,* and *cubes* have 1, 2, and 3 dimensions, respectively.

You have seen how the values of m and b affect the graph of $y = mx + b$ and how the values of a, b, and c affect the graph of $y = ax^2 + bx + c$.

You will now consider simple cubic equations of the form $y = ax^3 + d$ to see how changing the coefficient a and the constant d affect the graphs.

14. Complete Parts a–c for each of these three groups of equations.

Group I	**Group II**	**Group III**
$y = x^3$	$y = 2x^3$	$y = 3x^3 + 1$
$y = x^3 + 3$	$y = \frac{1}{2}x^3$	$y = 3x^3 - 1$
$y = x^3 - 3$	$y = -2x^3$	$y = -3x^3 - 1$

a. Graph the three equations in the same window of your calculator. Choose a window that shows all three graphs clearly. Sketch and label the graphs.

b. Write a sentence or two about how the graphs are the same and how they are different.

c. What does the coefficient of the x^3 term seem to tell you about the graph?

d. What does the constant tell you about the graph?

✅ *Develop & Understand: C*

Now consider what happens when a cubic equation has x^2 and x terms. Consider these three equations.

$$y = x^3 - x$$

$$y = -x^3 + 2x^2 + 5x - 6$$

$$y = 2x^3 - x^2 - 5x - 2$$

15. Graph the equations in the same window of your calculator. Sketch the graphs. Be sure to label the graphs and the axes.

16. Describe in words the general shapes of the graphs.

17. For each graph, find the points where the graph crosses the x-axis and y-axis.

18. Suppose you moved the graph of $y = x^3 - x$ up one unit.

   a. What is the equation of the new graph?

   b. Graph your new equation. How many x-intercepts does the graph have?

Share & Summarize

Look back at the graphs you made for this investigation. Describe what you observe about how graphs of linear, quadratic, and cubic relationships differ.

Real-World Link

Cubic equations are often involved in the design of the curved sections of sailboats.

Inquiry

Investigation **5** Graph Design Puzzles

Materials

• graphing calculator

These simple designs are made from the graphs of quadratic equations.

Design A

Design B

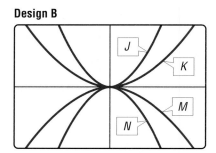

In this design, the parabolas are equally spaced and have their vertices on the *y*-axis.

Here each parabola has a vertex at the origin. Parabolas J and N are the same width. Parabolas K and M have the same width.

Design C

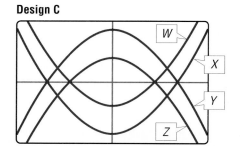

Design D

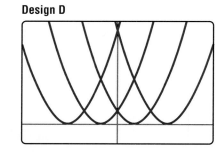

In this design, Parabolas W and Z are the same width, and their vertices are the same distance from the origin. Parabolas X and Y are the same width, and their vertices are the same distance from the origin.

In this design, the vertices of the four parabolas are equally spaced along the *x*-axis. Two of those points are to the left of the origin, and two are to the right.

Try It Out

With your group, choose Design A, B, or C. Try to create the design on your calculator. Use equations in the form $y = ax^2 + bx + c$, and experiment with different values of *a*, *b*, and *c*. You may need to adjust the viewing window to make the design look the way you want.

1. When you have created the design, make a sketch of the graph. Label each curve with its equation. Also label the axes, including the maximum and minimum values on each axis.

2. Different sets of equations and window settings can give the same design. Compare your results for Question 1 with the other members of your group or with other groups who chose your design. Did you record the same equations and window settings?

Try It Again

3. Now create each of the other three designs. For each design, make a sketch. Record the equations and window settings that you used. You may find Design D to be a challenge.

Take It Further

4. Work with your group to create a new design from the graphs of four quadratic equations.

   a. Make a sketch of your design. *On a separate sheet of paper,* record the equations and window settings that you used.

   b. Exchange designs with another group. Try to re-create the design.

What Did You Learn?

5. Write a report about the strategies that you used to re-create the designs. For each design, discuss the following points.

   a. How did you change the coefficients, *a* or *b*, or the constant, *c*, to create each design?

   b. Did any of the variables have a value of 0? If so, why was having that value equal to 0 necessary to create the design?

   c. Did you change the window settings to make any of the designs? If so, explain how changing either the range of *x* values or the range of *y* values affects the design.

Real-World Link

The design of satellite dishes is based on the parabola.

Practice & Apply

1. For each table of values, find an equation it may represent. Look for connections between the tables that may help you determine the equations. Explain how your found your solutions.

a.

x	y
−3	9
−2	4
−1	1
0	0
1	1
2	4
3	9

b.

x	y
−3	25
−2	16
−1	9
0	4
1	1
2	0
3	1

c.

x	y
−3	0
−2	1
−1	4
0	9
1	16
2	25
3	36

d.

x	y
−3	−18
−2	−8
−1	−2
0	0
1	−2
2	−8
3	−18

e.

x	y
−3	−6
−2	−4
−1	−2
0	0
1	2
2	4
3	6

f.

x	y
−3	−8
−2	2
−1	8
0	10
1	8
2	2
3	−8

In Exercises 2–5, match the equation with one of the graphs below. Explain your reasoning.

2. $y = x^2$

3. $y = (x - 2)^2$

4. $y = (x + 3)^2$

5. $y = -2x^2$

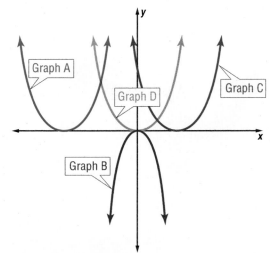

In Exercises 6 and 7, match the equation with one of the graphs below. Explain your reasoning.

6. $y = -2x^2 - 2x + 3$

7. $y = 2x^2 - 2x + 3$

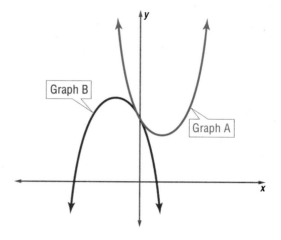

8. Could this be the graph of the equation $y = -x^2 + 1$? Explain.

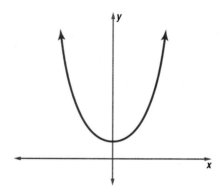

9. Could this be the graph of the equation $y = -x^2 - 1$? Explain.

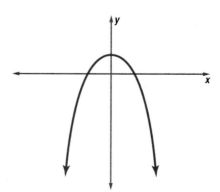

10. Could this be the graph of the equation $y = x^2 + 1$? Explain.

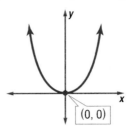

$(0, 0)$

11. Graph A is the graph of $y = x^2$.

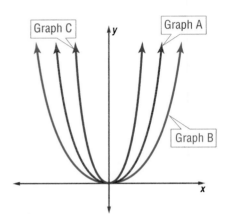

a. Graph B is the graph of either $y = 2x^2$ or $y = \dfrac{x^2}{2}$. Which equation is correct? Explain how you know.

b. Graph C is the graph of either $y = 3x^2$ or $y = \dfrac{x^2}{3}$. Which equation is correct? Explain how you know.

12. Consider these three graphs.

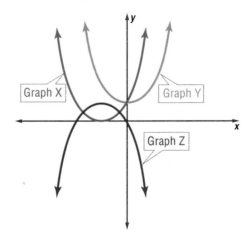

a. Could Graph X be the graph of $y = x^2 + 1$? Explain.

b. Could Graph Y be the graph of $y = (x + 1)^2$? Explain.

c. Could Graph Z be the graph of $y = -x^2 + 1$? Explain.

13. A place kicker on a football team attempted three field goals during practice. All three were kicked from the opponent's 40-yard line, which is 50 yards from the goalpost. For a field goal to-count, it must clear the crossbar, which is 10 feet high.

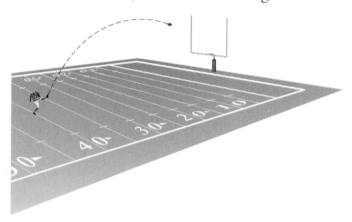

The football followed a different path through the air for each kick. These equations give the height of the kick in feet h for any distance from the kicker in yards d. Each kick was aimed directly at the center of the goalpost.

Kick 1: $h = 3.56d - 0.079d^2$

Kick 2: $h = 1.4d - 0.0246d^2$

Kick 3: $h = 2d - 0.033d^2$

a. For Kick 1, plot enough points to make a smooth curve. Check your graph by graphing Kick 1 using your calculator. Then graph Kicks 2 and 3 using your calculator. For window settings, choose x-values to represent the distance from the kicker in yards and y-values to represent the ball's height in feet.

b. Use your graphs to estimate the maximum height of each kick.

c. Use your graphs to estimate how many yards each kick traveled over the field before it struck the ground.

d. To make a field goal, the football must cross over the goalpost. That means it must be at least 10 feet high when it reaches the post, which is 50 yards from the kicker.

Use your graphs to estimate whether any of the kicks could have scored a field goal. Explain your reasoning.

Math Link

For Parts b-d, use the calculator's Trace feature. For Part d, press Trace, key in "50," and press enter to locate the point where $x = 50$. Also, try graphing $y = 10$ for a visual representation of the cross bar's height.

Tell whether each relationship is quadratic.

14. $y = 3x + 5$

15. $y = -(x + 1)^2$

16. $y = \dfrac{5}{x}$

17. $y = (x - 1)^2$

18. $y = -(x + 1)$

19. $y = \dfrac{5}{x^2}$

20. $y = (3x + 5)^2$

21. $y = 7x^3 + 3x^2 + 2$

22. $y = x^3 + 7x + 5$

23. $y = 2^x$

24. Consider these three equations.

$$y = x + 3 \qquad y = x^2 + 3 \qquad y = x^3 + 3$$

a. For each equation, draw a rough sketch showing the general shape of the graph. Put all three graphs on one set of axes.

b. Write a sentence or two about how the graphs are the same and how they are different.

25. Graph A is the graph of $y = x^3$.

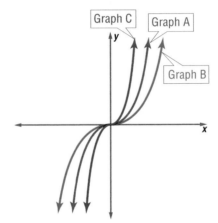

a. Graph B is the graph of either $y = \dfrac{x^3}{2}$ or $y = 2x^3$. Which equation is correct? Explain.

b. Graph C is the graph of either $y = \dfrac{x^3}{3}$ or $y = 3x^3$. Which equation is correct? Explain.

Connect & Extend

26. The quadratic equations below are more complicated than those with which you worked in Investigation 1. Just as with the simpler equations, you can make a table of values and plot a graph of the equation.

$$y = 3m^2 + 2m + 7$$

$$p = n^2 + n - 6$$

$$s = 2t^2 - 3t + 1$$

Here is a table of values and a graph for $y = 3m^2 + 2m + 7$.

m	−3	−2.5	−2	−1	−0.5	0	0.5	1	2	2.5
y	28	20.75	15	8	6.75	7	8.75	12	23	30.75

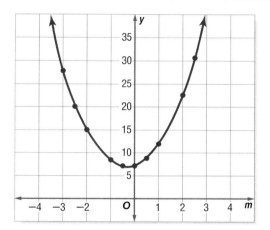

a. Prepare a table of ordered pairs for $p = n^2 + n - 6$. Plot the points on graph paper. Draw a smooth curve through them.

b. Prepare a table of ordered pairs for $s = 2t^2 - 3t + 1$. Plot the points on graph paper. Draw a smooth curve through them.

c. How are the graphs for the three equations alike?

d. How do the three graphs differ? In particular, where are their lowest points, and where are their lines of symmetry?

27. In Your Own Words Explain how the values of a and c affect the graph of the quadratic equation $y = ax^2 + c$.

28. Each of these four graphs represents either $y = x^2 + 1$ or $y = 2x^2 + 1$.

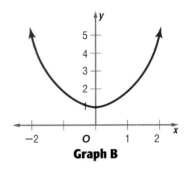

Graph A   **Graph B**

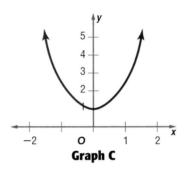

Graph C   **Graph D**

a. Match each graph to its equation. Explain your reasoning.

b. How can graphs that look different have the same equation?

29. Challenge Consider the equation $y = x^2 - 4x$.

a. Identify the values of a, b, and c in this quadratic equation.

b. Where does the graph of $y = x^2 - 4x$ cross the y-axis? How does this point relate to the value of c in the equation?

c. Graph $y = x^2 - 4x$ by making a table and plotting points. Make sure your graph shows both halves of the parabola.

d. Give the coordinates of the points where the graph crosses the x-axis.

e. Use the distributive property to write $x^2 - 4x$ as a product of two factors.

f. How do the points where the graph crosses the x-axis relate to this factored form?

Real-World Link

In May 1931, Auguste Piccard of Switzerland became the first person to reach the stratosphere when he ballooned to almost 52,000 feet. In October 1934, Jeanette Piccard became the first woman to reach the stratosphere when she and her husband (Auguste's twin brother Jean) ballooned to almost 58,000 feet.

· · · · · · · · · · · · · · · · · ·

30. Passengers in a hot air balloon can see greater and greater distances as the balloon rises. The table shows data relating the height of a hot air balloon with the distance the passengers can see, or the distance to the horizon.

Study the table to observe what happens to *d* as *h* increases by equal amounts.

Height (meters), *h*	Distance to Horizon (kilometers), *d*
0	5
10	11
20	16
30	20
40	23
50	25

a. Below are three graphs. Which is most likely to fit the data in the table? Explain.

In this graph, *d* increases by constant amounts as *h* increases.

In this graph, *d* increases by greater amounts as *h* increases.

In this graph, *d* increases by smaller amounts as *h* increases.

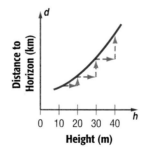

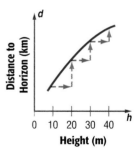

b. Check your answer to Part a by sketching a graph that represents the relationship between *d* and *h*.

c. Look at *d* in relation to *h*. Do you think these data could represent a quadratic relationship? Explain.

31. Consider this pattern of cube figures.

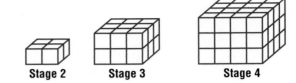

Stage 1   Stage 2   Stage 3   Stage 4

a. Copy and complete the table for the pattern.

Stage, n	1	2	3	4
Cubes, C				

b. How many cubes will be needed to build stage 5?

c. Write an equation for the number of cubes in stage n. Explain how you found your answer.

Real-World Link

On March 20, 1999, Bertrand Piccard (Auguste Piccard's grandson) and Brian Jones became the first aviators to circle Earth nonstop in a hot air balloon. Traveling the 42,810 kilometers took them 19 days, 1 hour, 49 minutes.

32. Sometimes people confuse the *quadratic* relationship $y = x^2$ with the *exponential* relationship $y = 2^x$.

a. Copy and complete the table of values for $y = x^2$ and $y = 2^x$.

x	-4	-3	-2	-1	0	1	2	3	4	5	6
$y = x^2$	16	9	4	1	0	1					
$y = 2^x$	$\frac{1}{16}$	$\frac{1}{8}$	$\frac{1}{4}$	$\frac{1}{2}$	1	2					

b. For which values of x shown are x^2 and 2^x equal?

c. For which values of x shown is x^2 greater than 2^x?

d. For which values of x shown is 2^x greater than x^2?

e. Compare the way the values of $y = x^2$ change as x increases by 1 to the way the values of $y = 2^x$ change as x increases by 1.

f. How do you think x^2 and 2^x compare for values of x greater than 6?

g. Use the table of values to graph $y = x^2$ and $y = 2^x$ on the same set of axes.

33. Preview Graph the following equations. Be sure to use positive and negative values for *x*.

How are the graphs similar? How are they different?

$$xy = 4 \qquad xy = -4$$

$$xy = 6 \qquad xy = -6$$

$$xy = 10 \qquad xy = -10$$

Mixed Review

34. Match each equation to a graph.

a. $x - \dfrac{3}{5} = 2y$ **b.** $3y - 4x - 1 = 8x - y$ **c.** $x + 5 = 4(y + 2)$

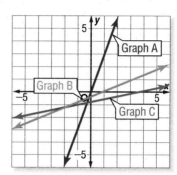

35. The coordinates of point *P* are (3, 5). Point *P* is translated 3 units up and then reflected about the *x*-axis. What are the coordinates of point *P'*?

36. The coordinates of point *P* are (−4, −6). Point *P* is translated 4 units down and reflected about the y-axis. What are the coordinates of *P'*?

For Exercises 37–39, give three values that satisfy each inequality. Then, graph each inequality on a number line.

37. $4 < k < 10$

38. $-3 \leq n < 7$

39. $-2 \leq t \leq 5$

LESSON 8.4

Inverse Variation

In this lesson, you will study a new type of relationship in which the product of two quantities is always the same. As with quadratics, you will discover some of the characteristics of this new type of relationship.

- graphs with a distinctive shape
- tables that show a recognizable pattern
- equations of a particular form

Explore

A group of students volunteered to paint a fence at a local park.

The more volunteers we get, the sooner we'll finish the job.

Assume all the volunteers work at the same rate. If you created a graph with the number of volunteers on the horizontal axis and the time needed to paint the fence on the vertical axis, what would the graph look like?

Discuss this question in your group and then sketch the graph. Just think about the general shape of the graph. Do not worry about exact numbers.

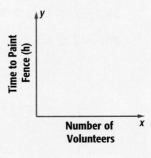

Now discuss the following questions with your class.

- Will the graph increase or decrease from left to right? Why?

- Will the graph be a straight line? Why or why not?

- Will the graph intersect the axes? Why or why not?

- Suppose one student starts at 9:00 A.M. and does not take a break. This student can finish the job by 5:00 P.M., and it would make sense to put the point (1, 8) on the graph. The point (2, 4) would also be on the graph. Explain what this point represents and how you know it would be on the graph.

- What would the point (4, 2) represent?

- Which of the following points might reasonably be on the graph that includes (1, 8), (2, 4), and (4, 2)?

$$(5, 1.6) \qquad (8, 1) \qquad (80, 0.1)$$

Explain your answers.

Suppose one volunteer can complete the job in 8 hours, two can finish in 4 hours, 3 in $2\frac{2}{3}$ hours, and so on. In each case, the number of *person-hours*, the number of people multiplied by the number of hours, is 8. So, one way to express the relationship between the time it takes to finish the job T and the number of volunteers V is

$$TV = 8.$$

In this lesson, you will explore several other situations in which the product of two variables is constant.

Investigation When *xy* Is Constant

Materials

• graph paper

As you work on these exercises, look for similarities in the equations, graphs, and tables.

✓ Develop & Understand: A

Ms. Fuentes is considering renting a house that has a large rectangular backyard. She wants to figure out if there will be room for her children's play equipment. The owner told her, "The backyard has an area of 2,000 square feet." Ms. Fuentes thought about what the owner said and tried to imagine the actual dimensions of the yard.

1. Use *x* to represent length, in feet, of one side of the yard and *y* to represent length, in feet, of an adjacent side. Select some possible values for *x* and *y* and complete a table like this one.

x	50	25	100				
y	40						
Area, *xy*	2,000						

2. Write an equation that shows how *x* and *y* are related.

3. Make a graph of the relationship between *x* and *y*. Start by plotting the points in your table. Add more points, if needed, until you can clearly see the shape of the graph. Draw a smooth curve through the points.

4. Think about how *y* changes as *x* changes for this situation.

   a. Copy and complete the table to show what happens.

x	20	30	40	50	60	70	80	90	100
y	100	66.7							
Decrease in *y*	—	33.3							

   b. As *x* increases by a fixed amount, such as 10, will *y* decrease by a fixed amount, an increasing amount, or a decreasing amount?

   c. If *x* doubles, what happens to *y*? If *x* triples, what happens to *y*? What happens to *y* if *x* is multiplied by *N*?

   d. As *x* gets very small and nears 0, what happens to *y*? As *x* grows very large, what happens to *y*?

✅ Develop & Understand: B

The owner of Ms. Fuentes' rented house decided to have the overgrown yard manicured. He contacted Rob, a high school student, and agreed to pay him $240 for the job.

If Rob works diligently, the job will be done quickly, and the hourly rate of pay will be high. If he works at a more leisurely pace, the job will take longer, and the hourly rate of pay will be lower.

5. Copy and complete the table.

Hours of Work	$\frac{1}{2}$		2		10	20	
Hourly Pay Rate		$240		$48			$8

6. Write an equation for this situation. Use h for number of hours to complete the job and d for number of dollars earned per hour.

7. Sketch a smooth graph by plotting enough number pairs (h, d) to see a pattern.

In all of the situations you have seen so far, the variables made sense for positive values only. The dimensions of Ms. Fuentes' yard, Rob's hourly pay rate, and the number of hours Rob works must all be greater than 0.

In the following exercises, you will look at another relationship in which a product of variables is constant. This time, you will consider both positive and negative values of the variables.

✅ Develop & Understand: C

Consider the relationship $xy = 3$, where x and y can be positive or negative.

8. If you start with values of x, it is often easier to find values of y by rewriting the equation so y is alone on one side. This is sometimes referred to as solving for y in terms of x. Solve the equation $xy = 3$ for y in terms of x.

9. Find values of y for five negative values and five positive values of x between -10 and 10. Record the (x, y) values in a table. Try some values that are not integers, including values between 0 and 1.

10. What is y when $x = 0$? What happens if you use your calculator to find y when $x = 0$?

11. Make a smooth graph of $xy = 3$ by first plotting the pairs in your table. Add more points as necessary to get a better sense of the graph's shape.

Think about what happens to the graph near the y-axis. Can the graph cross the y-axis?

12. How are this graph and equation like the yard work graph and equation from Exercises 5–7? How are they different?

> ### Share & Summarize
>
> **1.** How are the equations in Exercises 1–4, Exercises 5–7, and Exercises 8–10 alike? How are they different?
>
> **2.** How are the graphs alike? How are they different?

Investigation 2 Inverse Proportion

Vocabulary

hyperbola

inversely proportional

inverse variation

reciprocal relationship

Materials

• graph paper

You have been working with relationships in which multiplying two variables gives a constant product. That is,

$$xy = a$$

where x and y are the variables and a is a nonzero constant.

The graph of such a relationship is a curve like the one shown below.

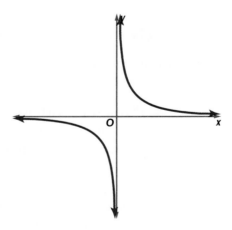

This type of curve is called a **hyperbola**.

Think about the three relationships you worked with in Investigation 1.

$$xy = 2{,}000 \qquad dh = 240 \qquad xy = 3$$

In each case, what happens to the value of one variable if you double the other variable?

What happens if you triple the value of one variable? What happens if you halve the value of one variable?

In Chapter 1, you learned that two variables are *directly proportional* if the values of the variables have a constant *ratio*. When two variables have a constant nonzero *product,* they are said to be **inversely proportional**. A relationship in which two variables are inversely proportional is called an **inverse variation**.

You will now explore the connection between inverse proportions and *reciprocals*.

✅ Develop & Understand: A

Math Link

The *reciprocal* of a number is what the number is multiplied by to get 1. For example, the reciprocal of 7 is $\frac{1}{7}$, since $7 \cdot \frac{1}{7} = 1$.

1. Find the reciprocal of each number. Express your answers as integers or fractions, not as mixed numbers or decimals.

a. 2

b. 5

c. -1

d. $\frac{1}{3}$

e. $\frac{3}{4}$

f. $-\frac{9}{7}$

g. $\frac{7}{3}$

h. -10

2. Look at your answers to Exercise 1.

a. Find the reciprocal of each answer.

b. What do you notice?

3. If x is not 0, what is the reciprocal of x?

4. What is the reciprocal of $\frac{1}{x}$?

5. What is the reciprocal of the reciprocal of x?

6. Find the reciprocal of each number. Express your answers as decimals.

   a. 8

   b. 100

   c. 0.2

   d. -0.25

   e. 12

   f. 7.5

   g. -1

   h. 0.0004

7. Look at your answers to Exercise 6.

   a. Find the reciprocal of each answer.

   b. What do you notice?

8. Consider the relationship $y = \frac{1}{x}$.

   a. What happens to y if you double x? If you triple x? If you quadruple x? If you halve x?

   b. Is y inversely proportional to x? In other words, is $y = \frac{1}{x}$ an inverse variation? Explain.

   c. Sketch a graph of $y = \frac{1}{x}$.

✅ Develop & Understand: B

Consider the relationship $y = \frac{5}{x}$.

9. What happens to y if you double x? If you triple x? If you quadruple x? If you halve x?

10. Is y inversely proportional to x? Explain.

11. What is the value of $y = \frac{5}{x}$ when $x = 0$? What happens if you calculate $\frac{5}{0}$ on your calculator?

12. Copy and complete the table. Write the entries as integers or fractions.

x	-5	-4	-3	-2	-1	$-\frac{1}{2}$	$-\frac{1}{4}$	0	$\frac{1}{4}$	$\frac{1}{2}$	1	2	3	4	5
$y = \frac{1}{x}$															
$y = \frac{5}{x}$															

13. Sketch a graph of $y = \frac{5}{x}$ on the same axes you used for the graph of $y = \frac{1}{x}$.

The relationships $y = \frac{1}{x}$ and $y = \frac{5}{x}$ are inverse variations. Notice that when the value of one variable is multiplied by a number, the value of the other variable is multiplied by the *reciprocal* of that number.

For example, for $y = \frac{5}{x}$, when $x = 1, y = 5$. But when you multiply x by 3, y becomes $\frac{5}{3}$, or $5 \cdot \frac{1}{3}$. For this reason, inverse variations are sometimes called **reciprocal relationships**.

✅ Develop & Understand: C

Tell whether each equation represents a reciprocal relationship. If the answer is no, tell what type of relationship the equation *does* represent.

14. $st = 6$

15. $s = 2t^2$

16. $s = 6t$

17. $y = \frac{x}{7}$

18. $y = \frac{7}{x}$

19. $x = \frac{2}{y}$

For Exercises 20–22, complete Parts a and b.

a. Tell which kind of relationship, linear, quadratic, or reciprocal, could exist between the variables.

b. Write an equation that relates the quantities.

20.

x	0.5	1	2	3	4	5	10
y	−2.5	−5	−10	−15	−20	−25	−50

21.

p	0.5	1	2	3	4	5	10
q	1.25	2	5	10	17	26	101

22.

t	0.5	1	2	3	4	5	10
r	3	1.5	0.75	0.5	0.375	0.3	0.15

Share & Summarize

1. If you are given a table of values for two variables, how can you tell whether the relationship between the variables could be an inverse variation?

2. If you are given an equation with two variables, how can you tell whether the relationship between the variables is an inverse variation?

Investigation ③ Relationships with Constant Products

Materials

• graphing calculator

You have been studying relationships in which the product of two quantities is a positive constant. All of these relationships can be written using equations in the form $xy = a$ or, equivalently, $y = \frac{a}{x}$, where a is a positive number. A group of equations of the same general form is sometimes called a *family* of equations.

Now you will explore other families of equations that describe relationships in which a product is constant. You will look at equations in the forms shown below.

• $xy = a$ or, equivalently, $y = \frac{a}{x}$, where a is a negative constant

• $(x + b)y = 1$ or, equivalently, $y = \frac{1}{x + b}$, where b is a constant

• $x(y - c) = 1$ or, equivalently, $y = \frac{1}{x} + c$, where c is a constant

Real-World Link
This type of relationship is used in the design of telescopic lenses, which collect and focus light in telescopes.

Think  Discuss

You have already studied the relationship $y = \frac{1}{x}$. Now think about $y = -\frac{1}{x}$. The table and graph below show this relationship.

x	−0.2	0.5	1	2	5
y	5	−2	−1	−0.5	−0.2

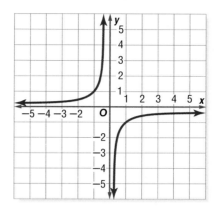

Think about parts of the graph that are not shown.

What is the y value for $x = 10$? For $x = 100$? For $x = 1,000$?

What happens as the x values grow even greater than 1,000? Does y ever equal 0?

What happens to the y values as positive x values get closer to 0? What happens when $x = 0$? What does this mean for the graph of $y = -\frac{1}{x}$?

What is the y value for $x = -10$? For $x = -100$? For $x = -1,000$?

What happens as the x values decrease beyond −1,000? Does y ever equal 0?

What happens to the y values as negative x values grow closer to 0? What happens when $x = 0$?

Math Link

A number's *absolute value* is its distance from 0 on the number line. The absolute value of -4 is 4. The absolute value of -4 is written as $|-4|$.

As you probably discovered for $y = \frac{1}{x}$ and $y = -\frac{1}{x}$, as the absolute value of x grows larger and larger, y approaches 0 without ever reaching it.

x	1	10	100	1,000	10,000	100,000	. . .
$\frac{1}{x}$	1	0.1	0.01	0.001	0.0001	0.00001	. . .

x	-1	-10	-100	$-1,000$	$-10,000$	$-100,000$	. . .
$\frac{1}{x}$	-1	-0.1	-0.01	-0.001	-0.0001	-0.00001	. . .

In the same way, as the absolute value of y grows larger and larger, x approaches 0 without ever reaching it. This is a characteristic of all inverse variations. There is always some value, not always 0, that x approaches but does not reach as the absolute value of y becomes very large. And, there is some value that y approaches but does not reach as the absolute value of x becomes very large.

✅ Develop & Understand: A

Work with your group to explore one of the families of equations described below. The variables a, b, and c are constants.

Family A
Equations of the form $y = \frac{a}{x}$, such as $y = \frac{3}{x}$ and $y = -\frac{7}{x}$

Family B
Equations of the form $y = \frac{1}{x + b}$, such as $y = \frac{1}{x + 2}$ and $y = \frac{1}{x - 1}$

Family C
Equations of the form $y = \frac{1}{x} + c$, such as $y = \frac{1}{x} + 5$ and $y = \frac{1}{x} - 2$

Your goal is to understand what the graphs of relationships in the family you choose look like and how the graphs change as a, b, or c changes. Your group should prepare a report describing its findings, including the following.

- a series of sketches, carefully drawn and labeled to show how the graphs change as the constant changes

- a written explanation of your findings, including answers to Exercises 1–6

1. How does the graph change as the constant changes?

2. How does changing the constant from positive to negative affect the graph?

3. What value does x approach as the absolute value of y grows very large? Does this change as the constant changes?

4. What value does y approach as the absolute value of x grows very large? Does this change as the constant changes?

5. For what value or values of x is there no value of y? Does this change as the constant changes?

6. How do the graphs compare to the graph of $y = \frac{1}{x}$?

> ## Share & Summarize
>
> **1.** Compare how the value of a affects graphs of equations of the forms $y = ax^2$ and $y = \frac{a}{x}$.
>
> **2.** Compare how the value of b affects graphs of equations of the forms $y = (x + b)^2$ and $y = \frac{1}{x + b}$.
>
> **3.** Compare how the value of c affects graphs of equations of the forms $y = x^2 + c$ and $y = \frac{1}{x} + c$.

Real-World Link

Equations involving constant products are used to describe the path that some comets take through the sky.

Halley's Comet

Practice & **Apply**

1. Trent gave his little sister Tia $3 to spend on stickers. The price per sticker varies depending on the type of sticker. Tia wants all of the stickers she buys to be the same type.

   a. Make a table of possible values for the price per sticker p and the number of stickers n. One pair of values is shown as an example.

Price per Sticker, p	$0.25						
Number Purchased, n	12						

   b. Make a graph of the relationship between p and n. Put price p on the horizontal axis. Start by plotting the points in your table. Add points as needed until you can clearly see the shape of the graph. Draw a smooth curve through the points.

   c. What happens to the value of n as the value of p doubles? As the value of p triples? As the value of p quadruples?

   d. What happens to the value of n as the value of p is halved? As the value of p is quartered?

   e. Write an equation for the relationship between p and n.

2. Orlando plans to serve a giant pizza at his birthday party. The amount of pizza each guest will be served depends on the number of guests.

   a. Copy and complete the table showing the fraction of the pizza each person will be served f for different numbers of people n. Assume each person gets the same amount of pizza and that the entire pizza is served.

n	1	2	3	4	5	6	7	8	9	10	11	12
f												

   b. Graph the relationship between n and f by plotting the points in your table and drawing a smooth curve through them.

   c. Write an equation for the relationship between n and f.

3. Consider the equation $xy = 10$, where x and y can be positive or negative.

    a. Solve for y in terms of x.

    b. Make a table of values for the equation. Choose x values between -10 and 10. Consider positive x values, negative x values, and x values between -1 and 1.

    c. What is the value of y when x is 0? Explain.

    d. Make a graph of this relationship by first plotting the points from your table. Add points as needed until you can see the shape. Then connect the points with a smooth curve. Think carefully about what happens to the graph near the y-axis.

In Exercises 4–9, decide whether the relationship is an inverse variation. If it is not, tell what type of relationship it is.

4. $st = \dfrac{1}{4}$ **5.** $s = \dfrac{6t}{8}$

6. $s = \dfrac{r^2}{4}$ **7.** $y = 0.25x$

8. $y = \dfrac{0.25}{x}$ **9.** $x = \dfrac{0.25}{y}$

For Exercises 10–12, do Parts a–c.

    a. Tell whether the variables could be inversely proportional. Explain how you know.

    b. Plot the points on a pair of axes. Put the variable in the top row on the horizontal axis.

    c. Write an equation for the relationship between the variables.

10.

m	0.5	1	2	3	4	5	10
n	60	30	15	10	7.5	6	3

11.

x	0.5	1	2	3	4	5	10
y	0.25	0.5	1	1.5	2	2.5	5

12.

t	0.25	0.5	0.75	1	1.26	1.5	2
r	0.5	0.25	0.1667	0.125	0.1	0.0833	0.0625

13. **Economics** Antoine wants to spend $4 on kiwi fruit at the farmer's market. The price per kiwi varies from stand to stand.

   a. Is the relationship between the price per kiwi in dollars p and the number of kiwi Antoine can purchase n an inverse variation? Explain.

   b. Write an equation for the relationship between p and n.

14. Mangos are very expensive at the farmer's market. They are $4 each, but Carmen cannot resist them.

   a. Is the relationship between the total price Carmen pays for mangos p and the number of mangos she buys n an inverse variation? Explain.

   b. Write an equation for the relationship between p and n.

15. Michelle spent $8 at the farmer's market. She bought mandarin oranges, which are $0.40 each, and nectarines, which are $0.80 each.

   a. Is the relationship between the number of mandarin oranges m and the number of nectarines n an inverse variation? Explain.

   b. Write an equation for the relationship between m and n.

16. Consider equations of the form $y = \frac{a}{x}$.

   a. On one set of axes, make rough sketches of the graphs for the three equations below. Use x and y values from −10 to 10.

   i. $y = -\frac{1}{x}$ **ii.** $y = -\frac{5}{x}$ **iii.** $y = -\frac{10}{x}$

   b. Describe how the graphs of $y = \frac{a}{x}$ change as a decreases.

17. Consider equations of the form $y = \frac{a}{x + b}$.

   a. On one set of axes, make rough sketches of the graphs for the three equations below. Use x and y values from −10 to 10.

   i. $y = \frac{1}{x + 1}$ **ii.** $y = \frac{1}{x + 3}$ **iii.** $y = \frac{1}{x + 5}$

   b. Describe how the graphs of $y = \frac{a}{x + b}$ change as b increases.

18. Consider equations of the form $y = \frac{a}{x} + c$.

   a. On one set of axes, make rough sketches of the graphs for the three equations below. Use x and y values from −10 to 10.

   i. $y = \frac{1}{x} + 2$ **ii.** $y = \frac{1}{x} + 4$ **iii.** $y = \frac{1}{x} + 6$

   b. Describe how the graphs change as the value of c increases.

Connect **&** *Extend*   **19. Physical Science** The time it takes to reach a destination depends on the speed of travel. Suppose it is 500 miles to a campsite you want to visit. Here are some ways you might get there.

Mode of Transportation	Average Speed (mph)
Subway Train	30
Car	50
Helicopter	85
Light Plane	
Rocket	1,100
Horseback	
Bicycle	

a. Make estimates for the missing speeds in the table. Do not worry about the accuracy of your estimates.

b. Copy and complete the table to show how long it would take to travel 500 miles using the estimated speed for each type of transportation.

Speed (mph)	30	50	85	1,100			
Travel Time (h)							

c. Write an equation stating the relationship between speed S and time T required to travel 500 miles.

d. Use the values in your table to sketch a smooth graph of the relationship between speed and time. Place speed on the horizontal axis.

e. What happens to the time values as the speed values increase?

f. Could the travel time for the 500-mile journey ever reach 0?

g. The speed of light is 186,000 miles per second. Determine how long the journey would take if you could ride on a light beam. Is this value shown on your graph?

h. Write an equation that gives the speed needed to travel 500 miles in T hours.

20. Compare the inverse relationship $y = \frac{3}{x}$ with the linear relationship $y = 3x$.

    a. Graph and label both equations on one set of axes.

    b. Consider the parts of the graphs in the first quadrant. What happens to each graph as x increases?

    c. Consider the parts of the graphs in the first quadrant. As x grows closer to 0, what happens to each graph?

    d. What happens when $x = 0$?

21. A large pipe organ has pipes with lengths ranging from a few centimeters to 4 meters. The table lists the pipe lengths for the E flats of all octaves, along with the frequencies of the sounds they produce. Higher sound frequencies correspond to higher pitches.

Pipe length (meters), *l*	4.0	2.0	1.0	0.5	0.25	0.125	0.0625
Frequency (cycles per second), *f*	39	78	156	312	622	1,244	2,488

    a. Plot the points, using a suitable scale for the axes. Put pipe length on the horizontal axis. Draw a smooth curve through the points.

    b. Use your graph to predict the pipe length that would produce an A note with a frequency of 440 cycles per second.

    c. Use your graph to predict the frequency produced by a pipe 3 meters long.

    d. What is the shape of the graph? What does this shape suggest about the relationship between frequency and pipe length?

    e. Find an equation that fits the data fairly well. Why do you think the data may not fit exactly?

22. You found that for the equation $y = \frac{1}{x}$, there is no y value corresponding to an x value of 0. We say that the equation $y = \frac{1}{x}$ is *undefined* for $x = 0$. In Parts a–f, list any x values for which the equation is undefined.

    a. $y = \frac{1}{x - 1}$ **b.** $y = \frac{2}{1 - 2x}$

    c. $y = \frac{1}{x + 2}$ **d.** $y = \frac{2}{x}$

    e. $y = \frac{x}{2}$ **f.** $y = \frac{4}{3x - 12}$

23. Consider the equation $y = \dfrac{1}{x^2}$.

   a. Make a table of values and a graph for this equation for x values between -10 and 10. Plot enough points to draw a smooth curve.

   b. How is this graph similar to the graph of $y = \dfrac{1}{x}$?

   c. How is this graph different from the graph of $y = \dfrac{1}{x}$?

24. A game company has spent $200,000 to develop new game software. The company estimates that manufacturing and shipping will be $4 for each unit it makes and sells.

   a. Suppose the company produces only 20 units of the program. What is the average cost of development for each of the 20 units?

   b. What is the average *total* cost, development plus manufacturing and shipping, for each of the 20 units?

   c. Write an expression for the average total cost per unit for n units.

   d. Complete the table for this situation.

Units	20	200	2,000	20,000	200,000	2,000,000
Average Total Cost per unit ($)	10,004					

   e. Suppose the company produces more and more units. The average cost will get closer and closer to a particular amount. What is that amount?

   f. How many units must the company produce for the average total cost to be less than $1 greater than the amount you answered in Part e? How many units are needed for the cost to be less than 1¢ greater?

25. **In Your Own Words** Describe an inverse variation. How can you recognize an inverse variation from a table, a graph, and a written description?

Mixed Review

Solve each equation.

26. $5x - 2 = 3x + 5$

27. $3(2n + 5) = 5n - 4$

28. $\dfrac{t - 6}{t + 2} = 5$

29. $\dfrac{x}{3} + \dfrac{x}{2} = 1$

Tell whether each figure has reflection symmetry, rotation symmetry, or both.

30. 　　**31.** 　　**32.**

33. Georgia brought a package containing 16 identical pieces of clay to school. She wanted to share the clay with some friends during recess.

a. Georgia wants to divide the clay evenly among her friends. If she invites three friends to join her, how many pieces will each friend receive, including Georgia?

b. If Georgia invites five friends to join her, how many pieces will each friend receive, including Georgia?

c. Write a formula giving the number of pieces n each friend will receive, including Georgia, if she invites f friends.

d. Use your expression in Part c to find n when f is 3 and when f is 5. If your answers do not agree with those for Parts a and b, find any mistakes you have made and correct them.

Conjectures

In the first four lessons of this chapter, you learned about quadratic and inverse relationships. This lesson will give you more practice with these types of relationships as you discover the process of making and proving *conjectures*.

A **conjecture** is an educated guess or generalization that has not yet been proven correct. Sometimes you have evidence that leads you to make a conjecture. Other times, you just "get a feeling" about what will happen and make a conjecture based on very little evidence.

Vocabulary

conjecture

Math Link

Consecutive integers follow one another, such as 1 and 2, 5 and 6, or −8 and −9.

> **Explore** ·
>
> Desiree and Yutaka were looking for patterns in the products of pairs of consecutive integers.
>
> $$5 \cdot 6 = 30$$
> $$7 \cdot 8 = 56$$
> $$10 \cdot 11 = 110$$
>
> Yutaka noticed a pattern relating the products to the square of the first number.
>
> Copy and complete the table.
>
> | First Integer | 5 | 7 | 10 | 90 | 0 | −4 | −1 |
> |---|---|---|---|---|---|---|---|
> | Second Integer | 6 | 8 | 11 | | | −3 | |
> | Product | 30 | 56 | 110 | | | | |
> | First Integer Squared | | | | | | | |
>
> Compare the third and fourth rows of your table. Try to make a conjecture about the relationship between the product of two consecutive integers and the square of the first integer. Test your conjecture on a few more examples of your own.
>
> Is there a way to show that your conjecture is true?

Investigation ① Make Conjectures

In this investigation, you will form conjectures about other situations.

✅ Develop & Understand: A

You know that in a table for a linear relationship, as x values increase by 1, the differences in consecutive y values are constant. For example, look at this table for $y = -2x + 10$.

x	1	2	3	4	5	6	7
y	8	6	4	2	0	−2	−4
Differences		−2	−2	−2	−2	−2	−2

Now look at the differences in y values for the quadratic relationship $y = -x^2 + 2x - 1$.

x	1	2	3	4	5	6	7
y	0	−1	−4	−9	−16	−25	−36
Differences		−1	−3	−5	−7	−9	−11

The differences in the y values are not constant. What happens if you look at *differences of the differences*?

1. Find the missing differences in this table for $y = -x^2 + 2x - 1$.

x	1	2	3	4	5	6	7
y	0	−1	−4	−9	−16	−25	−36
Differences of the y Values		−1	−3	−5	−7	−9	−11
Differences of the Differences			−2	?	?	?	?

To avoid confusion, the differences in the y values are called the *first differences*. The differences in the differences are called the *second differences*.

2. What do you notice about the second differences for the equation $y = -x^2 + 2x - 1$?

3. Work with a partner to explore the first and second differences for two more quadratic equations. Make a conjecture based on your findings.

Once you have made a conjecture, the next step is to find a convincing argument that explains why your conjecture is true or to find evidence that it is not true.

To show that a conjecture is true, it is not enough to show that it *usually* works. It is not even enough to say that the conjecture has worked in all the examples tested so far, unless you have tested *every* possible case. *You* may be convinced, but you need to be able to convince others as well.

And even if you are already convinced a conjecture is true, finding an argument to prove it is can help you understand *why* it is true.

✅ Develop & Understand: B

Consider the differences of squares of consecutive whole numbers.

Consecutive Numbers	Difference of Squares
1, 2	$2^2 - 1^2 = 3$
2, 3	$3^2 - 2^2 = 5$
3, 4	$4^2 - 3^2 = 7$
⋮	⋮
$n, n + 1$	$(n + 1)^2 - n^2 = D$

If n stands for any whole number, $n + 1$ is the next whole number. Below is an equation for the difference of the squares, D, of these numbers.

$$D = (n + 1)^2 - n^2$$

4. Copy and complete the table to show the value of D for the given values of n.

n	1	2	3	4	5	6	7	8
D	3	5	7					

5. Use what you know about constant differences to determine the type of relationship represented by $D = (n + 1)^2 - n^2$.

6. Use the table and your answer to Exercise 5 to make a conjecture about a simpler equation that relates D to n.

You have made a conjecture about a simpler equation for D. Now you can try to find a convincing argument to explain why your equation must be correct. One way to do this is to use a little geometry.

You can represent the square of a whole number as a square made from tiles. The tile pattern below represents the squares of consecutive whole numbers.

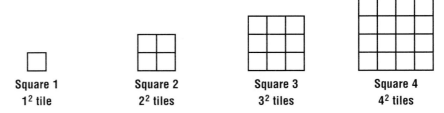

Square 1
1^2 tile

Square 2
2^2 tiles

Square 3
3^2 tiles

Square 4
4^2 tiles

The difference between the squares of two consecutive whole numbers is the number of tiles you must add to get from one square to the next. Be sure you understand why this is true.

7. Think about how you could add tiles to get from one square to the next in this pattern.

    a. Copy the pattern. Color the tiles that are added at each stage.

    b. How many tiles are added to go from square 1 to square 2? From square 2 to square 3? From square 3 to square 4?

    c. How many tiles are added to go from square n to square $n + 1$? Explain how you found your answer.

    d. If you did Part c correctly, your answer should prove your conjecture from Exercise 6. Explain.

Share & Summarize

1. Explain how finding first and second differences can help you determine whether a relationship is linear or quadratic.

2. How did you develop your conjecture for Exercise 6 on page 449?

3. How did you use geometry to prove your conjecture?

Investigation ② Detective Work

Making and proving a conjecture is a bit like being a detective. You might start with a hunch and then investigate further, looking for evidence to either support or disprove your hunch.

Good detectives do not search only for evidence to support their conjectures. They try to keep an open mind and look for evidence either way, even though it might show their hunch to be incorrect. Even one *counterexample,* an example for which a conjecture does not hold, will prove a conjecture incorrect.

✔ Develop & Understand: A

Prime numbers are important in cryptography, the study of codes. Many mathematicians have tried to find a rule that produces prime numbers.

Consider the expression $n^2 - n + 41$ as a possibility for such a rule.

1. Evaluate $n^2 - n + 41$ for $n = 1, 2, 3,$ and 4. Is each result a prime number? (Hint: If a number is *not* prime, it must have a prime factor that is equal to or less than its square root. For example, any nonprime number less than 100 must have 2, 3, 5, or 7 as a factor.)

2. Try some other values for n. What do you find? Compare your results with those of other students.

3. Do you think the expression will *always* give a prime number? Explain your thinking.

4. Without using your calculator, test the expression for $n = 41$. Is the result prime?

5. Explain how you could determine the answer for Exercise 4 just by looking at the expression $41^2 - 41 + 41$.

Here is how Dante and Kai reasoned about the sums of odd numbers.

✓ Develop & Understand: B

6. Convince yourself that Dante and Kai's argument is reasonable. Then try to answer Dante's question. How could you write the argument in a way that would convince someone else it must always be true?

7. Kai then wrote their argument.

   • If a number is even, it can be written $2k$, where k is a whole number.

   • If a number is odd, it is 1 more than an even number, so it can be written $2k + 1$.

   • So, odd + odd = $(2k + 1) + (2k + 1) = 4k + 2 = 2(2k + 1)$, which is an even number.

   Dante disagreed. He said, "What you've shown is that 2 times a particular odd number is an even number. That's already obvious because that's what *even* means."

   a. Discuss this with a partner. Who is correct? Explain.

   b. If you think Dante is correct, fix Kai's argument.

8. Make a conjecture about the sum of an even number and an odd number. Then write a convincing argument for why your conjecture must be true.

✅ Develop & Understand: C

Tate and Melinda were looking at data tables of the variables x, y, and z. Tate found that y was inversely proportional to x. Melinda found that z was also inversely proportional to x.

9. Make a conjecture about the relationship between y and z.

Now you will try to prove your conjecture.

10. Write equations for the relationships that you know.

    a. the relationship between x and y

    b. the relationship between x and z

11. Solve each of the equations that you wrote in Exercise 10 for x, if necessary.

12. You now have two expressions, each equal to x. What does that mean about the relationship between the two expressions? Write an equation to show this relationship.

13. Now solve for y in terms of z. What type of relationship exists between y and z? Was your conjecture correct?

Share & Summarize

1. What would you say to a friend who tells you that she can prove a general formula is correct and then shows you that her formula works only for a few particular examples?

2. Why is it important to discuss proofs of your conjectures with people who might not quickly agree with you?

Practice & Apply   In Exercises 1–4, make a conjecture about whether the relationship between x and y is linear, quadratic, or neither. Explain how you decided.

1.

x	1	2	3	4	5	6	7
y	−1	4	15	32	55	84	119

2.

x	1	2	3	4	5	6	7
y	4	16	64	256	1,024	4,096	16,384

3.

x	1	2	3	4	5	6	7
y	6	8	10	12	14	16	18

4.

x	1	2	3	4	5	6	7
y	4	12	24	40	60	84	112

5. Write an equation for the relationship in Exercise 3.

6. In Exercises 4–6 on page 449, you looked at $D = (n + 1)^2 - n^2$, the difference between squares of consecutive whole numbers. Now consider the following equation.

$$(m + 2)^2 - m^2 = d$$

In this case, d is the difference between the square of a whole number and the square of that whole number plus 2.

Numbers	Difference of Squares
1, 3	$3^2 - 1^2 = 8$
2, 4	$4^2 - 2^2 = 12$
3, 5	$5^2 - 3^2 = 16$
⋮	⋮
m, m + 2	$(m + 2)^2 - m^2 = d$

a. Copy and complete the table to show the value of d for consecutive values of m.

m	1	2	3	4	5	6
d	8	12	16			

b. Use what you know about constant differences to determine the type of relationship represented by $d = (m + 2)^2 - m^2$.

c. Make a conjecture about what a simpler equation for d might be. Check that your equation works for $m = 1$, $m = 2$, and $m = 3$.

d. You can use geometry to argue that your conjecture is true. Below are tile squares for 1^2 and 3^2. Think about how you add tiles to get from one square to the next. Copy the diagram. Color the tiles that you would add.

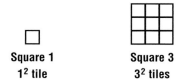

Square 1
1^2 tile

Square 3
3^2 tiles

e. Draw tile squares to represent 2^2 and 4^2. Color the tiles you would add to get from one to the other. Do the same for 3^2 and 5^2.

f. How many tiles do you add to go from the square for n^2 to the square for $(n + 2)^2$? Explain how you found your answer.

g. Does your answer from Part f prove your conjecture from Part c? Explain why or why not.

7. Santos conjectured that when you subtract an even number from an odd number, the result is odd. He tried to prove his conjecture.

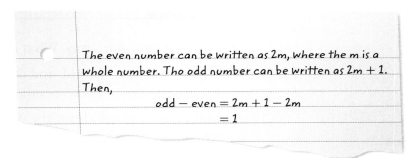

The even number can be written as $2m$, where the m is a whole number. The odd number can be written as $2m + 1$. Then,
$$\text{odd} - \text{even} = 2m + 1 - 2m$$
$$= 1$$

Santos said, "According to my proof, an odd number minus an even number is always 1. This is not true. What did I do wrong?"

a. What did Santos do incorrectly?

b. Give a correct proof of Santos' conjecture.

8. Make and prove a conjecture about the difference of two even numbers.

9. Make and prove a conjecture about the difference of two odd numbers.

10. You cannot prove a general conjecture by checking specific examples. However, just one false example, a counterexample, will disprove it. Show that the conjecture $(m + n)^2 = m^2 + n^2$ is false by giving values for m and n for which it does not work.

11. For each statement, provide a proof or a counterexample.

    a. The reciprocal of x times the reciprocal of y is the reciprocal of xy.

    b. The reciprocal of x plus the reciprocal of y is the reciprocal of $x + y$.

Connect & Extend

12. Below is a table of values for the cubic equation $y = x^3$. Notice that neither the first or second differences are constant.

x	1	2	3	4	5	6	7
y	1	8	27	64	125	216	343
First Difference		7	19	37	61	91	127
Second Difference			12	18	24	30	36

    a. Find the *third differences*. What do you notice?

    b. Explore the first, second, and third differences for two more cubic equations. Try to make a conjecture based on your findings.

    c. Make a conjecture about the differences for a *quartic equation,* which can be written in the form $y = ax^4 + bx^3 + cx^2 + dx + e$. Test your conjecture on $y = x^4$.

13. You may recall that the constant first difference for a linear equation is the value of m when the equation is written in $y = mx + b$ form. In this exercise, you will look for a relationship between a quadratic equation and the constant second difference.

    a. Find the constant second difference for this table of values for $y = \frac{1}{2}x^2 + 2x$.

x	1	2	3	4	5	6	7
y	2.5	6	10.5	16	22.5	30	38.5

b. Find the constant second difference for this table of values for $y = -x^2 + 2x - 1$.

x	1	2	3	4	5	6	7
y	0	−1	−4	−9	−16	−25	−36

c. Find the constant second difference for this table of values for $y = 3x^2 - 9$.

x	1	2	3	4	5	6	7
y	−6	3	18	39	66	99	138

d. In Parts a–c, look for a relationship between the constant second difference and the coefficient of x^2 in the quadratic equation. Try to make a conjecture about this relationship.

e. Test your conjecture on at least two more quadratic relationships.

Math Link

Goldbach's conjecture was made in 1742. Although it seems simple, no one has ever been able to prove that it is true.

14. Number Sense *Goldbach's conjecture* states that every even integer greater than 2 can be expressed as the sum of two prime numbers. Here are some examples.

$$4 = 2 + 2 \qquad 6 = 3 + 3 \qquad 8 = 3 + 5$$

a. Test Goldbach's conjecture on 10, 12, and 100.

b. Does the conjecture seem to be true?

c. What would you need to show in order to prove the conjecture?

In Exercises 15–17, decide whether the conjecture is true or false. Try to give a convincing proof of the conjectures that are true. For false conjectures, give a counterexample.

15. The value of $n^2 - n$ is always an even number.

16. The square of every even number is a multiple of 4.

17. The difference between any two square numbers is always an even number.

18. In Your Own Words Explain the difference between making a conjecture and proving something to someone.

Preview Writing an expression as the product of factors is called *factoring*.
For Exercises 19–22, match each trinomial with its binomial factors.

19. $x^2 + 7x + 10$ **A.** $(x + 6)(x - 6)$

20. $x^2 + 8x + 12$ **B.** $(x + 6)(x + 2)$

21. $x^2 + 10x + 25$ **C.** $(x + 2)(x + 5)$

22. $x^2 - 36$ **D.** $(x + 5)(x + 5)$

Mixed Review
Supply the missing exponent.

23. $24 \times 10^? = 24{,}000$

24. $2.8 \times 10^? = 28{,}000$

Simplify each expression as much as possible.

25. $a + 4(a - 2)$

26. $2b - (8 + 2b)$

27. $90 - (5c - 1)$

28. $4d(2 + e) - 2(3 - 2d)$

29. $-7(1 - f) + 2(7f + 2) - 9(f + 2)$

Solve each inequality. Graph the solution on the number line.

30. $5(9 - x) \le 4(x + 18)$

31. $3x - 9 < -4.5x + 6$

Review & Self-Assessment

Chapter Summary

You began this chapter by reviewing methods for solving equations in preparation for solving *quadratic equations*. You used graphs and tables to estimate solutions of equations.

Next, you focused on the characteristics of *quadratic relationships*. A quadratic relationship can be represented by equations of the form $y = ax^2 + bx + c$, where a, b, and c are constants, and a is not 0. The graph of a quadratic equation has a particular shape called a *parabola*.

Inverse or *reciprocal relationships* also have a particular shape called a hyperbola. You have seen inverse relationships with equations of the form $y = \frac{a}{x}$, $y = \frac{1}{x + b}$, and $y = \frac{1}{x} + c$. In addition, you saw what equations and graphs of *cubic* relationships look like.

You also made several *conjectures* in this chapter. *Proving* a conjecture not only allows you to convince others that it is true but can also help you understand *why* it is true.

Strategies and Applications

The questions in this section will help you review and apply the important ideas and strategies developed in this chapter.

Using graphs to estimate solutions of equations

1. Draw a graph of $y = 4x - 0.5x^2$. Find a value of c for which $4x - 0.5x^2 = c$ has only one solution.

Using tables to estimate solutions of equations

2. This table of values is for the equation $y = x^2 + x$. Between which two values of x would you expect to find a solution for $x^2 + x = \frac{11}{4}$?

x	y
1	2
1.1	2.31
1.2	2.64
1.3	2.99
1.4	3.36
1.5	3.75
1.6	4.16

Vocabulary

conjecture

cubic equation

hyperbola

inversely proportional

inverse variation

parabola

quadratic equation

reciprocal relationship

vertex

Recognizing quadratic relationships from graphs, equations, and tables

In Questions 3–10, determine whether the relationship between x and y could be quadratic, and explain how you know.

3.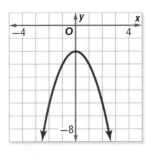

4.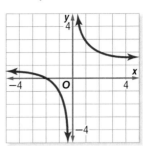

5.

x	−3	−2	−1	0	1	2	3
y	−10	−7	−4	−1	2	5	8

6.

x	−2	−1	0	1	2	3	4
y	−4	−5	−4	−1	4	11	20

7.

x	0	1	2	3	4	5	6
y	1	2	4	8	16	32	64

8. $y = 2x^3 + x^2$ **9.** $y = x(x + 2)$ **10.** $y = 2^x + 3$

Understanding the connections between quadratic equations and graphs

Match each equation with one of the graphs below.

11. $y = x^2$ **12.** $y = -x^2$

13. $y = x^2 - 4x + 4$ **14.** $y = x^2 + 4$

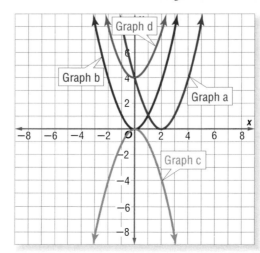

Recognizing inverse relationships from written descriptions, graphs, tables, and equations

In Questions 15–23, determine whether the relationship between x and y could be an inverse relationship, and explain how you know.

15.

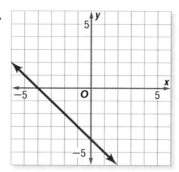

16.

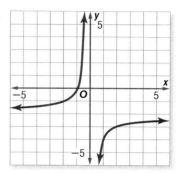

17. Carl sells hand-bound journals with blank pages. He notices that for every $5 that he increases the price x, his weekly sales y drop by 20 books.

18. Sandra has 60 square tiles. She tries to find as many rectangular arrangements of all 60 tiles as she can, recording the dimensions as length x and width y.

19.

x	1	2	4	5	6	9	10
y	60	30	15	12	10	$6\frac{2}{3}$	6

20.

x	0.5	1	1.5	2	2.5	3	3.5
y	840	420	280	210	168	140	120

21. $2xy = 7$

22. $y = \dfrac{x}{4}$

23. $x + 2 = \dfrac{1}{y}$

Solving real-world situations involving quadratic and inverse relationships

24. Jeanine was organizing an event to give her classmates a chance to read their poems before an audience. She scheduled the school library open for two hours. Jeanine wanted to determine how much time each poet would have. She realized it would depend on how many poets she invited.

a. Write an equation expressing the amount of time t in minutes available to each poet if there are n poets. Assume one poet starts immediately after another has finished.

b. If Jeanine invites eight poets, how much time will each have?

c. Suppose Jeanine wants to give each poet ten minutes. How many can she invite?

25. A radio station is broadcasting a live concert. When the audience applauds at the end of each song, the sound level rises, reaches a peak, and then falls in a way that can be approximated by a quadratic equation.

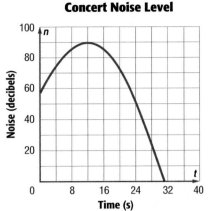

Concert Noise Level

The graph shows the relationship between the noise level n in decibels and the time t in seconds after a particular song ends.

a. At what time after the end of the song was the noise greatest? How loud did it get?

b. If a speaker tries to produce a noise that is too loud, it *distorts* the sound and produces static. The crew turns down the sound being broadcast when the noise level rises above 70 decibels and keeps it turned down until it is below 70 decibels once again. How long was the sound turned down following this song? How did you find your answer?

c. Noise levels of 50 decibels or lower are considered normal background noise. How long after the applause began did the sound subside to the level of normal background noise?

Making conjectures

26. Consider these sets of three consecutive whole numbers.

3, 4, 5	5, 6, 7
0, 1, 2	9, 10, 11

Compare the products of the first and last numbers of each set to the square of the middle number.

a. Make a conjecture about what you observe.

b. Suppose the middle number is x. What are the other two numbers? Rewrite your conjecture as a mathematical statement using x.

c. Find a counterexample or a proof for your conjecture. (Hint: You might want to look at this exercise geometrically.)

Demonstrating Skills

In Exercises 27–32, make a rough sketch showing the general shape and location of the graph of the equation.

27. $y = 3x^2$

28. $y = \dfrac{3}{x}$

29. $y = x^3$

30. $y = -x^2 + 3$

31. $y = -\dfrac{2}{x} - 1$

32. $y = (x + 1)^2$

Test-Taking Practice

SHORT RESPONSE

1 Draw the graph of the equation $y = \dfrac{1}{2}x^2 - 6$.

Show your work.

x	$y = \dfrac{1}{2}x^2 - 6$
−3	−1.5
−2	−4
−1	−5.5
0	−6
1	−5.5
2	−4
3	−1.5

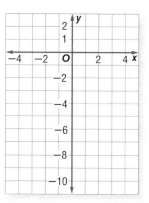

Answer _____

MULTIPLE CHOICE

2 Which of the following are quadratic functions?

    1. $xy = 3$

    2. $y = -4x^2 + 6$

    3. $f(x) = 7x^2 - 100$

A 1 only

B 3 only

C 1 and 3

D 2 and 3

3 Determine which relationship could be an inverse relationship.

F

x	1	2	3	4	5
y	20	10	5	$2\dfrac{1}{2}$	$1\dfrac{1}{4}$

G

x	1	2	3	5	6
y	15	$7\dfrac{1}{2}$	5	3	$2\dfrac{1}{2}$

H $y = -\dfrac{1}{2}x$

J $y = x^2 - 2^x$

Solve Quadratic Equations

Real-Life Math

Super Models Most computer programmers depend on mathematical equations and expressions in the software they design. Many computer and video games, for example, need to be able to model the paths of things flying through the air, such as balls for sports like basketball, baseball, football, soccer, and golf. Such paths are called *trajectories,* and quadratic equations can be used to model them.

Think About It Imagine the path of a ball that is shot for a basket. Can you describe the shape of its trajectory?

Math Online

Take the **Chapter Readiness Quiz** at glencoe.com.

Dear Family,

The next chapter deals with solving quadratic equations. Quadratic equations involve the square of the main variable and can be written in the form $ax^2 + bx + c = 0$, where a, b, and c are constants.

Quadratic equations are used to describe the movement of objects, such as the motion of basketballs, automobiles, and rockets. They are also used to determine the shapes of radar antennae and satellite dishes.

The class will learn three methods for solving quadratic equations. The first of these is **factoring**.

Key Concept—Factoring

Students have previously used the geometric model to the right to determine the product of $(m + 1)$ and $(m + 3)$.

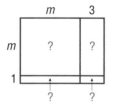

By determining the areas of the model's rectangles, students found the following product.

$$(m + 1)(m + 3) = m^2 + 3m + 1m + 3 = m^2 + 4m + 3$$

To factor, students reverse the process. They start with the rectangles and determine the polynomial factors.

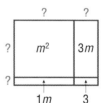

The trinomial $m^2 + 4m + 3$ equals $(m + 1)(m + 3)$.

Not all quadratic equations can be solved by factoring. The class will also complete the square and use the quadratic formula to solve quadratics.

Chapter Vocabulary

factoring

Home Activities

- Look for real-life situations that can be modeled by a quadratic equation.
- Encourage your student to share the methods he or she is using to solve quadratic equations. Ask him or her to explain each process.

LESSON
9.1

Backtracking

As you know, backtracking is a step-by-step process of undoing operations. To solve a linear equation by backtracking, you must know how to undo addition, subtraction, multiplication, and division. To use backtracking to solve nonlinear equations, however, you have to undo other operations as well.

Think & Discuss

How would you check to make sure that one operation really *does* undo another? Give an example to illustrate your thinking.

What would you do to undo each of these operations?
- finding the square roots of a number
- finding the reciprocal of a number
- changing the sign of a number
- raising a positive number to the nth power, such as 2^3

Real-World Link
Quadratic equations are used in many contexts. For example, $h = 1 + 2t - 4.9t^2$ might give the height in meters at time t seconds of someone jumping on a trampoline with an initial velocity of 2 m/s.

Investigation 1 Backtracking with New Operations

In this investigation, you will try out the ideas about undoing operations from Think & Discuss.

✅ *Develop & Understand: A*

Ben is solving the equation $\sqrt{2x - 11} = 5$.

Kai said Ben's flowchart should really be like this.

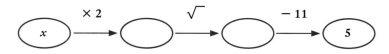

1. What is wrong with Kai's flowchart?

2. What equation would you solve by backtracking with Kai's flowchart?

3. Consider the equation $\sqrt{3x + 7} = 8$.

   a. Draw a flowchart for the equation.

   b. Use backtracking to find the solution. Check your answer by substituting it back into the equation.

Math Link

The $\sqrt{}$ sign refers to the nonnegative square root of a number, if one exists. This is also called the principal square root.

4. This flowchart is for the expression $\dfrac{24}{s-2}$.

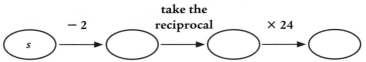

a. Try using the flowchart with a few numbers to see how it works. Record your results.

b. This flowchart is for the equation $\dfrac{24}{k-2}=8$.

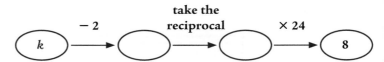

Backtrack to find the solution. Check that your solution is correct by substituting it into the equation.

5. This flowchart is for the equation $3-p=1$. The symbol $+/-$ means to take the opposite of the value; that is, to change its sign.

a. Solve the equation using backtracking.

b. Ben made this flowchart for $3-p=1$, but he got stuck when he tried to backtrack. Why can you not use his flowchart to solve the equation?

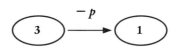

6. This flowchart is for the expression $\dfrac{2(3-t)}{4}$.

a. Try the flowchart with a few numbers to see how it works.

b. This flowchart is for the equation $\dfrac{2(3-t)}{4}=5$.

Backtrack to find the solution. Check that it is correct.

c. What operation undoes the "change sign" operation?

7. You can think of changing the sign of a number as multiplication by -1. For example, $-x = -1 \cdot x$.

   a. How do you usually undo multiplication by a number?

   b. What does your answer to Part c of Exercise 6 suggest about another way to undo multiplication by -1?

✅ Develop & Understand: B

Solve each equation by backtracking.

8. $\dfrac{4}{x} = 0.125$ **9.** $\dfrac{8 - z}{2} = 9$

10. $\dfrac{7 - m}{2} = 3$ **11.** $5\left(20 - \dfrac{a}{4}\right) = 85$

12. Consider this equation.

$$\frac{12}{3s - 1} = 6$$

   a. Draw a flowchart for the equation.

   b. Solve the equation by backtracking. Check your solution.

13. Katie drew this flowchart.

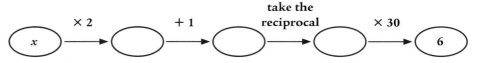

   a. What equation can be solved using Katie's flowchart?

   b. Solve the equation. Check your solution.

Share & Summarize

In this investigation, you learned how to undo taking the square root of a number, taking the reciprocal of a number, and changing the sign of a number.

1. Write an equation that can be solved by backtracking that uses all three of these operations. Find the solution of your equation.

2. Exchange equations with a partner. Try to solve your partner's equation. Check your answer by substitution.

Investigation 2 Backtracking with Powers

In this investigation, you will extend the types of equations for which you can use backtracking.

Think & Discuss

This flowchart is for the equation $x^2 = 9$. To solve this equation by backtracking, you must undo the "square" operation by taking the square root.

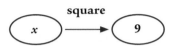

- How many solutions are there for $x^2 = 9$? How do you know?

Now consider the equation $(d - 2)^2 = 25$.
- Write the operations, in order, that you would use to evaluate the expression $(d - 2)^2$ for some value of d.
- Draw a flowchart for the equation $(d - 2)^2 = 25$.
- This equation has two solutions. Which step in your flowchart makes it possible for there to be two answers? Explain.
- As you backtrack beyond the step that makes two answers possible, there are two possible values for each oval. Think of a way you might show two values at each step. Then find both of the equation's solutions.

✓ Develop & Understand: A

1. Consider the equation $(a + 5)^2 = 25$.

   a. Draw a flowchart for the equation.

   b. Solve $(a + 5)^2 = 25$ by backtracking. Check your answer by substitution. Can you find more than one backtracking path (more than one solution)?

For the equations in Exercises 2 and 3, draw a flowchart. Solve the equation using backtracking. Check your answers.

2. $2(b - 4)^2 + 5 = 55$ 3. $3(c - 5)^2 - 5 = 7$

Solve each equation.

4. $(d - 2)^2 - 20 = 44$

5. $\sqrt{(2p - 3)^2 - 5} = 2$

6. $3(6 - t)^3 - 1 = 23$

7. $(e - 3)^4 = 81$

....

Real-World Link

The unit *foot-candle* measures the amount of light falling on one square foot of area. It was originally defined using a standardized candle burning one foot away from a surface.

....

8. When filmmakers make movies outside at night, they often use floodlights to brighten the actors and the scenery. The relationship between the brightness of an object *F*, measured in *foot-candles*, and its distance *d*, measured in feet, from the light source follows an *inverse square law*. For a particular 2,000-watt floodlight, the formula might be

$$F = \frac{360{,}000}{d^2}$$

a. Why do you think this is called an inverse square law?

b. Find the brightness for $d = 10$, 20, 30, and 50.

c. Explain the effect on an object's brightness of moving closer to or further from the light source.

d. Draw a flowchart for the brightness formula.

e. Use backtracking to solve this equation.

$$120 = \frac{360{,}000}{d^2}$$

What does the answer tell you?

✅ *Develop & Understand: B*

9. Consider the following equation.

$$(x - 3)^2 - 5 = 0$$

   a. Show that $3 + \sqrt{5}$ and $3 - \sqrt{5}$ are solutions of this equation.

   b. The values $3 + \sqrt{5}$ and $3 - \sqrt{5}$ are *exact* solutions of the equation. Because $\sqrt{5}$ is an irrational number, when you write it in a decimal form, you are giving an *approximation*, no matter how many decimal places you use.

      Write each solution from Part a as a decimal accurate to two places.

For each equation in Exercises 10–13, give the exact solutions and the approximate solutions correct to two decimal places.

10. $h^2 - 5 = 45$

11. $d^2 + 3 = 10$

12. $2m^2 + 5 = 25$

13. $5k^2 - 7 = 23$

14. $(2m - 3)^2 + 7 = 9$

15. $3(j + 5)^2 - 2 = 7$

16. $(4m + 1)^2 - 8 = 21$

17. $6(p + 2)^2 + 10 = 70$

> ## Share ❽ Summarize
>
> Use backtracking to solve this equation. Explain what you did at each step. Make note of any places you had to be particularly careful.
>
> $$2(3b - 4)^2 + 1 = 19$$

Practice & Apply

1. Consider the equation $-\sqrt{2x - 1} = -7$.

   a. Draw a flowchart for the equation.

   b. Solve the equation by backtracking.

Solve each equation.

2. $\sqrt{3x + 1} = 4$

3. $\dfrac{2}{3p - 1} = 5$

4. $\sqrt{a} = 1.5$

5. $\sqrt{2 - q} = 2.5$

6. $5\sqrt{\dfrac{z}{5} - 1} = 4$

7. $\dfrac{9}{4 - 7d} = 18$

8. $2(x - 4)^2 + 5 = 7$

9. $b^2 - 5 = 44$

10. $c^2 - 20 = 44$

11. $(l - 2)^2 - 5 = 44$

12. $(q - 2)^2 + 8 = 44$

13. $y^3 = 27$

14. $3(2w - 3)^2 - 5 = 70$

15. $(2t - 3)^2 - 20 = 44$

16. $y^3 = -27$

17. $(x + 2)^3 = 64$

Connect & Extend

18. A man controlling a robot with a camera eye has just sent it into a burning house to retrieve a safe full of money. He has given it this set of commands.

   • Forward 20 feet.
   • Right turn.
   • Forward 15 feet.
   • Left turn.
   • Forward 30 feet.
   • Right turn.
   • Forward 25 feet.
   • Pick up safe.

   The robot is now standing in the burning house holding the safe, but the robot controller has been overcome by smoke. You have been asked to get the robot back to its starting point.

   Use what you know about backtracking to write a set of commands that will bring the robot out of the burning house with the safe.

19. Mary Ann used her calculator to find $\sqrt{2}$, wrote down the result, and then cleared the calculator. She then used her calculator to square the number she had written down and obtained 1.99998.

Mary Ann concluded that squaring does not exactly undo taking the square root. Do you agree with her? What would you tell her?

20. Many equations cannot be solved directly by backtracking. Some have the variable stated more than once. Others involve variables as exponents.

Here are some equations that cannot be solved directly by backtracking.

$$f^2 = f + 1 \qquad x = \sqrt{x} + 1 \qquad k^2 + k = 0$$

$$1.1^B = 2 \qquad \frac{1}{x} = x^2 + 2$$

For each equation below, write *yes* if it can be solved directly by backtracking and *no* if it cannot.

a. $5 = \sqrt{x - 11}$

b. $4^d = 9$

c. $3g^2 = 5$

d. $\sqrt{x + 1} = x - 4$

21. Use backtracking to solve the equations in Parts a and b.

a. $(3x + 4)^2 = 25$

b. $(3x + 4)^2 = 0$

c. How many solutions did you find for each equation? Can you explain the difference?

22. Use backtracking to solve the equations in Parts a and b.

a. $(\sqrt{x})^2 = 5$

b. $\sqrt{x^2} = 5$

c. How many solutions did you find for each equation? Can you explain the difference?

23. In Your Own Words Explain how you can decide whether an equation can be directly solved by backtracking. Give an example of an equation that can be solved directly by backtracking and an example of one that cannot.

Mixed Review

Expand each expression.

24. $5(7x - 2)$

25. $-4m(6m - 9)$

26. $(t + 5)(t + 2)$

27. $(6k + 7)(6k - 7)$

Rewrite each equation in $y = mx + b$ form. Tell whether the relationship represented by the equation is increasing or decreasing.

28. $4(x - y) = -3$

29. $\dfrac{4 - 3x}{2y} = 1$

30. Order these numbers from least to greatest.

$$\sqrt[9]{-4} \qquad \sqrt[3]{-4} \qquad \sqrt[11]{-4} \qquad \sqrt[5]{-4} \qquad \sqrt[7]{-4}$$

Rewrite each expression using a single base and a single exponent.

31. $27x^3$

32. $a^{12} \cdot \left(a^2\right)^{-7}$

33. $\dfrac{32}{c^5}$

34. Graph the inequality $y > 3x + 2$.

Factoring

Some quadratic equations can be solved easily by backtracking. A second solution method called factoring can be used to solve other quadratic equations.

In this investigation, you will use rectangle models to think about factoring quadratic expressions.

The process of **factoring** means to write a number or algebraic expression as the product of factors.

Vocabulary

factoring

> ### Think & Discuss
>
> What two factors could make the product 24? How many factor pairs are there?
>
> What two factors could produce $12x$? Or x^2?

Investigation 1   A Geometric Model

Consider finding the binomial factors of $x^2 + 7x + 10$. It may be helpful to use a rectangular area model.

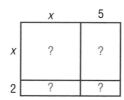

This model shows that you can find the area of the large rectangle by adding the areas of the smaller rectangles.

$$x \text{ by } x \longrightarrow x^2$$
$$x \text{ by } 5 \longrightarrow 5x$$
$$2 \text{ by } x \longrightarrow 2x$$
$$2 \text{ by } 5 \longrightarrow 10$$

The area of the large rectangle, and the product of $(x + 5)(x + 2)$, is $x^2 + 7x + 10$.

What if you start with the trinomial product and want to find the binomial factors? To do this, reverse the model.

Consider the following rectangle model for $x^2 + 7x + 10$

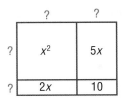

Use the left diagram to find the binomial factors. Ask yourself the following questions.

- What monomial times itself results in a product of x^2?
- What two monomials have a product of $5x$?
- What two monomials have a product of $2x$?
- What two positive integers have a product of 10?

As shown by the right diagram, these factors can be written alongside the top and left-side edges of the rectangle. The sums of the top factors, $x + 5$, and side factors, $x + 2$, are the binomial factors of $x^2 + 7x + 10$.

✅ Develop & Understand: A

Use the rectangle model to find the binomial factors of each trinomial.

1. $x^2 + 5x + 6$

	?	?
?	x^2	$3x$
?	$2x$	6

2. $m^2 + 6m + 8$

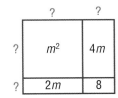

3. Draw a rectangle model to find the binomial factors of $y^2 + 4y + 3$.

4. Multiply two binomials to create a trinomial. Ask your partner to factor your trinomial using a rectangle model.

Share & Summarize

How can a rectangle diagram help you factor a quadratic expression? What number patterns did you notice that could help you with factoring?

Investigation ② Factor Quadratic Expressions

A quadratic equation may consist of a product of two factors on one side of the equals sign and 0 on the other side.

$$(x - 2)(x + 5) = 0$$

In this case, the solutions can be found exactly. This is because 0 has a special property.

Think & Discuss

If the product of two factors is 0, what must be true about the factors?

Find all the values of k that satisfy the equation $k(k - 3) = 0$. Explain how you know you have found them all.

Now find all the values of x that satisfy $(x - 2)(x + 5) = 0$. Explain how you found them.

You will now use the ideas from Think & Discuss to solve some equations written as a product equal to 0.

✅ Develop & Understand: A

Find all the solutions of each equation.

1. $(t - 1)(t - 3) = 0$ **2.** $(s + 1)(2s + 3) = 0$

3. $y(y - 4) = 0$ **4.** $(t - 3)(t - 3) = 0$

5. $x(3x + 7) = 0$ **6.** $(p + 4)(p + 4) = 0$

7. Using the same idea, find the solutions of this equation.
$$(2x + 1)(x + 8)(x - 1) = 0$$

You have seen how easily you can solve equations that are written as a product of factors equal to 0. Sometimes you can rewrite an equation in this form to make the solution easy to find.

Think & Discuss

The expression $(3x + 4)(3x - 4)$ can be rewritten as $9x^2 - 16$ when it is expanded. An expression such as $9x^2 - 16$ is called a *difference of two squares*. Can you explain why?

Now think about reversing the expansion. How would you factor $4a^2 - 25$? That is, how would you rewrite it as the product of two factors?

In Chapter 5, you learned how to square a binomial and rewrite it as a trinomial, an expression with three unlike terms.

$$(x - 5)^2 = x^2 - 10x + 25 \qquad (b + 5)^2 = b^2 + 10b + 25$$

Rewrite each trinomial below as the square of a binomial.

$$c^2 + 4c + 4 \qquad\qquad 16d^2 - 8d + 1$$

A trinomial such as $16d^2 - 8d + 1$ is called a *perfect square trinomial*. Can you explain why?

Which of these trinomials are perfect squares?

$$x^2 + 6x + 9 \qquad\qquad k^2 - 8k + 25$$
$$t^2 + 16t + 60 \qquad\qquad m^2 - 12m + 36$$
$$p^2 + 2p + 1 \qquad\qquad d^2 - 10d + 24$$
$$4y^2 + 4y + 4 \qquad\qquad 49s^2 - 28s + 4$$

Just by looking at the coefficients, how can you tell whether *any* trinomial in the form $ax^2 + bx + c$ is a perfect square?

Math Link

Subtracting a term means the coefficient is negative. For example, in the expression $16x^2 - 8x + 1$, the coefficient of x is -8.

✅ *Develop & Understand: B*

In Exercises 8–15, determine whether the quadratic expression to the left of the equal sign is the difference of two squares or a perfect square trinomial. If it is, rewrite it in factored form and solve the equation. If it is not in one of these special forms, explain how you know it is not.

8. $x^2 - 64 = 0$

9. $p^2 + 64 = 0$

10. $x^2 - 16x - 64 = 0$

11. $k^2 - 16k + 64 = 0$

12. $9y^2 - 1 = 0$

13. $9m^2 + 6m + 1 = 0$

14. $9g^2 - 4g - 1 = 0$

15. $y^2 + 9 = 0$

Each equation below has two variables. If the quadratic expression is the difference of two squares or a perfect square trinomial, rewrite the equation in factored form and solve for *a*. If it is in neither special form, explain how you know it is not.

16. $a^2 + 9b^2 = 0$

17. $a^2 - 4ab + 4b^2 = 0$

18. $4a^2 - b^2 = 0$

Share & Summarize

1. Give an example of a perfect square trinomial. Then give an example of a quadratic expression that is the difference of two squares. Explain how you know that your expressions are in the correct forms.

2. Explain why the only solutions of $4x^2 - 9 = 0$ are $x = 1.5$ and $x = -1.5$.

Real-World Link

Quadratic equations are often used to describe how the position of a moving object changes over time. For example, the equation $d = 25 - 3t^2$ might describe how many meters an accelerating hyena is from a rabbit after t seconds if it began running toward the rabbit from 25 meters away.

Investigation 3 Practice with Factoring

If a quadratic expression is equal to 0 and can be factored easily, finding its factors is an efficient way to solve the equation. In Investigation 1, you learned a technique for factoring that involved an area model. In this investigation, you will learn some new strategies for determining whether a quadratic expression can be easily factored and for factoring it when you can. For example, consider this expression.

$$x^2 + 8x + 12$$

If such an expression can be factored, it can be rewritten as the product of two linear expressions.

$$(x + m)(x + n)$$

Multiplying terms produces the following expression.

$$x^2 + (m + n)x + mn$$

You can use this idea to help factor any quadratic expression for which a, the coefficient of the squared variable, is equal to 1.

Example

Can $x^2 + 8x + 12$ be factored? If so, solve $x^2 + 8x + 12 = 0$.

First, compare the expanded form of $(x + m)(x + n)$ with the given expression.

$$x^2 + 8x + 12$$
$$x^2 + (m + n)x + mn$$

If $x^2 + 8x + 12$ can be factored into the form $(x + m)(x + n)$, the product of m and n must be 12. Their sum must be 8. The two numbers that fit these conditions 6 and 2. Therefore, the expression *can* be factored. The equation can be rewritten as follows.

$$(x + 2)(x + 6) = 0$$

So, the equation $x^2 + 8x + 12 = 0$ has two solutions, -2 and -6.

In this investigation, you will consider only cases in which m and n are integers.

✅ Develop & Understand: A

For Exercises 1–6, do the following.

- Think of the expression as a special case of $(x + m)(x + n)$. State the values of m and n.
- Use the fact that $(x + m)(x + n) = x^2 + (m + n)x + mn$ to expand the expression.

1. $(x + 7)(x + 1)$

2. $(x + 2)(x + 5)$

3. $(x - 4)(x - 5)$

4. $(x + 2)(x - 3)$

5. $(x - 2)(x + 3)$

6. $(x + 5)(x - 4)$

For Exercises 7–10, use $(x + m)(x + n) = x^2 + (m + n)x + mn$ to do the following.

- Determine what $m + n$ and mn equal. From this, find the values of m and n.
- Rewrite the expression as a product of two binomials. You may want to expand the product to check your result.

7. $x^2 + 7x + 6 = (x + \underline{})(x + \underline{})$

8. $x^2 - 7x + 6 = (x - \underline{})(x - \underline{})$

9. $x^2 - 4x - 12 = (x - \underline{})(x + \underline{})$

10. $x^2 + 4x - 12 = (x - \underline{})(x + \underline{})$

Use the method demonstrated in the example on the previous page to solve these equations.

11. $x^2 - 10x + 16 = 0$ **12.** $x^2 + 6x - 16 = 0$

Real-World Link

The quadratic equation $x = 10t + \frac{1}{2}(2.5)t^2$ describes the distance in meters traveled by a motorcycle that began from a certain point with velocity 10 m/s and increased its speed, or accelerated, at a rate of 2.5 m/s^2.

Think & Discuss

Kai organized the possibilities for factoring trinomials.

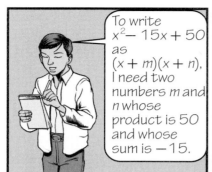
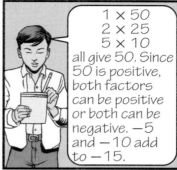

Use Kai's approach to factor these expressions.

$$x^2 + 11x + 10 \qquad x^2 - 7x + 10 \qquad x^2 - 3x - 10$$

Explain what happens if you use Kai's approach with $x^2 + 6x + 10$.

✓ Develop & Understand: B

Factor each quadratic expression using Kai's method, or state that it cannot be factored using his method.

13. $x^2 + 6x + 5$

14. $b^2 + 4b - 5$

15. $w^2 - 2w + 1$

16. $t^2 + 9t - 18$

17. $s^2 - 10s - 24$

18. $c^2 - 4c + 5$

19. Use Kai's approach to solve the equation $w^2 + 4w - 12 = 0$.

If every term of a quadratic expression has a common factor, rewriting it can make factoring easier. For example, $2x^2 + 12x + 10$ can be rewritten as $2(x^2 + 6x + 5)$, which factors to $2(x + 1)(x + 5)$.

Find the common factor for each expression. Then factor the expression as much as possible.

20. $3a^2 + 18a + 15$

21. $2b^2 + 8b - 10$

22. $4x^2 - 8x + 8$

23. $5t^2 + 25t - 70$

Challenge Sometimes a quadratic expression can be factored even though the coefficient of x^2 is not 1 and the terms do not have a common factor. For example, $2x^2 - 9x + 9$ can be factored as $(2x - 3)(x - 3)$. Use strategies like those you have used previously to factor these expressions.

24. $3x^2 - 11x - 4$

25. $8x^2 + 2x - 3$

If you are given a quadratic expression to factor, what steps would you follow to factor it or to determine that it cannot be factored using integers?

Investigation 4 Solve Quadratics

Sometimes you must rearrange the terms in a quadratic expression to see how or if the expression can be factored.

Think **&** *Discuss*

Mrs. Torres gave her class a number puzzle.

Why is Tamika's suggestion a good one? What quadratic equation should the class find when they finish rearranging?

Can the quadratic expression be factored? If so, what are the factors?

With what numbers could Mrs. Torres have started? Check your answer in her original number puzzle.

✅ *Develop & Understand: A*

Rearrange each equation so you can solve by factoring. Find the solutions.

1. $4a + 3 = 6a + a^2$ **2.** $b^2 - 12 = 4b$

3. $c(c + 4) + 3c + 12 = 0$ **4.** $d + \dfrac{6}{d} = 5$

5. $\dfrac{(x + 3)(x - 2)^2}{x - 2} = 3x - 3$ (Hint: Simplify the fraction first.)

6. Kenyon challenged his teacher, Ms. Hiroshi, with a number puzzle. "I'm thinking of a number. If you multiply my number by two more than the number, the result will be one less than four times my number."

     a. Write an equation for Kenyon's puzzle. Then use factoring to solve it. Check that your answer fits the puzzle.

     b. Kenyon expected his teacher to find two solutions to his puzzle. Why did she not find two?

7. A rectangular rug has an area of 15 square meters. Its length is two meters more than its width.

     a. Write an equation to show the relationship between the rug's area and its width.

     b. Solve your equation. Explain why only one of the solutions is useful for finding the rug's dimensions.

     c. What are the dimensions of the rug?

Math Link

When solving an equation, always check that the apparent solutions do not make any denominators in the original equation equal to 0.

8. When 20 is added to a number, the result is the square of the number. What could the number be? Show how you found your answer.

9. The sum of the squares of two consecutive integers is 145. Find all possibilities for the integers. Show how you found your answer.

10. Gabriela was trying to solve the equation $(x + 1)(x - 2) = 10$. This is how she reasoned.

> Two factors of 10 are 5 and 2.
> So, $x + 1 = 5$ must be one solution of the equation.
> That means $x = 4$.
> I'll check: $(4 + 1)(4 - 2) = 5 \cdot 2 = 10$.
> It checks!

a. What would have happened if Gabriela had guessed that $x - 2 = 5$?

b. Do you think Gabriela's method is an efficient way to solve quadratic equations? Explain.

c. Solve Gabriela's equation, $(x + 1)(x - 2) = 10$. Start by expanding and then rearranging. Check each solution.

d. Solve $(x + 5)(x - 2) = 30$.

Share & Summarize

1. Create a problem involving area that requires solving a quadratic equation.

2. Try to solve your problem by factoring. If you can, give the solutions of the equation and then answer the question. If not, explain why the expression cannot be factored.

Real-World Link

The quadratic equation $K = \frac{1}{2}(64)v^2$ gives the kinetic energy of a skydiver with a mass of 64 kg, or about 141 lb, falling through the sky with velocity v in m/s.

Practice & Apply

Use the rectangle model to find the binomial factors of each trinomial.

1. $p^2 + 6p + 9$

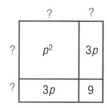

2. $t^2 + 7t + 12$

3. Draw a rectangle model to find the binomial factors of $b^2 + 2b + 1$.

Solve each equation.

4. $(x + 5)(x + 7) = 0$

5. $(x - 5)(x + 7) = 0$

6. $(x - 5)(x - 7) = 0$

7. $(x + 5)(x - 7) = 0$

In Exercises 8–17, determine whether the expression on the left of the equals sign is a difference of squares or a perfect square trinomial. If it is, indicate which and then factor the expression. Solve the equation for x. If the expression is in neither form, say so.

8. $x^2 - 49 = 0$

9. $x^2 + 49 = 0$

10. $x^2 + 14x - 49 = 0$

11. $x^2 - 14x + 49 = 0$

12. $49 - x^2 = 0$

13. $x^2 + 14x + 49 = 0$

14. $a^2x^2 + 4ab + b^2 = 0$

15. $a^2x^2 + 4abx + 4b^2 = 0$

16. $m^2x^2 - n^2 = 0$

17. $m^2x^2 + n^2 = 0$

18. In Your Own Words List the steps you would follow to solve a quadratic equation by factoring or to decide that it cannot be solved that way.

In Exercises 19-24, factor each quadratic expression that can be factored using integers. Identify those that cannot. Explain why they cannot be factored.

19. $d^2 - 15d + 54$

20. $g^2 - g - 6$

21. $z^2 + 2z - 6$

22. $h^2 - 3h - 28$

23. $2x^2 - 8x - 10$

24. $3c^2 - 9c + 6$

In Exercises 25–35, solve each equation by factoring using integers, if possible. If an equation cannot be solved in this way, explain why.

25. $k^2 + 15k + 30 = 0$ **26.** $n^2 - 17n + 42 = 0$

27. Challenge $2b^2 - 21b + 10 = 0$

28. Challenge $8r^2 + 5r - 3 = 0$

29. $4x + x^2 = 21$ **30.** $h^2 + 12 = 3h$

31. $14e = e^2 + 24$ **32.** $g^2 + 64 = 16g$

33. $u^2 + 5u = 36$ **34.** $(x + 3)(x - 4) = 30$

35. $\dfrac{(x + 1)^3}{x + 1} = 5x + 5$ (Hint: Simplify the fraction first.)

36. Carlos multiplied a number by itself and then added six. The result was five times the original number. Write and solve an equation to find his starting number.

Connect **&** *Extend*

37. Consider the following quadratic expressions.

$$x^2 + 5x + 2 \qquad z^2 + 10z + 6 \qquad n^2 + 8n + 3$$

     a. What happens when you try to draw rectangle models for these expressions?

     b. What do you think this means?

38. Because 7 is not a perfect square, the expression $4x^2 - 7$ does not look like the difference of two squares. But 7 *is* the square of *something*.

     a. Seven is the square of what number?

     b. How can you use your answer to Part a to factor $4x^2 - 7$ into a product of two binomials?

39. Geometry Each of these expressions represents one of the shaded areas below.

    i. $D^2 - d^2$ **ii.** $\pi(r + w)^2 - \pi r^2$

    iii. $(d + w)^2 - d^2$ **iv.** $4r^2 - \pi r^2$

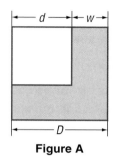

Figure A

Figure B

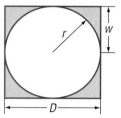

Figure C

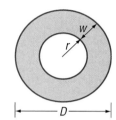

Figure D

    a. Match each expression with one of the shaded areas so that *every* figure is matched to a different expression.

    b. Write each expression in factored form. Factor out common factors, if possible.

In Exercises 40–45, factor the expression on the left side of each equation as much as possible. Find all the possible solutions. It will help to remember that $x^4 = (x^2)^2$, $x^8 = (x^4)^2$, and $x^3 = x(x^2)$.

40. $x^4 - 1 = 0$ **41.** $x^8 - 1 = 0$

42. $x^3 - 16x = 0$ **43.** $x^3 - 6x^2 + 9x = 0$

44. $x^4 - 2x^2 + 1 = 0$ **45.** $x^4 + 2x^2 + 1 = 0$

Solve each equation. Be sure to check your answers. (Hint: Try factoring the numerator first.)

46. $\dfrac{x^2 + 6x + 9}{x + 3} = 10$ **47.** $\dfrac{16x^2 - 81}{4x + 9} = 31$

48. The Numkenas built a small, square patio from square bricks with side lengths of one foot. They bought just enough bricks to build the patio. But after they built it, they decided it was too small.

To extend the length and width of the patio by d feet, they had to buy 24 more bricks. The original side length of the patio was 5 feet.

a. Draw a diagram to represent this situation. Be sure to show both the original patio and the new one.

b. Write an equation to represent this situation.

c. Simplify your equation. Solve it to find d, the amount by which the patio's length and width were increased. Check your answer.

49. Geometry The number of diagonals in an n-sided polygon is given by the equation $D = \dfrac{n^2 - 3n}{2}$. Some examples are shown below.

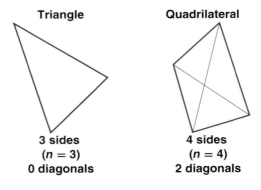

Triangle
3 sides
($n = 3$)
0 diagonals

Quadrilateral
4 sides
($n = 4$)
2 diagonals

Pentagon
5 sides
($n = 5$)
5 diagonals

Some of the following numbers are the number of diagonals in a polygon. For each possible value of D, set up an equation. Try to solve it for n by factoring. If an equation cannot be factored using integers, D cannot be the number of diagonals in a polygon. Indicate which values cannot be the number of diagonals in a polygon.

a. 20   **b.** 30   **c.** 35   **d.** 50   **e.** 54

50. The *triangular numbers* are a sequence of numbers that begins

$$1, 3, 6, 10, \ldots .$$

The numbers in this sequence represent the number of dots in a series of triangular shapes.

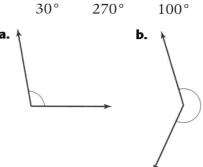

Triangle 1   Triangle 2   Triangle 3   Triangle 4

T, the number of dots in Triangle *n*, is given by this quadratic equation.

$$T = \frac{1}{2}(n^2 + n)$$

Some of the following numbers are triangular numbers. For each possible value of *T*, set up an equation. Try to solve it for *n* by factoring. If an equation cannot be factored using integers, *T* cannot be a triangular number. Indicate which numbers are not triangular.

a. 55   **b.** 120   **c.** 150   **d.** 200   **e.** 210

Mixed Review

51. Geometry There are 360° in a full circle. Without using a protractor, match each angle with one of the angle measurements below.

30°   270°   100°   50°   220°   80°

a.   **b.**   **c.**

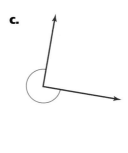

d.   **e.**   **f.**

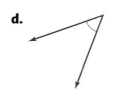

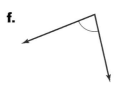

52. A rule for translating a figure on a coordinate grid changes the original coordinates (x, y) to the image coordinates $(x - 2, y + 3)$. On a coordinate grid, show the translation vector for this rule.

Simplify each expression as much as possible.

53. $\sqrt{34}$ **54.** $\sqrt{99x^4}$ **55.** $-\sqrt{60b}$

Completing the Square

Factoring is a very useful tool for solving equations. But factors and squares are not always easy to find. In Lessons 9.3 and 9.4, you will learn some techniques that will enable you to solve *every* quadratic equation.

You have used "doing the same thing to both sides" to solve linear equations. In this strategy, you write a series of equivalent equations that have the same solutions as the original equation but are easier to solve. You can also use this strategy with equations that contain square roots.

Example

Solve $\sqrt{3m + 7} = 5$.

$$3m + 7 = 25 \qquad \text{after squaring both sides}$$
$$3m = 18 \qquad \text{after subtracting 7 from both sides}$$
$$m = 6 \qquad \text{after dividing both sides by 3}$$

Think & Discuss

Why are both sides squared in the first solution step of the example?

In general, what kinds of "same things" do you know you can do to both sides to solve an equation?

Would you get an equivalent equation if you added 1 to the numerator of the fractions on both sides of an equation? Try it with $\frac{x}{2} = \frac{2x}{4}$.

What happens to an equation, such as $x = 2$, when you multiply both sides by x? Does the new equation have the same solutions?

What happens to the set of solutions when you multiply both sides of an equation, such as $x = 2$, by 0?

What effect would squaring both sides have on an equation? For example, begin with $x = 5$.

When you take the square root of both sides of an equation, what should you do to keep the same set of solutions? For example, begin with $w^2 = 36$.

Investigation ① Perfect Squares

If you can rearrange a quadratic equation into a form with a quadratic expression that is a perfect square on one side and a constant on the other side, you can solve the equation by taking the square root of both sides.

Example

Solve the equation $x^2 + 2x + 1 = 7$.

First, notice that $x^2 + 2x + 1$ is equal to $(x + 1)^2$. So, the equation $x^2 + 2x + 1 = 7$ can be solved by taking the square root of both sides.

$$(x + 1)^2 = 7$$

$$\sqrt{(x + 1)^2} = \sqrt{7} \text{ or } -\sqrt{7}$$

To write "$\sqrt{7}$ or $-\sqrt{7}$" more easily, use the $\pm$ symbol. The expression $\pm\sqrt{7}$ refers to both numbers, $\sqrt{7}$ and $-\sqrt{7}$.

$$\sqrt{(x + 1)^2} = \pm\sqrt{7}$$

$$x + 1 = \pm\sqrt{7}$$

$$x = -1 \pm \sqrt{7}$$

So, the solutions are $-1 + \sqrt{7}$ and $-1 - \sqrt{7}$.

✔ Develop & Understand: A

Find exact solutions of each equation, if possible, using any method you like.

Math Link
An *exact* solution does not involve approximations. For example, $x = \sqrt{2}$ is an exact solution of $x^2 = 2$, while $x = 1.414$ is an approximate solution to the nearest thousandth.

1. $(x - 3)^2 = 36$

2. $(k - 1)^2 - 25 = 0$

3. $2(r - 7)^2 = 32$

4. $(a - 4)^2 + 2 = 0$

5. $2(b - 3)^2 + 5 = 55$

6. $3(2c + 5)^2 - 63 = 300$

7. $(x - 4)^2 = 3$

8. $2(r - 3)^2 = -10$

9. $4(x + 2)^2 - 3 = 0$

10. Find approximate solutions of the equations in Exercises 7 and 9 to the nearest hundredth.

To use the solution method demonstrated in the example on the previous page, you need to be able to recognize quadratic expressions that can be rewritten as perfect squares. You worked with such *perfect square trinomials* in the last lesson.

✅ Develop & Understand: B

11. Which of these are perfect squares?

    a. $x^2 + 6x + 9$ b. $b^2 + 9$ c. $x^2 + 6x + 4$

    d. $m^2 + 12m - 36$ e. $m^2 - 12m + 36$ f. $y^2 + y + \frac{1}{4}$

    g. $r^2 - 16$ h. $1 + 2r + r^2$ i. $y^2 - 2y - 1$

12. Which of these are perfect squares?

    a. $4p^2 + 4p + 1$ b. $4q^2 + 4q + 4$ c. $4s^2 - 4s - 1$

    d. $4t^2 - 4t + 1$ e. $4v^2 + 9$ f. $4w^2 + 12w + 9$

13. Describe how you can tell whether an expression is a perfect square without factoring it.

Suppose you know the x^2 and x terms in a quadratic expression. You want to make it into a perfect square trinomial by adding a constant. How can you find the missing term?

Real-World Link

The quadratic equation $W = \frac{1}{2}kx^2$ describes the amount of work needed to stretch a spring x cm beyond its normal length in a unit called *Joules*. The value of k depends on the spring's strength.

. . . .

─ Example

If $x^2 + 20x + \underline{\hspace{1cm}}$ is a perfect square, then

$$x^2 + 20x + \underline{\hspace{1cm}} = (x + ?)^2.$$

The middle term of the expansion is twice the product of the coefficient of x and the constant term in the binomial being squared. What is the missing term?

$$20 = (2)(1)(?)$$

So, 10 is the missing term and the perfect square must be $(x + 10)^2$, or $x^2 + 20x + 100$.

✅ Develop & Understand: C

Complete each quadratic expression to make it a perfect square. Then write the completed expression in factored form.

14. $x^2 - 18x +$ _____

15. $x^2 + 22x +$ _____

16. $k^2 - 3k +$ _____

17. $25m^2 + 10m +$ _____

18. $16r^2 - 8r \square$ _____

19. $4z^2 - 12z \square$ _____

Share & Summarize

1. Why is it useful to look for perfect squares in quadratic expressions?

2. How can you recognize a perfect square trinomial?

Investigation 2 — Solve Quadratics by Completing the Square

In Exercises 1–10 on page 493, you learned that it is easy to solve a quadratic equation with a perfect square on one side and a constant on the other. You also solved equations that had a perfect square with a constant added or subtracted. A technique called *completing the square* can be used to rearrange quadratic equations into this form.

In Exercises 14–19 above, you found the constant that should be added to transform a quadratic expression into a perfect square. Using the same idea, some expressions that are not perfect squares can be rewritten as perfect squares with a constant added or subtracted.

Example

$x^2 + 6x + 10$ is not a perfect square. For $x^2 + 6x +$ _____ to be a perfect square, the added constant must be 9 because $x^2 + 6x + 9$ is a perfect square. Can you see why?

This means $x^2 + 6x + 10$ is 1 more than a perfect square. We can use this to rewrite the expression as a square plus 1.

$$x^2 + 6x + 10 = (x^2 + 6x + 9) + 1$$

$$= (x + 3)^2 + 1$$

✅ Develop & Understand: A

Rewrite each expression as a square with a constant added or subtracted.

1. $x^2 + 6x + 15 = x^2 + 6x + 9 + \underline{\quad} = (x + 3)^2 + \underline{\quad}$

2. $k^2 - 6k + 30 = k^2 - 6k + 9 + \underline{\quad} = (k - 3)^2 + \underline{\quad}$

3. $s^2 + 6s - 1 = s^2 + 6s + 9 - \underline{\quad} = (s + \underline{\quad})^2 - \underline{\quad}$

4. $r^2 - 6r - 21 = r^2 - 6r + 9\ \square\ \underline{\quad} = (r\ \square\ \underline{\quad})^2\ \square\ \underline{\quad}$

5. $m^2 + 12m + 30$

6. $h^2 - 5h$

7. $9r^2 + 18r - 20$

8. $9n^2 - 6n + 11$

Marcus and Lydia want to solve the equation $x^2 - 6x - 40 = 0$.

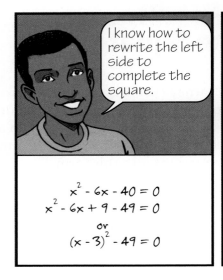

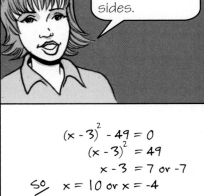

This method of solving equations is called *completing the square*.

✅ Develop & Understand: B

Find exact solutions of each equation by completing the square.

9. $x^2 - 8x - 9 = 0$

10. $w^2 - 8w + 6 = 0$

11. $9m^2 + 6m - 8 = 0$

What can you do if the coefficient of the squared variable is not a square? One approach is to first do the same thing to each side to produce an equivalent equation with 1, or some other square number, as the coefficient of the squared variable.

✅ Develop & Understand: C

12. To solve $2x^2 - 8x - 1 = 0$, you could divide both sides by 2, which gives the equivalent equation $x^2 - 4x - \frac{1}{2} = 0$. Complete the solution by solving this equivalent equation.

13. Use the method from Exercise 12 to solve $2m^2 - 12m + 7 = 0$.

14. Consider the equation $18x^2 - 12x - 3 = 0$.

    a. Try dividing the equation by 18 to make the coefficient of x^2 equal to 1.

    b. Now think about the coefficient 18. Find another number you could divide into 18 to get a perfect square. Divide the equation by that number.

    c. Use your answer to Part a or Part b to solve the equation.

15. Explain why $x^2 + 64 = 16x$ has only one solution.

16. Explain why $g^2 - 4g + 11 = 0$ has no solutions.

Share & Summarize

1. Give an example of a quadratic equation that is not a perfect square but that is easy to solve by completing the square. Solve your equation.

2. Suppose you have an equation in the form $y = ax^2 + bx + c$ for which the coefficient of x^2 is not a perfect square. How can you solve the equation? Illustrate your answer with an example.

Practice & Apply

Solve each equation.

1. $(x + 3)^2 = 25$

2. $(t + 7)^2 = 81$

3. $(s - 2)^2 - 5 = 31$

4. $(r - 8)^2 + 3 = 52$

5. $(2m + 1)^2 - 4 = 117$

6. $3(x - 3)^2 = 30$

7. $-2(y - 7)^2 + 4 = 0$

8. $4(2z + 3)^2 - 2 = -1$

Complete each quadratic expression so that it is a perfect square. Then write the completed expression in factored form.

9. $x^2 - 8x\ \square$ _____

10. $b^2 + 9b\ \square$ _____

11. $81d^2 - 90d\ \square$ _____

12. $9k^2 + 24k\ \square$ _____

Rewrite each expression as a square with a constant added or subtracted.

13. $r^2 - 6r + 1 = r^2 - 6r + 9 +$ _____ $= (r\ \square$ ____$)^2 -$ _____

14. $r^2 + 6r + 6 = (r\ \square$ ____$)^2\ \square$ _____

15. $p^2 - 16p + 60$

16. $g^2 - 3g - 1$

17. $a^2 + 10a + 101$

18. $4x^2 + 4x + 2$

Solve each equation by completing the square.

19. $t^2 + 6t + 2 = 0$

20. $p^2 + 4p + 1 = 0$

21. $v^2 - 2v - 6 = 0$

22. $m^2 + 2m - 11 = 0$

23. $n^2 + 14n = 2n + 3$

24. $b^2 - 3b = 3b + 7$

25. $x^2 - 6x = -5$

26. $a^2 + 10a + 26 = 0$

27. $2x^2 + 4x - 1 = 0$

28. $2u^2 + 3u - 2 = 0$

Real-World Link

The acceleration, in m/s^2, of a bicyclist coasting down a hill might be given by the quadratic equation $a = 0.12 - 0.0006v^2$, where v is the bike's velocity in m/s.

Connect & Extend

29. Stephen, Consuela, and Kwame each made up a number puzzle for their teacher, Mr. Karnowski.

- Stephen said, "I'm thinking of a number. If you subtract 1 from my number, square the result, and add 5, you will get 4."
- Consuela said, "I'm thinking of a number. If you subtract 1 from my number, square the result, and add 1, you will get 1."
- Kwame said, "I'm thinking of a number. If you double the number, subtract 5, square the result, and add 1, you will get 10."

After thinking about the puzzles, Mr. Karnowski said, "One of your puzzles has one solution, one of them has two solutions, and one doesn't have a solution."

Whose puzzle is which? Write an equation for each puzzle. Explain your answer.

30. Brianna and Lucita are playing tennis. On one volley, the height of the ball h, in feet, could have been described with the following equation, where t is the time in seconds since Brianna hit the ball.

$$h = -16(t - 1)^2 + 20$$

Assuming Lucita will let the ball bounce once, when will it hit the ground? Write and solve an equation to help you answer this question. Give your answer to the nearest hundreth of a second.

31. When you start the process of completing the square for an equation, you may be able to tell whether the equation has solutions without solving it.

a. Express each of these using a perfect square plus a constant. Without solving, decide whether the equation has a solution. Explain your answer.

   i. $x^2 + 6x + 15 = 0$

  ii. $x^2 + 6x + 5 = 0$

b. State a rule for determining whether an equation of the form $(x + a)^2 + c = 0$ has solutions. Explain your rule.

32. Geometry A rectangular painting has an area of 25 square feet. One side is two feet longer than the other.

    a. Quickly estimate approximate values for the lengths of the painting's sides. Do your estimates give an area that is too large or too small?

    b. Write an equation relating the sides and the area of the painting. Solve it exactly by completing the square.

    c. Compare your answer in Part b to your approximation in Part a.

33. History You may recall that when the famous German mathematician Gauss was a young boy, he amazed his teacher by rapidly computing the sum of the integers from 1 to 100. He realized that he could compute the sum without adding all the numbers by grouping the 100 numbers into pairs.

To see a shortcut for finding this sum, look at two lists of 1 to 100, one in reverse order.

1	2	3	4	5	6	7	...	50	...	94	95	96	97	98	99	100
100	99	98	97	96	95	94	...	51	...	7	6	5	4	3	2	1

    a. What is the sum of each pair?

    b. How many pairs are there?

    c. What is the sum of all these pairs?

    d. How many times is each of the integers from 1 to 100 counted in this sum?

    e. Consider your answers to Parts c and d. What is the sum of the integers from 1 to 100?

    f. Explain how you can use this same reasoning to find the sum of the integers from 1 to n for any value of n. Write a formula for s, the sum of the first n positive integers.

    g. Chloe added several consecutive numbers, starting at 1, and found a sum of 91. Write an equation you could use to find the numbers she added. Solve your equation by completing the square. Check your answer with the formula.

34. In Your Own Words Why are perfect squares useful in solving quadratic equations?

Mixed Review For the linear equations in Exercises 35–37, answer Parts a and b.

a. What is the constant difference between the y values as the x values increase by 1?

b. What is the constant difference between the y values as the x values decrease by 3?

35. $y = \dfrac{x}{2}$ **36.** $2x - 2 = y$ **37.** $\dfrac{y}{3x} = 3$

For Exercises 38–42, solve each inequality. Graph each solution on a number line.

38. $8y + 2 > 26$

39. $-4m + 9 \geq 19$

40. $7p + 12 < 10p - 21$

41. $3(2k + 5) \leq 5k + 9$

42. $0 > 5(7 - x) + 12x$

43. Geometry This figure has both reflection symmetry and rotation symmetry.

a. How many lines of symmetry does it have?

b. What is the angle of rotation?

For Exercises 44–49, simplify each expression. Answers should contain only positive exponents.

44. a^{-2} **45.** $3m^{-4}$

46. $\left(c^3 d^{-2}\right)\left(c^{-10} d^8\right)$ **47.** $\left(5m^2 n^{-3}\right)^4$

48. $\left(\dfrac{6a}{7b}\right)^{-2}$ **49.** $\left(\dfrac{r^{-3}}{2s^{-5}}\right)^4$

The Quadratic Formula

You have seen that some quadratic equations are easier to solve than others. Some can be solved quickly by factoring or by taking the square root of both sides. Any quadratic equation can be solved by completing the square, but it is not always obvious what has to be done.

Look again at the process of solving an equation by completing the square. The general quadratic equation, $ax^2 + bx + c = 0$, is solved below by completing the square.

To help you see each step more easily, a specific quadratic equation is solved alongside the general equation.

General Equation	**Specific Equation**
$ax^2 + bx + c = 0$	$2x^2 + 8x + \frac{1}{2} = 0$

Step 1. Divide by a.

$$x^2 + \frac{b}{a}x + \frac{c}{a} = 0 \qquad\qquad x^2 + 4x + \frac{1}{4} = 0$$

Step 2. Complete the square.

$$x^2 + \frac{b}{a}x + \frac{b^2}{4a^2} + \frac{c}{a} - \frac{b^2}{4a^2} = 0 \qquad x^2 + 4x + 4 + \frac{1}{4} - 4 = 0$$

Step 3. Rearrange terms.

$$x^2 + \frac{b}{a}x + \frac{b^2}{4a^2} = \frac{b^2}{4a^2} - \frac{c}{a} \qquad\qquad x^2 + 4x + 4 = 4 - \frac{1}{4}$$

Step 4. Factor the left side. Use a common denominator on the right side.

$$\left(x + \frac{b}{2a}\right)^2 = \frac{b^2 - 4ac}{4a^2} \qquad\qquad (x + 2)^2 = \frac{15}{4}$$

Step 5. Take the square root of both sides.

$$x + \frac{b}{2a} = \frac{\pm\sqrt{b^2 - 4ac}}{2a} \qquad\qquad x + 2 = \pm\frac{\sqrt{15}}{2}$$

Step 6. Subtract the constant added to x.

$$x = -\frac{b}{2a} \pm \frac{\sqrt{b^2 - 4ac}}{2a} \qquad\qquad x = -2 \pm \frac{\sqrt{15}}{2}$$

$$x = \frac{-b \pm \sqrt{b^2 - 4ac}}{2a} \qquad\qquad x = \frac{-4 \pm \sqrt{15}}{2}$$

This process gives us a formula that can be used to find the solutions of *any* quadratic equation.

The Quadratic Formula

The solutions of $ax^2 + bx + c = 0$ are

$$x = \frac{-b \pm \sqrt{b^2 - 4ac}}{2a}.$$

That is, the solutions are

$$x = \frac{-b + \sqrt{b^2 - 4ac}}{2a} \quad \text{and} \quad x = \frac{-b - \sqrt{b^2 - 4ac}}{2a}.$$

Think & Discuss

Identify the values a, b, and c. Use the quadratic formula to solve this equation.

$$x^2 + 3x - 5 = 0$$

Now solve the equation by completing the square. Do you get the same answer? Which method seems easier?

Investigation 1   Use the Quadratic Formula

This investigation will help you learn to use the quadratic formula.

Develop & Understand: A

For Exercise 1–8, do the following.

- Identify the values of a, b, and c referred to in the quadratic formula. You may need to rewrite the equation in the form $ax^2 + bx + c = 0$.
- Solve the equation using the quadratic formula.

1. $2x^2 + 3x = 0$

2. $7x^2 + x - 3 = 0$

3. $3 - x^2 + 2x = 0$

4. $6x + 2 = x^2$

5. $2x^2 = x - 5$

6. $x^2 - 12 = 0$

7. $x^2 = 5x$

8. $x(x - 6) = 3$

9. Consider the equation $x^2 + 3x + 2 = 0$.

    a. Solve the equation by factoring.

    b. Now use the quadratic formula to solve the equation.

    c. Which method seems easier?

    d. Which of Exercises 1–8 could you have solved by factoring?

✓ Develop & Understand: B

For each problem, do the following.

- Solve the equation using any method you like. Check your answers.
- If you did not solve the equation by factoring, decide whether you could have used factoring to solve it.

10. $x^2 - 5x + 6 = 0$ 11. $w^2 - 6w + 9 = 0$

12. $t^2 + 4t + 1 = 0$ 13. $x^2 - x + 2 = 0$

14. $k^2 + 4k + 2 = 0$ 15. $3g^2 - 2g - 2 = 0$

16. $z^2 - 12z + 36 = 0$ 17. $2e^2 + 7e + 6 = 0$

18. $x^2 + x = 15 - x$ 19. $3n^2 + 14 = 8n^2 + 3n$

Share & Summarize

1. What is the connection between the quadratic formula and the process of completing the square?

2. You have learned several methods for solving quadratic equations. These are backtracking, factoring, completing the square, and the quadratic formula.

    a. Which of these can be used with only some quadratic equations?

    b. Which can be used with all quadratic equations?

3. When you are given a quadratic equation to solve, how do you choose a solution method?

Investigation 2 Apply the Quadratic Formula

In Chapter 8, you examined quadratic equations in specific situations and estimated solutions using a graphing calculator. Now you will apply the quadratic formula to solve them exactly.

✅ Develop & Understand: A

In some of these situations, the quadratic formula will give two solutions. Make sure your answers make sense within the given context.

1. Josefina, a tapestry maker, has a client who wants a handwoven design with an area of four square meters. Josefina decides it will have a rectangular shape and a length one meter longer than its width.

   a. Write an equation representing the client's requirements and Josefina's decision.

   b. Use the quadratic formula to find the width and the length of the tapestry. Express your answer in two ways, exactly and to the nearest centimeter.

2. Another client wants a rectangular tapestry with an area of 6 square meters, but she insists the length be exactly 2 meters longer than the width.

   Write and solve an equation that represents this situation. Give exact dimensions of the tapestry. Then estimate the dimensions to the nearest centimeter.

3. Jesse threw a superball to the ground. It bounced straight up with an initial speed of 30 feet per second. The height h in feet t seconds after the ball left the ground is given by the formula $h = 30t - 16t^2$. Write and solve an equation to find the value of t when the ball returns to the ground.

4. When an object is thrown straight upward, its height h in feet after t seconds can be estimated using the formula $h = s + vt - 16t^2$, where s is the initial height (at $t = 0$) and v is the initial velocity. In Exercise 3, s was 0 feet and v was 30 feet per second. So, the ball's height was estimated by $h = 30t - 16t^2$.

Suppose Jesse threw the ball upward instead of bouncing it so that the ball's height, when $t = 0$, was 5 feet above the ground. But the initial velocity was still 30 ft/s.

   a. Write an equation describing the height h of Jesse's ball after t seconds.

   b. Write and solve an equation to find how long it takes the ball to reach the ground.

5. If an object is dropped with an initial velocity of 0, its height can be approximated by adding its starting height to $-16t^2$, which represents the effect of gravity. For example, the height of a rock dropped from 20 feet above the ground can be approximated by the formula $h = -16t^2 + 20$.

   Write and solve an equation to determine how many seconds pass until a rock dropped from 100 ft hits the ground.

✅ Develop & Understand: B

The town of Seaside is considering allowing a large tourist resort to be built along the oceanfront. Some residents are in favor of the plan because it will bring income to the community. Others are against it, saying it will disrupt their lifestyle.

The state tourism board has a formula for computing the overall tourism rating T of an area based on two factors: the uniqueness rating U and the amenities rating A.

6. The amenities rating scale A is used to assess the attractiveness of a tourist destination, including how easy it is to find a place to stay. Seaside currently has a rating of 5. It is estimated that for every 100 beds the resort opens to tourists, A will increase by 2 points.

   If the resort has p hundreds of beds, what is the estimate for the new amenities rating?

7. The uniqueness rating scale U is used to assess the special features that will attract tourists. Seaside currently has a high uniqueness rating, 20, because dolphins are often sighted close to the local beaches.

A committee has gathered evidence that an increase in tourists will keep dolphins from coming near shore. They estimate that for every 100 beds in the resort, U will drop by 2 points.

If the resort has p hundreds of beds, what is the estimate for the new uniqueness rating?

8. The overall tourism rating T is computed by multiplying A and U.

   a. What is the town's current tourism rating?

   b. Write an expression in terms of p for the estimated tourism rating if p hundreds of beds are added.

   c. Use your expression from Part b to decide for what values of p there would be no change in the tourism rating. (Hint: Write and solve an equation.)

   d. For what numbers of beds would the resort create a decrease in the tourism rating?

   e. Seaside's town council believes that the disruption to the town's lifestyle could not be justified unless the development resulted in an increase in the tourism rating to at least 140 points.

      i. Use your expression from Part b to decide for what values of p you would expect to achieve a tourism rating of 140 points.

      ii. What values of p would give a tourism rating *over* 140 points?

   f. What would you advise the council to do?

Share & Summarize

Explain the steps involved in using the quadratic formula to solve an equation.

Investigation (3) What Does $b^2 - 4ac$ Tell You?

In some situations, you might be more interested in knowing *how many* solutions a quadratic equation has than exactly what the solutions *are*. For example, suppose you are thinking about the height of a thrown ball at various times. You could solve a quadratic equation to find at what time the ball reaches a certain height.

However, if you want to know only *whether* it reaches that height and do not care about *when* it does, you want the answer to the following question.

Does this equation have any real solutions?

You need only part of the quadratic formula, the expression $b^2 - 4ac$, to answer this question.

Math Link

The quadratic formula is
$$x = \frac{-b \pm \sqrt{b^2 - 4ac}}{2a}.$$

Think & Discuss

You have seen examples of quadratic equations that have no real solutions. Sometimes this is easy to tell without using the quadratic formula. Give an example of such an equation. Explain how you know that it does not have real solutions.

Some quadratic equations have exactly one real solution. Give an example.

Of course, many quadratic equations have two real solutions. Give an example.

✓ Develop & Understand: A

You will now investigate the relationship between the value of $b^2 - 4ac$ and the number of real solutions of $ax^2 + bx + c = 0$.

1. The equation $x^2 + 1 = 0$ has no real solutions.

   a. Explain why this is true.

   b. What is the value of $b^2 - 4ac$ for this equation? Is it positive, negative, or 0?

   c. Give another example of a quadratic equation that you know has no real solutions. Find the value of $b^2 - 4ac$ for your example. Is it positive, negative, or 0?

Real-World Link

The acceleration of the passengers on a carnival ride might be given by the quadratic equation $a = 0.2v^2$, where a is the acceleration toward the center of the ride in m/s² and v is the velocity in m/s.

2. The equation $(x - 3)(x + 5) = 0$ has two solutions.

   a. Express this equation in the form $ax^2 + bx + c = 0$. Find the value of $b^2 - 4ac$. Is it positive, negative, or 0?

   b. Give another example of a quadratic equation with two solutions. Find the value of $b^2 - 4ac$ for your equation.

3. The expression $x^2 + 2x + 1$ is a perfect square trinomial since it is equal to $(x + 1)^2$. The equation $x^2 + 2x + 1 = 0$ has one solution.

   a. Is the value of $b^2 - 4ac$ for this equation positive, negative, or 0?

   b. Give another example of a quadratic equation with one solution. Is the value of $b^2 - 4ac$ for your equation positive, negative, or 0?

4. As you know, a quadratic equation can have zero, one, or two solutions. This exercise will help you explain the connection between the value of $b^2 - 4ac$ and the number of solutions an equation has.

   a. Where does the expression $b^2 - 4ac$ occur in the quadratic formula?

   b. What value or values must $b^2 - 4ac$ have for the quadratic formula to give no solutions? Explain.

   c. What value or values must $b^2 - 4ac$ have for the quadratic formula to give one solution? Explain.

   d. What value or values must $b^2 - 4ac$ have for the quadratic formula to give two solutions? Explain.

✅ Develop & Understand: B

Find the number of solutions each equation has.

5. $2x^2 - 9x + 5 = 0$

6. $3x^2 - 7x + 9 = 0$

In Exercises 3–5 on pages 505 and 506, Jesse bounced a superball with a velocity of 30 feet per second as it left the ground. The ball's height is given by the formula $h = 30t - 16t^2$, where t is time in seconds since the ball left the ground.

7. You can use your knowledge of the quadratic formula to find how high the ball will travel. First, look at whether the ball will reach 100 feet.

   a. What equation would you solve to find if and when the height of the ball reaches 100 feet?

   b. Write your equation in the form $at^2 + bt + c = 0$.

   c. What is the value of $b^2 - 4ac$ for your equation?

   d. Will the ball reach a height of 100 feet? Explain.

8. **Challenge** This graph of $h = 30t - 16t^2$ can help you determine just how high the ball will go.

Superball Bounce Height

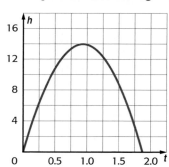

a. Suppose M is the maximum height reached by the ball. Write an equation to represent when the ball is at this height.

b. Write your equation in the form $at^2 + bt + c = 0$.

c. How many solutions will this equation have? Hint: Look at the graph.

d. What do you know about the value of $b^2 - 4ac$ for a quadratic equation with the number of solutions that this equation has?

e. Use your answer to Part d to help you find the value of M. Show how you found your answer.

f. How high does the ball travel?

g. Write and solve an equation to find how long it takes the ball to reach this height.

Share & Summarize

1. Without actually solving it, how can you tell whether a quadratic equation has zero, one, or two solutions?

2. For a quadratic relationship in the form $y = ax^2 + bx + c$, how can you tell whether y ever has a certain value d?

Inquiry

Investigation 4 The Golden Ratio

Materials

- ruler
- graph paper (optional)

Real-World Link

The golden ratio was used in the construction of many buildings in ancient Greece, including the Parthenon. The dimensions of the front of the Parthenon form a perfect golden retangle.

In this investigation, you will work with a ratio that has been important since the time of the early Greeks. The ratio arises in many surprising places, including mathematics, art, music, architecture, and genetics.

What Do You Like?

1. Here are several rectangles. Which do you think is the most "appealing to the eye"? You do not need reasons for your answers. Simply state which one you like.

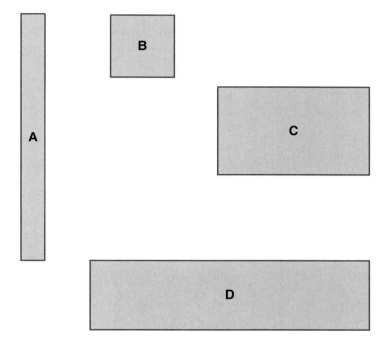

2. Draw some other rectangles that have a shape that is pleasing to you. Explain why you think one shape is more pleasing than another.

Many people think Rectangle C, and rectangles geometrically similar to it, is the most pleasing to the eye. It is called a *golden rectangle*. The ratio of its sides, the ratio of the long side to the short side, is the *golden ratio*.

Go on

One special property of a golden rectangle is that, when you add a square to its longer side to form a new rectangle, the new shape is similar to the original. So, the new rectangle is a golden rectangle and its sides are in the golden ratio.

Math Link

In similar figures, corresponding sides have lengths that share a common ratio and corresponding angles are congruent.

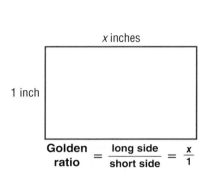

$$\text{Golden ratio} = \frac{\text{long side}}{\text{short side}} = \frac{x}{1}$$

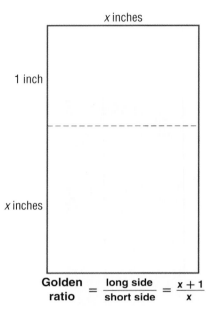

$$\text{Golden ratio} = \frac{\text{long side}}{\text{short side}} = \frac{x+1}{x}$$

Try It Out

3. Measure the dimensions of the two rectangles above. Determine whether they have the same $\dfrac{\text{long side}}{\text{short side}}$ ratio. What is the ratio?

4. Now find the $\dfrac{\text{long side}}{\text{short side}}$ ratio of each rectangle you drew in Question 2.

Solve It

5. The ratios of the two rectangles above must be equal. Write an equation setting the ratios equal to each other. That is, complete this equation.

$$\frac{x}{1} = \text{_____}$$

6. Now write your equation in the form $ax^2 + bx + c = 0$. (Hint: You will need to think about how to get x out of the denominator.)

7. Find the exact solutions of your equation. Then express the solutions to the nearest thousandth.

8. Do both of your solutions make sense? Explain.

9. What is the value of the golden ratio?

10. Compare the value of the golden ratio with the ratios you measured in Question 3. What do you notice?

11. Now compare the value of the golden ratio to the measurements you made of your own rectangles in Question 4. What do you notice?

Going Further

12. Using graph paper or ordinary paper and a ruler, you can draw a rectangle that is almost a golden rectangle.

   a. *Step 1:* Start with a square with side lengths of one unit in the top left corner of your page. What is the ratio of the sides?

   b. *Step 2:* Add another square next to the first to make a larger rectangle. What is the ratio of the long side to the short side of this new rectangle?

   c. *Step 3:* Add a square next to the longer side of your rectangle to make an even larger rectangle. What is the $\dfrac{\text{long side}}{\text{short side}}$ ratio of this new rectangle?

   d. Repeat Step 3 as many times as you can on your paper. What is the ratio for the final rectangle you make? Compare this value to the golden ratio you calculated in Question 9.

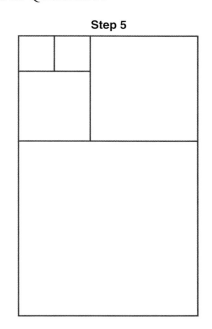

Step 4   **Step 5**

Inquiry

Math Link

The Fibonacci sequence is named for its discoverer Leonardo Fibonacci, also known as Leonardo Pisano, who was born about 1170 A.D. in the city of Pisa, Italy. He was one of the first people to introduce the Hindu-Arabic number system, which uses the digits 0 to 9 and a decimal point, into Europe.

13. Look at the dimensions of the rectangles you made in Question 12.
- The first is $1 \cdot 1$.
- The next four are $1 \cdot 2$, $2 \cdot 3$, $3 \cdot 5$, and $5 \cdot 8$.
- Listing only the smaller dimension in each rectangle gives the *Fibonacci sequence*.

$$1, 1, 2, 3, 5, 8, \ldots .$$

Look for a pattern in the Fibonacci sequence. What are the next two numbers? How did you find them?

14. Compute the sequence of ratios of Fibonacci numbers up to the ratio $\dfrac{\text{tenth Fibonacci number}}{\text{ninth Fibonacci number}}$, or $\dfrac{F_{10}}{F_9}$. The first two are computed below.

$$\frac{F_2}{F_1} = \frac{1}{1} = 1 \qquad \frac{F_3}{F_2} = \frac{2}{1} = 2$$

15. Compare the ratios to the golden ratio. What do you notice?

Finding Out More

16. The golden ratio and the Fibonacci numbers appear in many contexts both inside and outside mathematics. For example, pineapples have scales in sets of 8, 13, and 21 rows.

Look for answers to some of these questions at the library or on the Internet.
- How do the golden ratio and the Fibonacci sequence appear in the natural world?
- How has the golden ratio been used by Leonardo da Vinci and other artists?
- How is the golden ratio applied in architecture? In music?

Practice & Apply

For Exercises 1–4, solve each equation using the quadratic formula, if possible.

1. $2x^2 + 5x = 0$

2. $5x^2 + 7x + 4 = 0$

3. $c^2 - 10 = 0$

4. $b^2 + 10 = 0$

5. Solve the equation $9x^2 - 16 = 0$ by factoring and by using the quadratic formula.

6. Geometry The area of a photograph is 320 square centimeters. Its length is 2 cm more than twice its width. Write and solve an equation to find its dimensions.

7. Physical Science Suppose that, at some point into its flight, a particular rocket's height h, in meters, above sea level t seconds after launching is represented by the formula $h = 2t(60 - t)$.

    a. How many seconds after launching will the rocket return to sea level?

    b. Write and solve an equation to find when the rocket will be 1,200 m above sea level.

For Exercises 8–10, find the number of solutions to each quadratic equation without actually solving the equation. Explain how you know your answers are correct.

8. $x^2 + 2x + 3 = 0$

9. $x^2 - 2x - 3 = 0$

10. $9x^2 + 12x + 4 = 0$

11. A ball is thrown upward with a starting velocity of 40 feet per second from 5 feet above the ground. The equation describing the height h of the ball after t seconds is $h = 40t - 16t^2 + 5$.

    a. Will the ball travel as high as 100 feet? Explain.

    b. Will it travel as high as 15 feet? Explain.

    c. Challenge Find the ball's maximum height.

Connect & Extend

12. When Lourdes solved the equation $2x^2 - 13x = 24$, she was surprised to find that the solutions were exactly 8 and -1.5. Victor said he thought this meant the equation could have been solved by factoring.

    a. Write a quadratic equation in factored form that has the solutions 8 and -1.5.

    b. Expand the factors to write an equation without parentheses. Was Victor correct? (Hint: If your equation contains a fraction, try multiplying by its denominator to get only integers for coefficients.)

    c. Write one advantage and one disadvantage of using the quadratic formula to solve the equation $2x^2 - 13x = 24$.

13. Consider the equation $3x + \frac{1}{x} = 4$.

    a. Do you see any obvious solutions to this equation?

    b. Now solve the equation using the quadratic formula. (Hint: First, write an equivalent quadratic equation.) Check your solutions in the original equation.

Challenge Although these equations are not quadratic, the quadratic formula can help you solve them. Try to solve them. Explain your reasoning.

14. $(x^2 - 2x - 2)^2 = 0$

15. $x^3 - 2x^2 - 2x = 0$

16. History Here is a problem posed by the 12th-century Indian mathematician Bhaskara:*

The eighth part of a troop of monkeys, squared, was skipping in a grove and delighted with their sport. Twelve remaining monkeys were seen on the hill, amused with chattering to each other. How many were there in all?

That is, take $\frac{1}{8}$ of the entire troop and square the result. That number of monkeys, along with the 12 on the hill, form the entire troop. How many monkeys are there in the troop? Show your work.

17. In Chapter 7, you solved inequalities involving linear relationships. Use the same ideas here to solve inequalities involving quadratic relationships.

    a. First, use the quadratic formula to solve $x^2 - 3x - 7 = 0$.

    b. Use the information from Part a to help graph $y = x^2 - 3x - 7$. You may want to plot some additional points.

    Use your solutions and graph to solve each inequality.

    c. $x^2 - 3x - 7 < 0$

    d. $x^2 - 3x \geq 7$

    e. $x^2 - 3x \leq 7$

18. Consider the quadratic relationship $y = x(x - 1)$.

    a. For what values of x is $y = 0$?

    b. Is y positive or negative for x values between those you listed in Part a?

    c. Can y ever be equal to -1? Explain.

    d. Can y ever be equal to 1? Explain.

    e. Sketch a graph of this relationship.

    f. Challenge Use your knowledge of the quadratic formula and your graph to find the *minimum* value of y.

19. Challenge Use the quadratic formula to solve this exercise.

Jermaine wants to construct a large picture frame using a 20-foot strip of wood.

    a. Express the height and area of the frame in terms of its width.

    b. Sketch a graph of the relationship between area and width. Is there a maximum area or a minimum area?

    c. Use the quadratic formula to find the maximum or minimum area for Jermaine's frame. Explain how you found your answer.

    d. What dimensions give this area?

20. In Your Own Words Describe the relationship between the graph of $y = ax^2 + bx + c$ and solutions of the equation $ax^2 + bx + c = d$.

If the equation $ax^2 + bx + c = d$ has no solutions, what does that mean about the graph?

Mixed Review

Write each expression in the form 7^b.

21. $\dfrac{7^{23}}{7^{15}}$

22. $\left(7^3\right)^{10}$

23. $\left(\dfrac{1}{7}\right)^{11}$

Recall that for linear equations, first differences are constant. For quadratic equations, second differences are constant. Determine whether the relationship in each table could be linear, quadratic, or neither.

24.

x	y
−3	−4
−2	1
−1	4
0	5
1	4
2	1

25.

x	y
−3	7
−2	2
−1	−1
0	−2
1	−2
2	−1

26.

x	y
−3	0
−2	−0.5
−1	−1
0	−1.5
1	−2
2	−2.5

27.

x	y
−3	23
−2	13
−1	7
0	5
1	13
2	23

Rewrite each equation in slope-intercept form.

28. $2(y + x) + 1 = 3x - 2y + 3$

29. $6y + \dfrac{3}{7}x - 2 = 0$

30. $8 = -(3x + 4) + (4 - y) - (2y + 10)$

Review & Self-Assessment

Chapter Summary

Quadratic equations can be solved with several methods. In this chapter, you began with backtracking to solve quadratics of a particular form as well as equations requiring finding reciprocals, taking square roots, and changing signs.

You used rectangle models to find binomial factors. You then learned how to solve some quadratic equations by *factoring* and using the fact that when a product is equal to 0, at least one of the factors must be equal to 0.

As these methods do not work well for all quadratic equations, you also learned how to *complete the square* and to use the *quadratic formula*.

$$x = \frac{-b \pm \sqrt{b^2 - 4ac}}{2a}$$

Strategies and Applications

The questions in this section will help you review and apply the important ideas and strategies developed in this chapter.

Using backtracking to undo square roots, squares, reciprocals, and changes of sign

1. Identify the operation that undoes each given operation. Note the cautions, if any, you must take when undoing the given operation.

   a. taking the square root   **b.** taking the reciprocal

   c. changing sign   **d.** squaring

Indicate whether you can solve each equation directly by backtracking. If so, draw a flowchart and find the solution. If not, explain why not.

2. $\sqrt{2x + 3} - 4 = 7$ 3. $\dfrac{24}{y - 7} = 4$

4. $3a - \sqrt{2a + 3} - 4 = 7$ 5. $(4n + 5)^2 - 3 = 6$

Solving quadratic equations by factoring

Tell whether you can solve each equation by factoring using integers. If you can, do so. Show your work. If not, explain why not.

6. $g^2 + 3g = -6$ 7. $81x^2 + 1 = -18x$

8. $3k^2 - 5k - 12 = 12 + 2k^2$ 9. $4w^2 - 9 = 0$

10. $(x + 5)(x - 1) = -8$ 11. $2s^2 - 4s + 2 = 0$

Solving quadratic equations by completing the square

12. Explain what it means to solve by "completing the square." Use the equation $4x^2 + 20x - 8 = 0$ to illustrate your explanation.

13. Give an example of a quadratic equation that is possible, but not easy, to solve by completing the square.

Understanding and applying the quadratic formula

14. How was the quadratic formula derived? That is, what technique or method was used and on what equation?

15. Suppose a, b, c, and d are all numbers not equal to 0. Explain why the solutions of $ax^2 + bx + c = d$ are not $x = \dfrac{-b \pm \sqrt{b^2 - 4ac}}{2a}$.

16. How can you determine the number of solutions of a quadratic equation in the form $ax^2 + bx + c = 0$ using the value of $b^2 - 4ac$?

Demonstrating Skills

Factor each expression.

17. $a^2 + 3a$

18. $2b^2 - 2$

19. $c + 14c + 49$

20. $8d^2 - 8d + 2$

21. $e^2 + 8e - 9$

22. $f^2 + 7f + 10$

Write an expression equivalent to the given expression by completing the square.

23. $4g^2 + 12g - 3$

24. $h^2 - 10h + 7$

25. $2j^2 + 24j$

Tell how many solutions each equation has. Do not solve.

26. $k^2 + 10 = 20k - 90$

27. $2m^2 + 3m + 3 = -5$

Solve each equation, if possible.

28. $\sqrt{3n + 1} = 13$

29. $\dfrac{60}{-(2p - 3)} = 12$

30. $(7q + 3)(q - 8) = 0$

31. $(10r + 4)(5r + 4) = -2$

32. $4s^2 + 3s - 40 = 3s - 41$

33. $t^2 - 100 = 0$

34. $2u^2 - 4u = 14$

35. $9v^2 - 3 = 4v^2 + 32$

36. $5w^2 = 8w$

37. $3 - 9x - x^2 = 17$

Test-Taking Practice

SHORT RESPONSE

1 Find two solutions to the equation $0 = x^2 + 2x - 8$.

Show your work.

Answer _____

MULTIPLE CHOICE

2 Which equation can be solved by factoring using integers?
A $y = x^2 - 4x + 24$
B $y = x^2 + 3x + 16$
C $y = x^2 - 5x - 7$
D $y = x^2 - x - 12$

3 Which equation has solutions -4 and -2?
F $y = x^2 - 6x - 8$
G $y = x^2 + 6x + 8$
H $y = x^2 - 4x - 2$
J $y = x^2 + 4x - 8$

4 Which expression is equivalent to $x^2 - 4x - 2$?
A $(x + 2)^2 + 2$
B $(x - 2)^2 - 6$
C $(x - 4)^2 - 2$
D $(x - 2)^2 + 4$

5 Solve the equation below.
$$0 = 2x^2 - 8x + 6$$
F $-1, -3$
G $1, 3$
H $\frac{1}{4}, \frac{9}{4}$
J $2 \pm 4\sqrt{7}$

6 Solve the equation below.
$$\sqrt{5x - 4} = 11$$
A $x = 5$
B $x = 5.2$
C $x = 25$
D $x = 120$

7 Which trinomial is *not* a perfect square trinomial?
F $x^2 + 4x + 4$
G $x^2 + 8x + 16$
H $x^2 - 10x + 25$
J $x^2 + 10x - 8$

Functions and Their Graphs

Real-Life Math

Flattening the Globe Creating an accurate map of the world is difficult to do because you must show a three-dimensional surface using only two dimensions. Mathematical functions called *projections* help cartographers create maps. A projection assigns every point on a three-dimensional globe to a point on a two-dimensional surface, in effect, *flattening* the globe.

Some projections can create very interesting maps. The Mercator projection exaggerates the areas of landmasses farthest from the equator, such as Greenland and Antarctica. On this type of map, Greenland looks like it is almost the size of Africa. In fact, it has only about 7% of Africa's area. Goode's interrupted projection reduces this distortion but breaks the oceans and Antarctica into pieces.

Think About It In Geometry, you studied nets. How are nets of geometric solids similar to projections?

Math Online
Take the **Chapter Readiness Quiz** at glencoe.com.

Dear Family,

Chapter 10 is about functions and their graphs. The concept of functions is central to Algebra and has been a major thread throughout this course.

Key Concept—Functions

One useful way to think about a function is as a machine that takes some input, a number or something else, and produces an output.

The output must be *unique*, meaning you get only one output for a particular input. Also, the output must be *consistent* in that you get that output every time you use the same input. For example, if 3 is the input, it goes into the machine and is multiplied by 5. So, the output is 15. Every time you input 3, you will get the same answer, 15.

Functions are often expressed as mathematical sentences. Each rule below describes the function represented by the machine above, multiply by 5.

$$y = 5x \qquad f(x) = 5x \qquad g(t) = 5t$$

Once the class has looked at functions using input-output machines, students will go on to use graphs for finding the maximum and minimum values of functions. The class will also use functions to solve exercises.

Chapter Vocabulary

domain   range

function   *x*-intercepts

Home Activities

Help your student think of some situations that can be represented as functions, like the situations below.

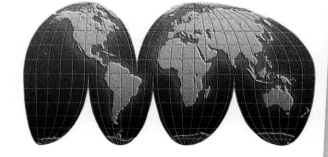

- Input: the total restaurant bill
  Output: the tip at 15%

- Input: the side length of a square
  Output: the area of that square

- Input: number of adult moviegoers
  Output: the total cost for tickets at $8.00 each

LESSON 10.1

Functions

In your study of Algebra, you have analyzed many relationships between variables.

A car traveling along a highway at 55 miles per hour for t hours will cover a distance of $55t$ miles. This can be represented by the equation $d = 55t$.

When a quarterback throws a football, the height of the ball in yards when it has traveled d yards might be described by the equation $h = 2 + 0.8d - 0.02d^2$.

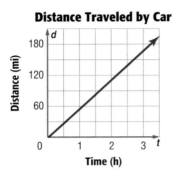

Distance Traveled by Car

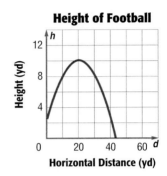

Height of Football

Vocabulary

function

Many of the relationships you have studied have a special name called *functions*. In mathematics, a **function** is a relationship between an input variable and an output variable in which there is only one output for each input.

- In the car example, the input variable is the time spent on the highway. The output variable is the distance traveled. Since there can be only one distance traveled for any given time, the relationship is a function. In this case, the distance traveled is a *function of* the time.

- In the football example, the input variable is the horizontal distance the ball has traveled. The output variable is the ball's height. Since there can be only one height for any given horizontal distance, the relationship is a function. In this case, the height is a *function of* the horizontal distance.

One way to think about a function is to imagine a machine that takes some input and produces an output. Input can be a number, a word, or something else, depending on the function.

For example, suppose you put 10 into a function machine for the football example. Since the machine is a function, the output must be *unique*. If you put 10 into the machine, it can give an output of 8, but it cannot give both 8 and some other number.

For a function machine, the output must be consistent. That is, the machine will always give the same output for the same input. If you get an output of 8 for an input of 10, then every time you put 10 into the machine, the output will be 8.

It *is* possible that two or more inputs will produce the same output. For example, the football-height function machine will produce 8 when you put 10 or 30 into it. Try it.

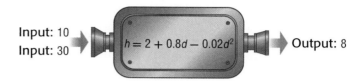

Input: 10
Input: 30
$h = 2 + 0.8d - 0.02d^2$
Output: 8

If more than one output is possible for a given input, the relationship is *not* a function. For example, a machine that outputs the square roots of a positive number cannot be a function because every positive number has *two* square roots.

Think & Discuss

Here are some examples of functions. For each function, explain why there is only one possible output for each input.

- Input: a number
  Output: twice that number

- Input: the name of a state
  Output: the state's capital

× 2

Capital?

- Input: an integer
  Output: classification as even or odd

- Input: the side length of a square
  Output: the area of that square

- Input: a person's Social Security number
  Output: that person's birth date

- Input: a word
  Output: the first letter of that word

Which function gives the same outputs for different inputs? Explain.

The following relationships are *not* functions. For each, explain why there might be more than one output for some inputs.

- Input: a number
  Output: a number less than that number

- Input: a city name
  Output: the state in which that city can be found

- Input: a whole number
  Output: a factor of that number

- Input: a word
  Output: that word with the letters rearranged

Investigation 1 Function Machines

You can describe a function in various ways, such as using words, symbols, graphs, or machines. In this investigation, you will think about functions as machines.

✅ Develop & Understand: A

Two machines that each perform one operation have been connected to form a more complicated function called Function A. Function A takes an input, doubles it, and then produces 7 more than that result as an output.

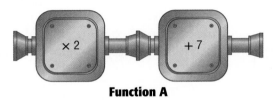

Function A

1. If the input is 5, what is the output?

2. If the input is −4, what is the output?

3. If the output is −10, what could have been the input?

4. Is there more than one answer to Exercise 3? Explain why or why not.

5. If the input is some number *x*, what is the output?

6. Function A is called a *linear function*. Explain why that makes sense.

7. Function B is represented by this machine connection. Is it the same as Function A? Explain.

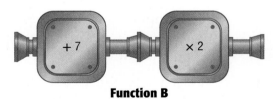

Function B

8. Describe a connection that would "undo" Function A. That is, create a connection so that if you put a number into Function A and then put the output into your connection, you *always* get back your original number.

✅ Develop & Understand: B

The "Prime?" machine takes positive whole numbers as inputs and outputs *yes* if a number is prime and *no* if a number is not prime.

9. If the input is 3, what is the output?

10. If the input is 2, what is the output?

11. If the input is 100, what is the output?

12. If the input is 1, what is the output?

13. If the output is *yes,* what could have been the input?

14. Is there more than one answer to Exercise 13? Explain why or why not.

15. If possible, describe a machine that would undo the "Prime?" machine. That is, create a machine that takes the output from the "Prime?" machine and always produces the original number. If it is not possible, explain why not.

✅ Develop & Understand: C

The "3" machine takes numbers as inputs and always outputs the number 3.

16. If the input is 17, what is the output?

17. If the input is −2, what is the output?

18. If the output is 3, what could have been the input?

19. Is there more than one answer to Exercise 18? Explain why or why not.

20. Explain why "3" is a function.

21. The function "3" is a *constant function.* Explain why that name makes sense.

22. If possible, describe a machine that would undo the "3" machine. That is, create a machine that takes the output from the "3" machine and always produces your original number. If it is not possible, explain why not.

Share & Summarize

Lucita and Ben are trying to decide whether $y = x^4$ is a function.

Who is correct, Ben or Lucita? Is $y = x^4$ a function? Explain how you know.

<table>
<tr><td>Investigation 2</td><td colspan="2">Describe Functions with Rules and Graphs</td></tr>
</table>

Investigation 2 — Describe Functions with Rules and Graphs

Vocabulary

domain

Materials

• graphing calculator

Functions, like the one described by this connection, are a type of rule that assigns one output value to each input value. You can often write such rules as algebraic equations. This is easier than drawing machines.

For example, each of these equations describes the same function as the one shown by the connection. Multiply the input by 5 and then add 1.

$$y = 5x + 1 \qquad f(x) = 5x + 1 \qquad g(t) = 5t + 1$$

Letters like f and g are often used to name functions. In the second rule above, the variable x represents the input, f is the name of the function, and $f(x)$ represents the output. The symbol $f(x)$ is read "f of x." It does *not* mean "f times x." Instead, it means "apply rule f to the value x." For example, $f(2) = 5(2) + 1 = 11$. This is illustrated below.

✅ *Develop & Understand: A*

Kenneth is thinking about a rule to change one number into another number. He is wondering whether his rule is a function.

1. Make an input/output table for Kenneth's rule. Show outputs for at least four inputs.

2. Is Kenneth's rule a function? How can you tell?

3. For Parts a–c, decide which functions describe Kenneth's rule.

   a. $y = (2x + 1)^2$
   $y = 2x^2 + 1$
   $y = 2(x + 1)^2$
   $y = (2x)^2 + 1$

   b. $m = (2n + 1)^2$
   $a = (2b + 1)^2$
   $p = (2t + 1)^2$

   c. $f(z) = 2(z + 1)^2$
   $g(x) = (2x + 1)^2$
   $p(t) = (2t + 1)^2$
   $j(k) = 1 + (2k)^2$

✅ *Develop & Understand: B*

You can graph a function with the input variable on the horizontal axis, or the *x*-axis, and the output variable on the vertical axis, or the *y*-axis.

4. Graph Kenneth's rule from Exercise 3 on your calculator.

   a. What did you enter into the calculator for the rule?

   b. Sketch the graph. Remember to label the minimum and maximum values on each axis.

5. Decide which graphs below represent functions. Explain how you decided.

a.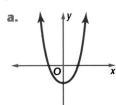

b.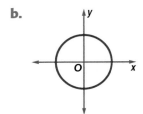

c.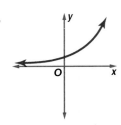

d.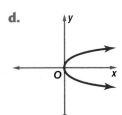

e.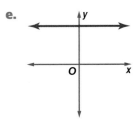

f.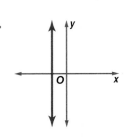

g.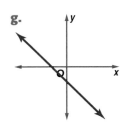

h.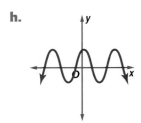

i.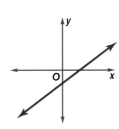

When you have a function such as $f(x) = x^2$, you may want to find the value of the function for different values of x.

Example

Consider the function $f(x) = x^2$.

If $x = 3$, then $f(3) = 3^2 = 9$. Finding $f(3)$ is like putting 3 into this machine.

If $x = -10$, then $f(-10) = (-10)^2 = 100$.

Remember, $f(2)$ does not mean "f times 2." It means "use 2 as the input to machine f" or "evaluate the function f with the input 2."

✅ *Develop & Understand: C*

The distance fallen by skydivers before they open their parachutes is a function of the time since they fell from the aircraft. The function is approximated by $f(t) = 4.9t^2$, where t is the time in seconds and $f(t)$ is the distance in meters.

Math Link

From previous chapters, you might recall that the equation $d = 16t^2$ can also be used to describe the distance traveled by a falling object. The equations $d = 16t^2$ and $d = 4.9t^2$ describe the same relationship, only distance is in feet in the first equation and meters in the second.

6. What does $f(2)$ represent in the skydiving situation? What is the numerical value of $f(2)$?

7. How far has a skydiver fallen after 10 seconds?

8. In the context of this situation, would it make sense to find the value of $f(-3)$? Explain your answer.

Some functions can have only certain inputs. In the skydiver example, only positive numbers make sense as inputs because the function measures how far a skydiver has fallen *after* jumping.

As another example, here is a function you considered earlier.

- Input: an integer

  Output: classification as even or odd

The input is described as "an integer" because non-integers, such as $\frac{3}{4}$ and -12.92, do not make sense as inputs. It is not reasonable to ask whether such numbers are even or odd.

The set of allowable inputs to a function is called the **domain** of that function. If some numbers are not allowed as inputs, we say they *are not in the domain* of the function.

> **Think & Discuss**
>
> Consider the function $r(x) = \frac{1}{x}$.
>
> What numbers are not in the domain of this function? Why?

✅ Develop & Understand: D

In Exercises 9–13, describe the domain of the function.

9. $f(x) = x^2$

10. $g(t) = \sqrt{t}$

11. $R(x) = \dfrac{1}{1 - x}$

12. $e(n)$ is the number of factors of n.

13. $q(p)$ is *yes* if p is evenly divisible by 3 and *no* if p is not evenly divisible by 3.

Real-World Link

Ants are found all over the world except in the polar regions. It is estimated that there are 10,000 different species of ants and 10 million billion individual ants.

14. The number of legs in an ant farm is a function of the number of ants in the farm. Specifically, the number of legs is 6 times the number of ants.

   a. If there are 2,523 ants in the farm, how many legs are there?

   b. What numbers cannot be inputs to this "number of legs" function? Explain your answer.

   c. You can describe this "number of legs" function using algebraic symbols. Let a be the number of ants, and write a function g so that $g(a)$ is the number of legs.

> **Share & Summarize**
>
> A particular function can be described in several ways, including using words, equations, tables, graphs, and machines.
>
> **1.** Describe, write, or draw three representations of this function.
>
> $$g(x) = 7 - 3x$$
>
> **2.** Are any numbers not in the domain of $g(x) = 7 - 3x$? If so, which numbers?

Investigation 3 · Maximum Values of Functions

Graphs are very useful for finding approximate maximum and minimum values of functions. For example, you might consider the maximum height a thrown or bounced ball might reach. A manufacturer might use a function to predict the price that will give the maximum profit for a product.

✓ Develop & Understand: A

Elise threw a stone vertically up from the edge of a pier. The height of the stone above the water level is a function of t, where t is the number of seconds after the stone is thrown. The function, which measures height in meters, is below.

$$h(t) = 15t - 4.9t^2 + 6$$

At the right is a graph of this relationship.

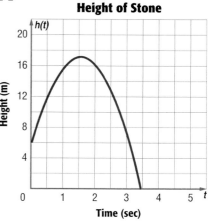

Height of Stone

1. When the stone first leaves Elise's hand, about how high is it above the water? Explain how you can find the answer from the equation or the graph.

2. About how high is the pier? Explain why your answer is reasonable.

3. When is the stone at a height of 15 meters?

4. Between what times is the stone more than 15 meters above the water?

5. To the nearest meter, what is the maximum height the stone reaches?

6. About how long after it is thrown does the stone reach its maximum height?

7. Use your calculator's Trace and Zoom features to better approximate the stone's maximum height. Find the maximum height to the nearest hundredth of a meter.

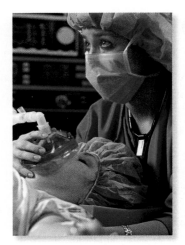

✅ Develop & Understand: B

A company that manufactures medicine has researched the concentration of a local anesthetic in a patient's bloodstream. The company found that the concentration can be approximately calculated with the function

$$C(t) = \frac{21t}{t^2 + 1.3t + 2.9}$$

where t is the number of minutes after the anesthetic is administered and $C(t)$ is the concentration of the anesthetic, measured in grams per liter. A higher concentration means the patient is less likely to feel pain.

8. Find $C(1)$, $C(6)$, and $C(10)$. What does each of these values represent in terms of this situation?

9. The graph shows the relationship between $C(t)$ and t. Use it to estimate the maximum concentration reached by the anesthetic.

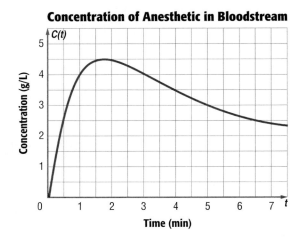

Concentration of Anesthetic in Bloodstream

10. About how long does it take the anesthetic to reach the maximum concentration?

11. Using the equation, draw your own graph on your calculator. Use Zoom and Trace to find the answers to Exercises 9 and 10 to the nearest hundredth of the given units, g/L and min.

12. Tests have shown that when the concentration reaches 2 g/L, patients report feeling numbness. About how long after the injection does this happen?

13. A doctor wants to stitch a cut in Julia's hand. She expects the stitching to take about 3 minutes. How long after she injects Julia with anesthetic should she wait before she starts? Explain.

Share & Summarize

Decide whether each of these functions has a maximum value. If so, approximate the maximum value and the input that produces it.

1.

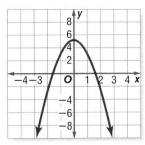

2.

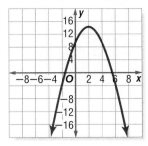

3.

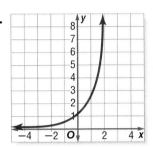

4.

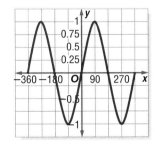

Investigation ④ Maximum Areas, Minimum Lengths

Materials

• graphing calculator

These shapes have the same perimeter but different areas.

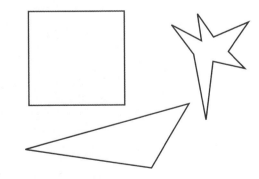

Real-World Link
Geometers are mathematicians who specialize in geometry.

Farmers, builders, and geometers often want to maximize the area of a shape for a given perimeter. In this investigation, you will consider the maximum area for rectangular shapes with a given perimeter. You will also consider the minimum perimeter for a given area.

✅ Develop & Understand: A

Keisha and her twin sister, Monifa, have bought some guinea pigs. They are building a fenced pen for the animals. They have six meters of fencing. They want to give their pets as much space as possible.

Keisha drew some possible rectangular shapes for the pen.

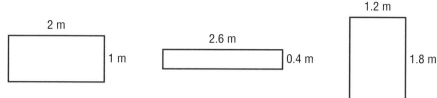

Real-World Link

Guinea pigs are native to South America and live an average of eight years.

. . . .

1. The twins need to consider two dimensions for a rectangular pen, length and width. Copy and complete the table, which relates possible lengths and widths, both measured in meters. The total perimeter must be six meters in each case.

Length	0.5	1	1.5	2	2.5	3
Width	2.5	2				
Perimeter		6				

2. If the length of the rectangle increases by a certain amount, what happens to its width?

3. Write an equation that gives the width *W* for any length *L*.

Your equation shows the relationship between one dimension of the rectangular pen and the other. Keisha and Monifa want to find the greatest rectangular area that they can enclose using six meters of fencing. The mathematical relationship that they need is between one of the dimensions, such as length, and the area.

4. Complete this table, showing dimensions and area of some possible rectangles. All measurements are in meters.

Length	0.5	1	1.5	2	2.5	3
Width	2.5					
Area	1.25					

5. Write an equation for the function *A* giving the area for length *L*.

6. Use your calculator to graph the length of the pen versus its area, using your function from Exercise 5. Sketch your graph. Remember to label the minimum and maximum values on each axis.

7. What length and width should the pen be to produce the greatest area from 6 meters of fencing? Use the graph that you drew to approximate your answer.

✅ Develop & Understand: B

Members of a family want to build a rectangular chicken pen. They will use an existing stone wall as one side and fencing for the other three sides. They want to know what dimensions will give an area of 40 m^2 using a minimum length of fencing.

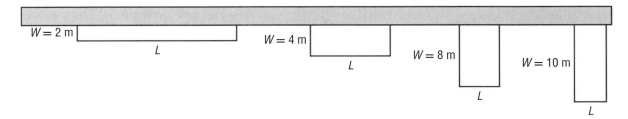

8. Copy and complete the table. Try additional width values, if necessary, to determine the least amount of fencing needed.

Width, W	2	4	8	10
Length, L				
Amount of Fencing				

9. Express the length of the pen in terms of W.

10. Use your expression from Exercise 9 to write the amount of fencing as a function of W. Name your function F.

$$F(W) = \underline{\hspace{2cm}}$$

11. Use one of the following methods to find the width that requires the least amount of fencing.

   • Graph the amount of fencing versus width, using your function from Exercise 10. Approximate the least value for the length.

   • Use a calculator to guess-check-and-improve.

Share & Summarize

Hernando was experimenting with his calculator, adding positive numbers and their reciprocals. Here are some examples.

$$5 + \frac{1}{5} = 5.2 \qquad 0.1 + \frac{1}{0.1} = 10.1 \qquad 1.25 + \frac{1}{1.25} = 2.05$$

1. Do you think there is a minimum total he can produce doing this? If so, what is it? If not, explain why not. (Hint: Let x stand for the number, and write an equation to express what Hernando is doing.)

2. Do you think there is a maximum total he can produce? If so, what is it? If not, explain why not.

Inquiry

Investigation Maximize Volume

Materials

- 5-inch-by-8-inch cards
- ruler
- scissors
- graphing calculator
- tape

Your teacher will give you 5-inch-by-8-inch cards. You can cut squares out of the corners of a card and then fold the sides to make an open box, one without a top.

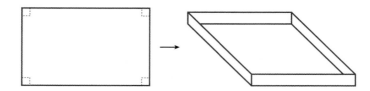

Your challenge is to create the box with the greatest possible volume.

Try It Out

The volume of your box will depend on the side length of the squares that you cut from it.

1. Using the method above, try to create the box with the greatest possible volume. Be careful to cut squares of the *same size* from each corner. Record the side lengths of the squares you cut. You can refer to them later.

2. Compare the greatest volume that you found with the greatest volume found by others in the class. Record the side length of the cutout squares for the boxes with the greatest volume.

Analyze the Situation

3. Each dimension of your box depends on the side length of the squares that you cut. Copy and complete the table for squares of different side lengths. All measurements are in inches.

Side Length of Square	0	0.5	1	1.5	2	2.5
Height of Box	0					
Length of Box	8					
Width of Box	5					

4. Add a row to your table, calculating the volume of the box for each side length of the square. Of those boxes listed in the table, which has the greatest volume?

Of course, there are more possible square sizes than the six listed in the table. You can use functions and graphs to help you check *all* the possibilities.

Math Link

For a rectangular prism such as a box, volume is the area of the base times the height.

---

5. If the side length of the square you cut out is *x*, find each of the following in terms of *x*.

    a. the height of the box   **b.** the length of the box

    c. the width of the box   **d.** the volume of the box

6. Based on your answer to Part d of Exercise 5, write an equation for the function relating the box's volume to the side length of the square that you cut. Name your function *v*.

7. Use your calculator to graph the volume function, and sketch the graph. Then use Zoom and Trace to estimate the value of *x* that gives the maximum volume.

What Have You Learned?

You estimated the maximum volume of the open box that you can make from a 5-inch-by-8-inch card. Instead, start with a standard sheet of paper, 8.5 inches by 11 inches.

8. Use what you learned in this investigation to answer these questions. Show your work, including sketches of any graphs you make.

    a. What cutout size will maximize the volume for an open box made from a standard sheet of paper?

    b. What is the greatest possible volume?

9. Use an ordinary sheet of paper and your answers to Question 8 to create the box that you think has the greatest volume. Tape the corners to make it strong.

Practice & Apply

1. Consider this function machine.

    a. If the input is 10, what is the output?

    b. If the input is $-\frac{2}{3}$, what is the output?

    c. If the input is 1.5, what is the output?

    d. If the input is some number x, what is the output?

    e. If the output is -9, what was the input?

    f. Suppose you want a function machine that will undo this machine. That is, if you put a number first through the "÷ 2" machine and then through your new machine, it *always* returns your original number.

       What function machine would accomplish this?

2. Consider this function machine, which squares the input.

    a. If the input is $\frac{4}{3}$, what is the output?

    b. If the input is $-\frac{4}{3}$, what is the output?

    c. If the output is 9, what was the input?

    d. Suppose you want a function machine that will undo this machine. That is, if you put a number first through the "Square" machine and then through your new machine, it *always* returns your original number.

       What type of function machine would accomplish this?

3. Consider this connection, Function F.

Function F

a. If the input is 1.5, what is the output?

b. If the input is −3, what is the output?

c. If the input is 11, what is the output?

d. If the input is some number *x,* what is the output?

e. If the output is −8, what was the input?

f. Suppose you want a function machine that will undo this machine. That is, if you put a number through the Function F machine and then through your new connection, it will *always* return your original number.

What function machine would accomplish this?

Tell whether each example below is a function. Explain how you decided.

4. Input: a circle
Output: the ratio of the circumference to the diameter

5. Input: a rugby team
Output: a member of the team

6. Input: a CD
Output: a song from the CD

7. Input: an insect
Output: six legs

8. Input: a phone book
Output: a phone number

Real-World Link
Rugby is played in more than 100 countries by several million people.

Determine if the relationship represented by each input/output table could be a function.

9.

Input	Output
−3	4
−2	3
−1	2
0	1
1	0
2	−1
3	−2

10.

Input	Output
−3	0
−2	−2.828 and 2.828
−1	−2.236 and 2.236
0	−3 and 3
1	−2.236 and 2.236
2	−2.828 and 2.828
3	0

11.

Input	Output
−3	$\frac{1}{3}$
−2	$\frac{1}{2}$
−1	1
0	undefined
1	1
2	$\frac{1}{2}$
3	$\frac{1}{2}$

12. Consider this rule. *Square a number, subtract 2, and then divide by 2.*

   a. Copy and complete the table using this rule.

   b. Sketch a graph of the relationship shown in your table.

   c. Is this rule a function? How do you know?

Input	Output
−3	
−2	
−1	
0	
1	
2	
3	

13. When Kyle entered Math class, the table and functions below were on the board. Kyle thought the values in the first column of the table were function inputs and that the values in the second column were outputs.

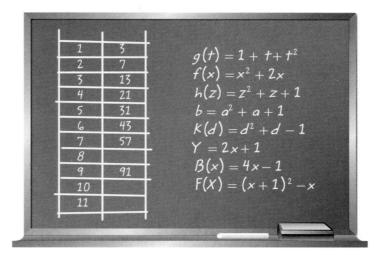

1	3
2	7
3	13
4	21
5	31
6	43
7	57
8	
9	91
10	
11	

$g(t) = 1 + t + t^2$
$f(x) = x^2 + 2x$
$h(z) = z^2 + z + 1$
$b = a^2 + a + 1$
$k(d) = d^2 + d - 1$
$Y = 2x + 1$
$B(x) = 4x - 1$
$F(X) = (x + 1)^2 - x$

a. Which of the functions might be shown in the table? Explain.

b. Complete the table by finding the missing values of the function.

14. Physical Science A rock falls over the edge of a cliff 600 meters high. The distance in meters the rock falls is a function of time in seconds and can be approximated by the function $s(t) = 4.9t^2$.

a. Find the value of $s(8)$. In this situation, what does $s(8)$ represent?

b. How far has the rock fallen after 9 seconds? After 10 seconds?

c. When does the rock hit the ground?

d. What is the domain of the function $s(t) = 4.9t^2$ in this context?

Describe the domain of each function.

15. $f(x) = 2^x$

16. $g(x) = \dfrac{1}{x + 1}$

17. $h(x) = \dfrac{1}{x + 1} + \dfrac{1}{x - 1}$

18. Which of the following are not graphs of functions? Explain how you know.

a.
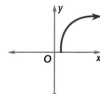

b.
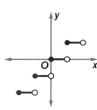

c.
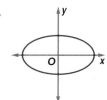

d.
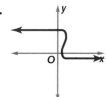

19. Suppose a person throws a stone straight upward so that its height h in meters is given by the function $h(t) = 6 + 20t - 4.9t^2$, where t represents the time in seconds since the stone was released.

   a. Find $h(4)$. What does it represent in this situation?

   b. Find the height of the stone after 3 seconds.

   c. Sketch a graph of the stone's height over time.

   d. Use your graph to approximate the stone's maximum height. How long does it take the stone to reach this height?

20. **Economics** ABC Deli sells several kinds of sandwiches, all at the same price. The weekly profit of this small business is a function of the price of its sandwiches. This relationship between profit P in hundreds of dollars and the price per sandwich s in dollars is given by the following equation.

$$P(s) = -s(s - 7)$$

   a. Complete the table for this function.

   b. Explain the meaning of $(7, 0)$ in terms of the deli's profit.

   c. Extend your table to search for the sandwich price that will yield the maximum profit.

   d. What is the maximum profit this business can expect in a week?

s	P(s)
0	0
1	
2	
3	
4	
5	
6	
7	

For Exercises 21 and 22, find the maximum value of each function. Then determine the input value that yields that maximum value.

21. $f(t) = 200t - 5t^2$

22. $k(t) = 4 + 4t - 4t^2$

23. Meagan gave her little brother an 8-meter strip of cardboard to make a rectangular fort for his toy soldiers.

  a. Copy and complete the table, which relates possible lengths and widths for the fort.

Length (m)	0.5	1	1.5	2	2.5	3	3.5
Width (m)							
Perimeter (m)							

  b. Write an equation for the function that gives the width for any length L. Name the function W.

  c. Add a row to your table showing the area of some possible rectangles.

  d. Write an equation for the function A giving the area for length L.

  e. Use your function from Part d to sketch a graph of the fort's area in terms of its length.

  f. What dimensions give the greatest area for the fort?

24. In Your Own Words Give one example of a function that can be described by an algebraic equation. Explain how you know that your example is a function.

25. **Geometry** Roof gutters are designed to channel rainwater away from the roof of a house, protecting the house from excess moisture.

    If you cut through a gutter and look at its side view, you see a *cross section*. Here are some cross sections of gutters.

    Nick's Metalworks wants to produce some gutters from a roll of metal that is 39 cm wide. The gutters should have vertical sides. Nick has drawn some possible cross sections.

    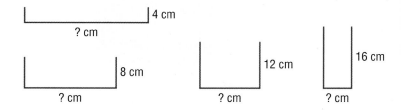

    a. To keep the gutters from overflowing during heavy rainfall, the company wants them to have the greatest cross-sectional area possible. Copy and complete the table to show the widths and areas for gutters of various heights.

Height (cm), h	4	8	12	16
Width (cm), w				
Area (cm²), A				

    b. Find a formula for width *w* in terms of height *h*.

    c. Write an equation for the cross-sectional area *A* as a function of *h*. What sort of function is it?

    d. Sketch a graph of the area function.

    e. Estimate the gutter height that gives the greatest area.

Connect & Extend

26. Create a function machine that produces three more than twice every input as an output.

27. Create a function machine that produces one less than one-third every input as an output.

28. Create a function machine that returns an odd number for any whole-number input.

29. Physical Science Think about the relationship between the temperature of a hot cup of coffee and the time in minutes since the coffee was poured.

    a. Sketch a graph of how you think the relationship between temperature and time might look. (Hint: Think about the rate at which the coffee cools. Does it cool more quickly at first?)

    b. Is this relationship a function? If so, explain why.

30. Geometry The sum of the interior angles of a polygon is a function of the number of its sides. For example, the sum of the interior angles of a triangle is 180°, of a square is 360°, of a pentagon is 540°, and of a hexagon is 720°.

    a. What is the sum of the interior angles of a polygon with 12 sides, which is also called a dodecagon? Use the pattern in the angle sums for the above-mentioned polygons.

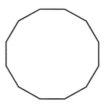

    b. Write an equation for the function relating the number of sides to the angle sum. Name the function g, and use s to represent the number of sides.

    c. What is the domain of this function? Explain your answer.

Challenge In Exercises 31–33, write an equation for a function f that does not have the given numbers in its domain.

31. 3 and -3

32. negative numbers

33. positive numbers

Use the given graph and a table of values to find the minimum value of each function and the input that produces it.

34. $f(x) = x + x^2$

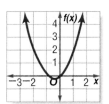

35. $f(x) = 1 - x + x^2$

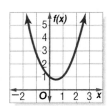

36. You can think of a *sequence* as a function for which the input variables are the counting numbers (1, 2, 3, 4, ...). For example, the sequence of even whole numbers greater than zero can be given by the function $f(n) = 2n$, where 1, 2, 3, 4, ... are the inputs.

a. List the first seven terms of the sequence described by the function $g(n) = \dfrac{1}{2^n}$, for n starting at 1.

b. Add the first five terms of this sequence.

c. Add the first six terms of this sequence.

d. Add the first seven terms of this sequence.

e. Suppose you were to add *all* the terms of this sequence for some large value of n, such as 100 terms. Do you think the sum of this sequence approaches a particular value? Do you think it increases indefinitely?

37. Two numbers add to 1. What is the maximum value of their *product*? Explain.

38. Economics Designers at a company that manufactures charcoal pencils for artists have decided to redesign the shipping boxes for the pencils. The pencils are in the shape of rectangular prisms with a 0.25-inch-by-0.25-inch base and a length of 8 inches. The company plans to package a dozen pencils in each box.

a. Calculate the volume of a single pencil. Then find the volume each box must contain. That is, find the volume of 12 pencils.

b. One dimension of the box must be 8 in., which is the length of the pencils. Using x, y, and 8 for the dimensions of the box, write a formula for the volume a box can hold.

c. Use the total volume of the 12 pencils, along with your formula from Part b, to write an equation for y in terms of x.

The company wants to use as little cardboard as possible in making the boxes.

d. Write a formula for the surface area S of the box, using only x as the input variable. Ignore the area of the flaps that hold the box together. (Hint: You may want to write it using x and y first and then replace y with an expression in terms of x.)

e. Make a table of values giving the surface area of the box for different values of x. Since the pencils are 0.25 in. wide, the dimensions of the box must be multiples of 0.25 in. For example, 0.25 in., 0.5 in., and 0.75 in. are multiples of 0.25 in.

f. What dimensions should the box be so that it uses the least amount of cardboard?

Mixed Review

Make a rough sketch showing the general shape and location of the graph of each equation.

39. $y = x^2 + 3$ **40.** $y = \frac{2}{x}$ **41.** $y = x^3 - 1$

Write an equation to represent the value of B in terms of r.

42.

r	0	1	2	3	4
B	3	0.6	0.12	0.024	0.0048

43.

r	0	1	2	3	4
B	12	4.8	1.92	0.768	0.3072

44. Astronomy A lightyear, the distance light travels in one year, is 5.88×10^{12} miles. Answer these questions without using your calculator.

a. The star Alpha Centauri is about 4 lightyears from Earth. Write this distance in miles, using scientific notation.

b. The star Betelgeuse is about 500 lightyears from Earth. Write this distance in miles, using scientific notation.

c. Suppose a light beam went from Earth to Betelgeuse, and another light beam went from Earth to Alpha Centauri. How much further did the beam going to Betelgeuse have to travel?

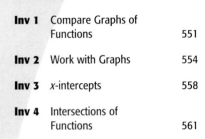

Graphs of Functions

Ms. Torres drew the graphs and table below to show her class that the graphs of $f(x) = x^2$ and $g(x) = (x - 3)^2$ are related.

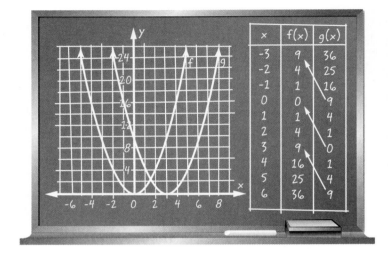

Ms. Torres then asked the class why the graphs look as they do.

Math Link

A *translation* is a transformation that moves a figure a specific amount in a specific direction. It does not change the figure's size or orientation.

Think & Discuss

Which of the four students' comments do you find the most helpful in understanding why the graphs look as they do? Explain.

Describe in your own words why it seems reasonable for the graph of $g(x) = (x - 3)^2$ to be 3 units to the right of the graph of $f(x) = x^2$.

Trace the graph of $f(x) = x^2$, and place it on top of the graph of $g(x) = (x - 3)^2$. Line up the parabolas. Are the parabolas congruent?

The graph of $g(x) = (x - 3)^2$ is related to that of $f(x) = x^2$ by a translation. What are the direction and distance of the translation?

Predict what the graph of $h(x) = (x + 4)^2$ looks like. Test your prediction by graphing with your calculator.

Is the graph of h related to the graph of f by a translation? If so, specify the direction and the distance of the translation.

Investigation ⓵ Compare Graphs of Functions

Materials

- graphing calculator

In this investigation, you will explore sets of related functions.

✅ Develop & Understand: A

For Exercises 1 and 2, do Parts a–c. Work in pairs or groups of four. Your group will need two graphing calculators, one for each exercise.

a. Graph the four equations in the same window. Make a quick sketch of the graphs. Do not erase your graphs for Exercise 1 when you go on to Exercise 2. You will need both sets for Exercise 3.

b. Describe how the four graphs in the set are alike and different. Use the concept of translation in your comparisons.

Math Link

When you make sketches of graphs from your calculator, label the graphs with their names, such as *j, f, g, h*. Also, label the minimum and maximum values on each axis.

c. Write equations for two more functions that belong in the set.

1. $j(x) = (x + 1)^2$

   $f(x) = x^2$

   $g(x) = (x - 1)^2$

   $h(x) = (x - 2)^2$

2. $j(x) = \dfrac{1}{x + 1}$

   $f(x) = \dfrac{1}{x}$

   $g(x) = \dfrac{1}{x - 1}$

   $h(x) = \dfrac{1}{x - 2}$

3. Describe how the two sets of graphs are alike and different.

4. On which graph would you find the point (4, 9)? Explain.

✅ Develop & Understand: B

Work in pairs or groups of four for Exercises 5–7.

5. Consider the function $f(x) = 2^x$.

   a. Write equations for three functions, g, h, and j, so that their graphs have the same shape as the graph of f, but:

   i. g is translated 2 units to the right of f.

   ii. h is translated 3 units to the right of f.

   iii. j is translated 3 units to the left of f.

   b. Graph the four functions in the same window. Make a quick sketch of the graphs.

6. Consider the function $f(x) = 2x^2$.

   a. Write equations for three functions, g, h, and j, so that their graphs have the same shape as the graph of f, but:

   i. g is translated 1 unit to the right of f.

   ii. h is translated 2 units to the left of f.

   iii. j is translated 3 units to the right of f.

   b. Graph the four functions in the same window. Make a quick sketch of the graphs.

7. On which graph from Exercises 5 and 6 would you find the point $(-3, 1)$? Explain how you found your answer.

✅ Develop & Understand: C

Work in pairs or groups of four with two graphing calculators. Do Parts a–c for Exercises 8 and 9.

   a. Graph the four equations in the same window. Make a quick sketch of the graphs, remembering to label them. Do not erase your graphs for Exercise 8 when you go on to Exercise 9.

   b. Describe how the four graphs in the set are alike and different. Use the concept of translation in your comparisons.

   c. Write two more functions that belong in the set.

8. $j(x) = 2^x - 1$

   $f(x) = 2^x$

   $g(x) = 2^x + 1$

   $h(x) = 2^x + 2$

9. $j(x) = \frac{1}{x} - 1$

$f(x) = \frac{1}{x}$

$g(x) = \frac{1}{x} + 1$

$h(x) = \frac{1}{x} + 2$

10. Describe how the two sets of graphs are alike and how they are different.

11. On which graph would you find the point $\left(3, \frac{4}{3}\right)$? Explain.

Share & Summarize

1. Suppose you have the graph of a function f. You create a new function g by using the rule for f but replacing the variable x by the expression $x + h$, for some constant h.

If $f(x) = 2x$, for example, you might replace x with $x + 3$ to get $g(x) = 2(x + 3)$.

If $f(x) = 3x^2 - 2$, you might replace x with $x - 5$ to get $g(x) = 3(x - 5)^2 - 2$.

Write a sentence or two describing the differences and similarities between the graphs of f and g. You may want to draw sketches to help you explain.

2. Suppose you create a function g by adding a constant h to f.

For example, $f(x) = 2x$ and $g(x) = 2x + 3$, or $f(x) = 3x^2 - 2$ and $g(x) = 3x^2 - 7$.

Describe the differences and similarities between the graphs of f and g. You may want to include sketches.

3. Suppose you want to know whether the point (a, b) is on the graph of a function. How could you find out this information?

Investigation ② Work with Graphs

Vocabulary

range

Materials

• graphing calculator
  (optional)

Earlier, you saw that the maximum or minimum value of a quadratic function can be found by looking at the vertex of its graph, which is a parabola. You also learned that parabolas are symmetric. They can be folded on a line of symmetry so that the two sides match.

You will now examine connections between the graph and the equation of a quadratic function. You will also learn what the range of a function is and how it relates to the maximum or minimum point.

> ### Think & Discuss
>
> Look at this graph of $f(x) = (x - 2)^2 + 1$.
>
> Find the values of x for $f(x) = 1$, $f(x) = 2$, and $f(x) = 5$.
>
> Now find values of x for $f(x) = 0$, $f(x) = -1$, and $f(x) = -5$.
>
> Describe all possible values for $f(x)$.
>
> Describe all values $f(x)$ can never be.

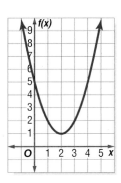

All the possible *output* values of a function f are the **range** of the function. For the function graphed above, the range is $f(x) \geq 1$. No matter what you substitute for x, the value of $f(x)$ will always be greater than or equal to 1. Every value greater than or equal to 1 has an input value. Numbers less than 1 are not in the range of $f(x) = (x - 2)^2 + 1$.

Real-World Link

A biologist might need to know the range of a function modeling the temperature of a body of water over time to study how water temperature affects an organism living there.

✅ *Develop & Understand: A*

For each function, specify the domain, or possible inputs, and range, or possible outputs. For some functions, it may help to make a graph with a calculator.

1. $g(x) = 4^{x+2}$ (Hint: Can $g(x)$ be negative? Zero?)

2. $h(s) = \dfrac{1}{s+3}$

3. $c(x) = -3x + 4$

4. Input: a state
   Output: the capital of that state

5. Input: a number
   Output: the integer part of that number
           For example, if the input is 4.5, the output is 4.
           If the input is -3.2, the output is -3.

It is not always easy to determine the range of a function. You can test a few logical input values and draw some conclusions from the outputs, or you can graph the function. You will see how the range of a quadratic function is related to its vertex. To begin, consider how to find the vertex.

⎡*Example*

This graph is of the function $f(x) = (x - 2)^2 + 1$.

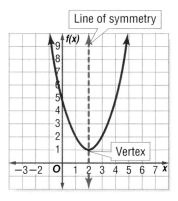

> ### Math Link
> A parabola has its maximum or minimum value at the vertex.

For this parabola, the line of symmetry is the line $x = 2$.

The turning point, or *vertex*, is the point on the graph where $x = 2$. When $x = 2, f(x) = (2 - 2)^2 + 1 = 1$, so the vertex has coordinates $(2, 1)$.

✅ *Develop & Understand: B*

6. Below are the graphs of these functions.

$$f(x) = 3x^2 \qquad\qquad g(x) = 3(x-2)^2 + 4$$

i.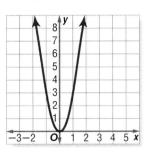

ii.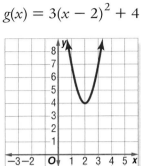

a. Without using your calculator, decide which graph represents which function.

b. Sketch the graphs. Draw the line of symmetry for each.

c. What is the vertex of each graph?

d. Specify the range of each function.

e. How is the range of a function related to the vertex?

7. Consider this graph.

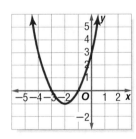

a. What is the graph's line of symmetry? What is its vertex?

b. Which of these functions does the graph represent? Explain how you know.

$$f(x) = (x + 2)^2 + 1 \qquad\qquad g(x) = (x + 2)^2 - 1$$
$$h(x) = -(x - 2)^2 + 1 \qquad\qquad i(x) = (x - 2)^2 - 1$$

c. For each function in Part b, specify the range and the vertex.

d. How is the range related to the maximum or minimum point of a function?

8. Answer these questions about the function $f(x) = (x - 3)^2 - 1$ without drawing the graph.

   a. What is the line of symmetry?

   b. What is the vertex?

9. A parabola has a vertex at the point (3, 4).

   a. Write an equation for a quadratic function whose graph has this vertex. Check your answer by graphing.

   b. Are there other parabolas with this vertex? If so, state two more. How many are there?

10. Suppose you have graphs of these quadratic functions.

$$f(x) = (x - h)^2 + k \qquad g(x) = x^2$$

   a. How is the graph of f related to the graph of g?

   b. What is the vertex of g? What is the vertex of f?

You can predict the line of symmetry and the vertex of the parabola without drawing a graph when a quadratic function is written in a form like $f(x) = 2(x - 3)^2 + 1$.

It is much harder to visualize the graph when a quadratic function is written in a form like $f(x) = 2x^2 - 12x + 19$. In this case, it is helpful to rewrite the function by completing the square.

✅ *Develop & Understand: C*

For each function, do Parts a–d.

- **a.** Complete the square to rewrite $f(x) = ax^2 + bx + c$ in the form $f(x) = a(x - h)^2 + k$.

- **b.** Find the line of symmetry of the graph of f.

- **c.** Find the coordinates of the vertex of the parabola.

- **d.** Use the rewritten form of the function to sketch its graph. Check with a graphing calculator.

11. $f(x) = x^2 + 8x + 7$

12. $f(x) = -x^2 + 4x + 1$ (Hint: Factor out -1 first.)

13. $f(x) = x^2 - 6x - 3$

Share & Summarize

Describe the relationship between a range of a quadratic function and the vertex of the related parabola.

Investigation 3 x-intercepts

Vocabulary

x-intercepts

Materials

- GeoMirror

Recall that the y value at which a graph crosses the y-axis is called the y-intercept. In the same way, the x values at which a graph crosses the x-axis are called the **x-intercepts**.

Think & Discuss

How are the x-intercepts of the graph of a function f related to the solutions of $f(x) = 0$? For example, how is the x-intercept of $f(x) = 3x + 7$ related to the solution of $3x + 7 = 0$?

Without making a graph, find the x-intercepts of these functions.

$$h(x) = (3x + 1)(x - 4) \qquad j(x) = x^2 - 7x - 18$$

You will now explore how the x-intercepts of a quadratic function are related to each other and to a parabola's line of symmetry.

✓ Develop & Understand: A

The graph shows the function
$f(x) = 3x^2 - 3x - 6$.

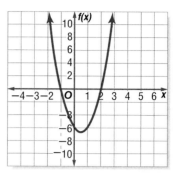

1. Estimate the x-intercepts of f.

2. Find the line of symmetry.

3. On each side of the line of symmetry is an x-intercept. Use your GeoMirror to find a reflection of the x-intercept on the left side over the line of symmetry. What do you notice?

4. Find the distance between each estimated x-intercept and the line of symmetry.

5. Find the exact values of the x-intercepts by solving the equation $3x^2 - 3x - 6 = 0$.

6. Check that the distance between each x-intercept that you found in Exercise 5 and the line of symmetry is equal to your answer from Exercise 4.

You can find the x-intercepts and the vertex of the graph of a quadratic function and use them to make a quick sketch of the graph.

─ Example

Sketch a graph of the function $g(x) = (3x - 7)(x + 1)$ without using a calculator.

The x-intercepts of g are the solutions of $(3x - 7)(x + 1) = 0$. Since one factor must be 0, the solutions are $\frac{7}{3}$ and -1.

The vertex must be halfway between these x-intercepts.

Its x value is the mean of the solutions, $\dfrac{\frac{7}{3} + (-1)}{2} = \dfrac{2}{3}$.

Since $g\left(\frac{2}{3}\right) = -\frac{25}{3}$, the vertex is $\left(\frac{2}{3}, -\frac{25}{3}\right.$.$)$

Now plot the vertex and the points where the graph of g crosses the x-axis $\left(\frac{7}{3}, 0\right)$ and $(-1, 0)$. Then draw a parabola through the three points.

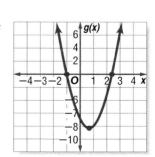

✓ *Develop & Understand: B*

For the quadratic equations given in Exercises 7–9, do Parts a–c.

    a. Find the *x*-intercepts of *f*.

    b. Use the *x*-intercepts to find the vertex of the parabola.

    c. Plot the three points from Parts a and b. Then draw a parabola.

7. $f(x) = (x - 2)(x + 3)$

8. $f(x) = -(x - 5)(2x + 1)$

9. $f(x) = x^2 - 3x - 40$

10. Angelo drew this parabola on a graphing calculator and made a sketch of it to take home for homework. By the time he got home, he had forgotten what function had generated the parabola. He did remember that he was trying to solve an equation like $(t + \underline{\hphantom{xx}})(t - \underline{\hphantom{xx}}) = 0$.

What function had he used? What are the solutions of his equation?

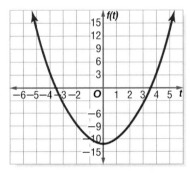

Share & Summarize

Explain how to find the *x*-intercepts and the vertex of the graph of this function.

$$f(x) = 3x^2 - 44x - 15$$

Investigation ④ Intersections of Functions

Materials

- graphing calculator

Suppose you encounter an equation that you do not exactly know how to solve. There *are* ways to find approximate solutions.

One nice method for finding an approximate solution follows.

- Think of each side of the equation as a function.
- Graph the two functions.
- Find the point or points where the functions intersect.
- Check that the x value, or input, at each intersection point gives you approximately the same y value, or output, for both functions. That is, check that the two sides of the original equation are approximately equal at that input.

Example

Solve $x^3 = 5x + 10$.

- First, think of each side of the equation as a function.
  $$f(x) = x^3 \text{ and } g(x) = 5x + 10.$$
- Graph the two functions.

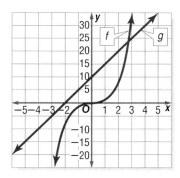

- Find the point or points where the functions intersect. In this case, there is only one point, near $x = 2.9$.
- Check: At $x = 2.9$, $f(x) = 24.389$ and $g(x) = 24.5$. The two sides of the original equation are approximately equal.

✓ Develop & Understand: A

Suppose you are given two options for getting paid at a babysitting job.

- You can earn $10 per hour for each hour worked.
- You can earn $2 if you stay 1 hour, $4 if you stay 2 hours, $8 if you stay 3 hours, and so on. The amount you earn doubles for each additional hour that you stay.

Is there some number of hours for which you would earn the same amount using either payment plan? In this exercise set, you will explore this question.

Math Link

Although you have evaluated exponential expressions for integer powers only, it is possible for exponents to be fractions or decimals.

1. Write an equation for a function L that describes earning $10 per hour. Use h for the input variable.

2. Write an equation for a function D that describes doubling the amount that you earn for each hour that you stay. Use h for the input variable.

3. Graph your two functions in a single window. Sketch the graphs. Label which graph matches which function.

4. How many solutions can you find to $L(h) = D(h)$? Use Zoom and Trace to approximate the solutions.

5. If you had a babysitting job, how would you decide which payment plan to choose? Explain.

✓ Develop & Understand: B

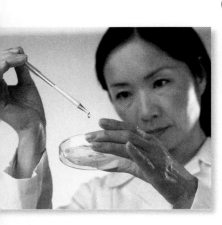

In a culture dish, a population of bacteria is growing at a rate of 10% each hour. There were 1,000 bacteria in the dish at the beginning of the experiment. An hour later, there will be 10% of 1,000, or 100, more bacteria, for a total of 1,100 bacteria in the dish.

6. Make a table to show the bacteria population after 1 hour, 2 hours, and 3 hours.

7. Write an equation for a function that represents how many bacteria there will be after x hours. Name the function p.

8. After some amount of time, the bacteria will double in number. That is, $p(x) = 2,000$. After how many hours will this be? Use your graphing calculator to find an approximate solution.

9. Explain how you found your answer to Exercise 8.

10. What equation would you solve to find how many hours are needed for the number of bacteria in the dish to triple? Find an approximate solution.

✅ Develop & Understand: C

Consider how you might solve the cubic equation $x^3 = 2x - 0.5$.

11. If you use the method of graphing two functions and finding their intersections, what two functions would you graph?

12. Graph your functions. How many solutions does the function $x^3 = 2x - 0.5$ have?

13. Approximate all the solutions of the equation. Check your solutions by substituting.

14. Sasha suggested doing the same thing to both sides of the equation to obtain $x^3 - 2x + 0.5 = 0$ and then graphing this function

$$h(x) = x^3 - 2x + 0.5.$$

   a. Where would you look to find the solutions of $x^3 - 2x + 0.5 = 0$?

   b. Graph the function. Estimate the solutions using your graph.

   c. Did you prefer Sasha's method or graphing two functions and finding their intersections?

Share & Summarize

For each equation, state whether you can solve it exactly or approximately. Then solve it the best way you can.

1. $400k + 10 = 500k$

2. $3^G = 1.5G + 5$

3. $x^2 = \sqrt{x + 1}$

Practice & Apply

1. Sketch a graph of each function in Part a on a single set of axes. Using a different set of axes, do the same for the functions in Part b. Then answer the questions that follow.

    a. $y = 2x$
    $y = 2(x + 1)$
    $y = 2(x + 2)$
    $y = 2(x + 3)$

    b. $y = 2(x - 1)$
    $y = 2(x - 1) + 1$
    $y = 2(x - 1) + 2$
    $y = 2(x - 1) + 3$

    c. Describe how the four graphs in Part a are like those in Part b.

    d. Describe how the four graphs in Part a are different from Part b.

    e. Find another function that belongs to the set of functions in Part a and another that belongs to the set in Part b.

    f. Which of the eight graphs contains the point $(3, 7)$? Explain how you found your answer.

2. Below is a graph of $f(x) = x^2 + 3x - 2$.

    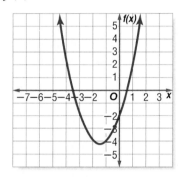

    a. Sketch a graph of $g(x) = (x - 2)^2 + 3(x - 2) - 2$ and one of $h(x) = (x + 2)^2 + 3(x + 2) - 2$.

    b. How are the graphs of g and h related to the graph of f?

    In Exercises 3–6, write an equation for the function g so that the graph of g has the same shape as the graph of f.

3. The graph of g is translated 5 units to the right of the graph of $f(x) = \frac{1}{x}$.

4. The graph of g is translated 3 units up from the graph of $f(x) = x^2 + x - 2$.

5. The graph of g is translated 1 unit to the left of the graph of $f(x) = \frac{1}{x - 3}$.

6. The graph of g is translated 4 units to the left of $f(x) = (x + 1)^2$.

Sketch a graph of each function. State the domain and the range.

7. $f(x) = 4 - (x - 2)^2$ **8.** $g(x) = 5 - (x - 1)$ **9.** $f(x) = \dfrac{1}{x - 10}$

10. Consider these functions.

$$f(x) = 2x^2 + 1 \qquad g(x) = 2(x - 1)^2 + 2$$

  a. What are the coordinates of the vertices for the graphs of these two functions?

  b. What is the line of symmetry for each?

  c. Sketch the graphs of both functions on one set of axes.

11. Consider this graph.

  a. Identify the line of symmetry and the vertex.

  b. Which of these functions does the graph represent? Explain how you know.

$$f(x) = -3 + (x - 2)^2$$

$$g(x) = 3 - (x + 2)^2$$

$$h(x) = 3 - (x - 2)^2$$

Identify the vertex, the line of symmetry, and the range of each function.

12. $f(x) = 4 - (x + 3)^2$

13. $g(p) = (p - 5)^2 - 9$

14. $h(x) = (x + 6)^2 - 3$

15. Consider a parabola with its vertex at $(-2, 6)$.

  a. Write an equation for a quadratic function for a parabola with this vertex.

  b. Are there other parabolas with this vertex? If so, state two more. How many more are there?

In Exercises 16–18, do Parts a–c.

a. Complete the square to rewrite $f(x) = ax^2 + bx + c$ in the form $f(x) = a(x - h)^2 + k$.

b. Find the line of symmetry of the graph of f.

c. Find the coordinates of the vertex of the parabola.

16. $f(x) = x^2 - 2x - 6$

17. $f(x) = 3 + 4x - x^2$

18. $f(x) = x^2 + 8x - 1$

19. Consider this graph of $f(x) = -x^2 + 2x + 5$.

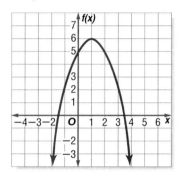

a. Use the graph to find approximate solutions of $f(x) = 0$. Explain how you found your answer.

b. Use the quadratic formula or complete the square to solve $f(x) = 0$ exactly. How close are your approximations?

c. Find the vertex of the graph of f.

d. What is the line of symmetry of the graph of f?

e. Find the distance between each solution of $f(x) = 0$ and the line of symmetry.

20. Consider this graph of $f(x) = x^2 - \dfrac{1}{2}x - \dfrac{3}{16}$.

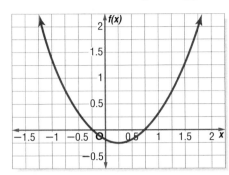

a. Use the graph to find approximate solutions of $f(x) = 0$.

b. Use the quadratic formula or complete the square to solve $f(x) = 0$ exactly. How close are your approximations?

c. Find the vertex of the graph of f.

d. What is the line of symmetry of the graph of f?

e. Find the distance between each solution of $f(x) = 0$ and the line of symmetry.

Real-World Link

An engineer might need to know the range of a function modeling the increase in the length of steel railroad sections versus temperature to determine how far apart the sections should be laid.

. . . .

In Exercises 21–23, the graph of the function is a parabola. Do Parts a–c for each exercise.

a. Find the x-intercepts of the parabola.

b. Use the x-intercepts to find the line of symmetry and the vertex.

c. Use the x-intercepts and the vertex to sketch the parabola.

21. $g(x) = (x - 3)(x + 0.5)$

22. $h(x) = (2x + 3)(x - 1)$

23. $f(x) = -x^2 - 4x + 5$

In Exercises 24–26, use graphs to determine how many solutions the equation has.

24. $x^2 - 2x = 4 - x - x^2$

25. $x^2 - x - 2 = 1 - 2x - x^2$

26. $2 - x^2 = 1 - 2x^2$

Sketch graphs to find approximate solutions of each equation.

27. $x^3 - 4x - 1 = x - 1$

28. $x^3 - 4x - 1 = x + 4$

29. $x^3 - 4x - 1 = 5 - x^2$

Connect & Extend   **30.** Sketch a graph of each function in Part a on a single set of axes. Do the same for Part b, using a new set of axes. Then answer the questions that follow. (Hint: Sketch the graph of $y = \dfrac{1}{x^2}$ by plotting points, and use this graph to help sketch the others.)

a. $y = \dfrac{1}{x^2}$

$y = \dfrac{1}{(x-2)^2}$

$y = \dfrac{1}{(x+2)^2}$

b. $y = -\dfrac{1}{x^2}$

$y = 3 - \dfrac{1}{x^2}$

$y = 3 - \dfrac{1}{(x+2)^2}$

c. Describe how the graphs in Part a are like those in Part b.

d. Describe how the graphs in Part a are different from those in Part b.

e. Find another function that belongs to the set of functions in Part a and another that belongs to the set in Part b.

31. Physical Science A launcher positioned 6 feet above ground level fires a rubber ball vertically with an initial velocity of 60 feet per second. The equation relating the height of the ball over time t is $h(t) = 6 + 60t - 16t^2$, where h is in feet and t is in seconds.

a. Sketch a graph of h.

Another rubber ball is launched two seconds later with the same direction and initial velocity.

b. Suppose you graph the height of the second ball with time since the *first* ball was launched on the horizontal axis. How will the second graph be related to the first?

c. Write an equation for the height of the second ball over time.

d. Will the second ball collide with the first ball when the first ball is on its way up or on its way down? Explain how you could tell from the graphs of the two functions.

For each function f, write a new function g translated 2 units down and 4 units to the left of f.

32. $f(x) = 2^{x+1} - 1$

33. $f(x) = 2(x - 3)^2 + 1$

34. $f(x) = (x - 1)^3 - x + 1$

35. $f(x) = 1 + \dfrac{1}{x^2 + 1}$

36. Geometry A piece of wire 20 cm long is used to make a rectangle.

    a. Call the length of the rectangle L. Write a formula for the width W of the rectangle in terms of its length.

    b. Write a function for the area A of the rectangle in terms of the length L.

    c. Complete the square of the quadratic expression you wrote for Part b. Use your rewritten expression to find the coordinates of the vertex of the graph of Function A.

    d. What are the dimensions of the sides of the rectangle with the maximum area? What is the area of that rectangle?

37. The expression $\dfrac{2x^2}{x}$ is equivalent to $2x$ for all x values except 0. $\dfrac{2x^2}{x}$ is undefined for $x = 0$. The graph of $g(x) = \dfrac{2x^2}{x}$ looks like the graph of $h(x) = 2x$ with a hole at $x = 0$.

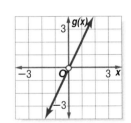

The domain of g is all real numbers except 0. The range of g is also all real numbers except 0.

Now consider the function $f(x) = \dfrac{(x - 2)(x + 1)}{x - 2}$. Sketch a graph of f, and give its domain and range.

In Exercises 38–40, do Parts a–c.

    a. Write the equation in the form $f(x) = ax^2 + bx + c$. Then complete the square to rewrite it in the form $f(x) = a(x - h)^2 + k$.

    b. Find the line of symmetry of the graph of f.

    c. Find the coordinates of the parabola's vertex.

38. $f(x) = 2x^2 - 8x + 2x^2 - 1$

39. $f(x) = 1 + 4x - 2x^2$

40. $f(x) = -x^2 - x - 1 - x - (2 + x)$

41. Sketch a graph and use it to explain why the equation $3^x + 2 = 0$ has no solutions.

42. The cubic function $c(x) = x^3 + 2x^2 - x - 2$ can be rewritten as $c(x) = (x + 2)(x + 1)(x - 1)$.

   a. Find the x-intercepts of c.

   b. Find the y-intercept of c.

   c. Use the intercepts to draw a rough sketch of c.

43. Consider the equation $(x + 2)^2 - 2 = -x^2 + 4$.

   a. Use the method of graphing two functions to estimate solutions of the equation.

   b. Use the quadratic formula to find the exact solutions of this equation. How close were your estimates?

44. Sketch a graph of $y = \frac{1}{x}$. Use your sketch to think about these questions.

   a. How many solutions of $\frac{1}{x} = 5$ are there?

   b. How many solutions of $\frac{1}{x - 5} = 5$ are there?

   c. How many solutions of $\frac{1}{x} = x$ are there? Of $\frac{1}{x} = -x$?

   d. How many solutions of $\frac{1}{x} = x^2$ are there?

   e. Use the method of graphing two functions to show the solutions of $\frac{1}{x} = (x - 3)^2$. Use your graph to estimate those solutions.

45. Use the method of graphing two functions and locating the points of intersection to find at least four values of x for which each inequality is satisfied.

   a. $x^2 - 2x - 7 < 2x - 3$

   b. $x^2 - 2x - 7 > 2x - 3$

46. Geometry The radius of the cylindrical container is one unit less than the side length of the square base of the rectangular container.

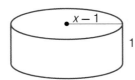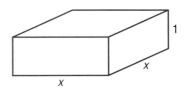

a. For each container, write a function for the volume.

b. Use the volume functions to make a graph comparing the volumes of the two containers. Keep in mind that the value of *x* must be greater than one, or the cylindrical container would not exist.

c. For what value of *x* do the containers hold the same amount?

47. In Your Own Words Explain how graphing can help you solve equations. How would you decide when to find approximate solutions with a graph and when to find exact solutions using algebra?

Mixed Review

Write an equation to represent the value of *y* in terms of *x*.

48.

x	0	1	2	3	4
y	1.2	2.4	4.8	9.6	19.2

49.

x	1	2	3	4	5
y	21	63	189	567	1,701

Evaluate or simplify without using a calculator.

50. $2\sqrt[4]{81a^6}$, where *a* is nonnegative

51. $\sqrt[3]{\left(\frac{8}{125}\right)^2}$

Simplify each expression as much as possible.

52. $a + 4(a - 2)$

53. $2b - (8 + 2b)$

54. $90 - (5c - 1)$

55. $4d(2 + e) - 2(3 + 2d)$

56. $-7(1 - f) + 2(7f + 2) - 9(f + 2)$

Review & Self-Assessment

Chapter Summary

This chapter focused on a particular type of mathematical relationship called a *function*. A mathematical function produces a single output for each input and can be described with a graph or an equation.

You worked with graphs and equations to find the maximum and minimum values of functions. You used these extreme values to identify the range of a function and to solve exercises involving maximum height or maximum area.

You studied in depth the graphs of quadratic functions. You found the line of symmetry and the coordinates of the vertex by inspecting graphs and by completing the square of quadratic expressions. You also solved equations of the form $f(x) = 0$ to find the x-intercepts of quadratic functions.

Finally, you solved equations of the form $f(x) = g(x)$ by locating the points where the graphs of f and g intersect.

Strategies and Applications

The questions in this section will help you review and apply the important ideas and strategies developed in this chapter.

Understanding functions and describing the domain and range of a function

1. Explain how you can determine whether a relationship could be a function by examining a table of inputs and outputs.

2. Give an example of a relationship that is *not* a function.

3. Explain how you can tell whether a relationship is a function by looking at its graph. Give an example of a graph that is not a function.

4. Give an example of a function in which negative numbers do not make sense as part of the domain.

5. Describe the range of the function $k(n) = 3n^2 - 4$.

Finding the maximum and minimum values of quadratic functions

6. Explain two ways to find the maximum or minimum value of a quadratic function.

Vocabulary

domain

function

range

x-intercepts

7. Consider all the possible rectangles with a perimeter of 22 centimeters.

    a. If the length of one such rectangle is x cm, write an equation for a function A for the area of the rectangle.

    b. Use your answer to Part a to find the maximum possible area of the rectangle.

    c. What dimensions give the maximum area?

8. Assume the function $H(t) = 100t - 4.9t^2$ gives the height in meters of a rocket launched vertically from ground level, where t is time in seconds. Estimate the maximum height of the rocket. Tell how many seconds after its launch it attains this maximum height.

Understanding and using graphs of quadratic functions

9. Explain how the graphs of g and h are related to the graph of $f(t) = 2t^2$.

$$g(t) = 10 + 2(t + 2)^2 \qquad h(t) = 2(t - 2)^2 - 3$$

10. Which of these quadratic functions has its vertex at $(3, -3)$?

$$g(t) = 3(t + 3)^2 - 3 \qquad h(t) = 4(t - 3)^2 - 3 \qquad k(t) = 3 - 3(t - 3)^2$$

11. Write the equation for a quadratic function with vertex $(-6, 1)$.

12. Explain how the range of a quadratic function is related to the vertex of its parabola.

13. Explain two methods for finding the vertex and the line of symmetry for the graph of $g(x) = (x + 2)(x + 4)$. Give the vertex and the line of symmetry.

14. Explain how you can use the x-intercepts of a quadratic function f to find its vertex.

Solving equations involving two functions

15. Explain how to use the method of graphing two functions to solve the equation $x^3 = 2x^2 - 1$.

16. Determine how many solutions this equation has, and explain how you found your answer.

$$x^2 + 2x - 3 = x - 2$$

Demonstrating Skills

Copy and complete each table for the given function.

17. $g(x) = x^2 + 3x - 1$

Input	−2	−1	0	1	2
Output					

18. $h(x) = \dfrac{1}{4 - x}$

Input	−4	−2	0	2	4
Output					

19. This is a graph of $f(x) = 2^x$.

  a. Sketch a graph of $g(x) = 2^{x-2}$.

  b. Sketch a graph of $h(x) = 2^{x+3}$.

  c. Sketch a graph of $j(x) = 2^x - 2$.

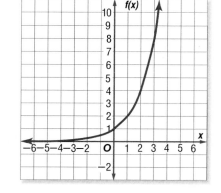

Tell whether each function has a minimum or maximum value, and give the coordinates of this point.

20. $f(x) = -x^2 + 2x - 2$

21. $j(x) = -5 + x + x^2$

22. $k(x) = 3 - 4(1 - x)^2$

Write the equation for a function g that is the same shape as f but translated 2 units to the left and 1 unit down.

23. $f(x) = -1 + \dfrac{1}{x^3 + 1}$

24. $f(x) = 3^{x+1} - 2$

25. $f(x) = x(x - 2)$

26. Determine the vertex and the line of symmetry of $f(x) = (x + 5)^2 + 9$ without graphing.

For each quadratic function, complete the square and find the vertex and the line of symmetry of its parabola without graphing.

27. $Q(x) = 2x^2 + 2x - 6$

28. $m(x) = -x^2 + \dfrac{7}{2}x - 3$

29. $r(x) = x(x + 3)$

30. Consider the function $f(x) = -x^2 + 8x - 7$.

  a. Find the x-intercepts of the graph of f.

  b. What is the line of symmetry and the vertex of the graph of f?

  c. Use the x-intercepts and the vertex to sketch a graph of f.

31. Solve this equation by graphing.

$$x^2 + 1 = 0.5x + 2.5$$

32. Explain how to solve the equation $x^2 + x = -x - 1$ without graphing. Solve for x.

33. Start with a 4-inch-by-6-inch card. What cut-out size will maximize the volume for an open box made from the card?

34. Start with a 4-inch-by-6-inch card. What is the greatest possible volume?

35. Start with a 10-inch-by-16-inch sheet of paper. What cut-out size will maximize the volume for an open box made from the card?

36. Start with a 10-inch-by-16-inch card. What is the greatest possible volume?

37. Start with a 15-inch-by-24-inch sheet of paper. What cut-out size will maximize the volume for an open box made from the card?

38. Start with a 15-inch-by-24-inch card. What is the greatest possible volume?

Test-Taking Practice

SHORT RESPONSE

1 The height of a ball thrown into the air is given by the function $y = -t^2 + 4t + 3$, where y is the height of the ball in feet and t is time in seconds. Find the maximum height of the ball and the time it takes to reach this height.

Show your work.

Answer _____

MULTIPLE CHOICE

2 Which equation represents a function g that is the same shape as $f(x) = x(x + 1)$ but translated 1 unit to the right and 2 units down?
A $g(x) = (x - 1)x - 2$
B $g(x) = (x - 1)x + 2$
C $g(x) = (x - 2)x + 1$
D $g(x) = (x + 1)(x + 2) - 2$

3 Determine the vertex and the line of symmetry of $f(x) = (x - 3)^2 + 5$.
F $(3, 5); x = 3$
G $(-3, 5); x = 3$
H $(3, 5); x = 5$
J $(3, -5); x = 3$

4 Which coordinates are the minimum value of the function $f(x) = x^2 + 2x - 3$?
A $(-3, 0)$
B $(-3, 1)$
C $(-1, -4)$
D $(1, 0)$

5 Find the x-intercepts of the graph of $f(x) = (x + 2)^2 - 4$.
F $(-4, 0), (0, 0)$
G $(-2, 4), (2, 12)$
H $(0, 0), (4, 0)$
J the graph has no x-intercepts

CHAPTER 111

Data and Probability

Real-Life Math

Old West Action Crosswords, coded messages, and hidden words are all types of word puzzles. Another type of word puzzle is an *anagram*. Anagrams are particularly fun because you do not need someone to create a puzzle for you. Choose a word or phrase, and try to rearrange the letters to form another word or phrase. For example, *Old West action* is an anagram of *Clint Eastwood,* a star of many western films.

While a short word such as *star* has 24 combinations, including *rats, tars,* and *tsar,* a word does not have to be much longer to have thousands or millions of possible rearrangements. For example, the letters in *Clint Eastwood* can be arranged in 1,556,755,200 different ways.

Think About It
List as many anagrams of the word *stop* as you can.

Math Online
Take the **Chapter Readiness Quiz** at glencoe.com.

Dear Family,

In the next few weeks, the class will be looking at many new situations that involve probability. In finding the probability that something will occur, you must first find all the *possible* outcomes. For example, if you are drawing blocks from a bag containing three blue, two green, and five white blocks, there are ten possible outcomes: the ten blocks. The probability of drawing a green block is the ratio of the number of green blocks, two, to the number of possible outcomes, $\frac{2}{10}$.

Key Concept—Probability

The class will consider more complicated situations soon. For example, imagine that you draw a block and keep it and then draw another block. The class will learn how to calculate such probabilities as the probability of drawing a green block first and a blue block second, using methods from the mathematical field called *combinatorics*. This method includes finding the possible combinations of items.

One such method is to use a tree diagram, like the one below, to record all ten possible first draws, and then for each first draw, the nine possible second draws. This diagram shows the branches for a tree diagram when green is the result of the first draw.

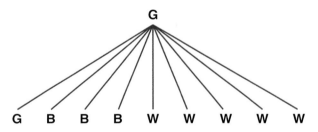

At the end of this chapter, the class will apply what it has learned to analyze the fairness and probabilities of complicated games, such as sports playoffs and state lotteries. We will answer questions like the following.

- What are the chances of winning any of several lotteries?
- Is one team favored by a particular playoff structure?
- Which playoff structure is the fairest in a given situation?

Chapter Vocabulary

quartile   sample space

Home Activities

- Help your student think about common occurrences of this topic such as lotteries.
- Play a game with your student that involves the use of dice or spinners and probability.

LESSON
11.1

Counting Strategies

A *probability* is a number between 0 and 1 that indicates how likely something is to happen. Often the key to determining the probability that something will occur is to first find all the possible *outcomes*.

When you toss a coin, the two possible outcomes, heads and tails, are equally likely. So, the probability of getting heads is 1 out of 2, or $\frac{1}{2}$, or 0.5, or 50%.

Think & Discuss

Suppose you turn the two spinners below. Each spinner has an equal chance of landing on white, blue, or orange.

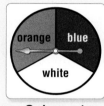 

Spinner 1   **Spinner 2**

List all the possible outcomes. For example, one outcome is spinner 1 landing on white and spinner 2 landing on orange. You can use the notation white/orange, or WO, to represent this outcome.

How many possible outcomes are there?

In how many outcomes does spinner 1 land on blue and spinner 2 land on orange? What is the probability of this happening?

In how many outcomes does one spinner land on blue and the other land on orange? What is the probability of this happening?

In how many outcomes does spinner 2 land on blue? What is the probability of this happening?

In some situations, counting outcomes is not as easy as it sounds. In this lesson, you will investigate some counting strategies.

Inquiry

Investigation 1 Pizza Toppings

Materials

- cubes in 5 colors
- circles to represent pizzas

Paula's Pizza Place offers four toppings for its vegetarian cheese pizzas.

Customers can order a pizza with any combination of toppings, from cheese to all four toppings. However, a topping can be used only once. A customer cannot order a pizza with two helpings of artichokes, for example.

Make a Prediction

1. How many different pizzas do you think can be made using these four toppings?

Try It Out

You can explore this situation by making a model. Use colored cubes to represent the toppings and circles to represent pizzas. Create a few pizzas by placing the cubes on the circles. Remember, use only one helping of a topping on any one pizza.

See whether you can make all the possible pizzas. Try to find a systematic way to organize the pizzas so you can be sure you found them all.

2. List all the different pizzas that are possible. How many possibilities are there?

Go on ▶

Try It Again

You will now solve the situation again but with fewer toppings available to choose. As you work, look for a pattern in your results that may confirm your answer about the four-topping pizza.

3. How many different pizzas can you make with one topping?

4. How many different pizzas can you make with two toppings?

5. How many different pizzas can you make with three toppings?

6. Organize your results from Questions 3–5 in a table.

Toppings	1	2	3
Different Pizzas			

7. Do you see a pattern in your results? If so, does the number of different pizzas you made from four toppings fit the pattern? If not, check your work for Questions 3–5.

8. Suppose Paula adds pineapple to the list of toppings. How many combinations do you predict are now possible? Use a fifth cube color to represent pineapple. Make enough pizzas to see a pattern. Check your prediction.

What Did You Learn?

Review your results for all the pizza combinations. You should see a pattern in them. If you do not, check your work.

9. Use the pattern in your results to extend your table up to at least 12 toppings. You may use a calculator, but try to complete the table without using paper-and-cube pizza models.

10. Make a list of all the pizza toppings you can recall. In a short report, explain how to determine the number of combinations that can be made from your list of toppings. In your report, also answer this question.

If you order a different type of pizza every day, how many days, weeks, or months will pass before you will have ordered all of the possibilities?

Investigation One-on-One Basketball

Vocabulary

sample space

Materials

- 4 identical slips of paper
- container

Ally, Brad, Carol, and Doug are playing one-on-one basketball. To decide the two players for each game, they put their names into a hat and pull out two at random.

To find the probability that Brad and Carol will play the next one-on-one game, you might start by first listing all the possible pairs of the four friends. Each pair is an *outcome* in this situation. The set of all possible outcomes, in this case, the set of all possible pairs, is called the **sample space**.

There are many ways to find the sample space for a particular situation, but you need to be careful. If there are numerous possible outcomes, it can be difficult to determine whether you have listed all of them or have listed an outcome more than once.

✅ Develop & Understand: A

You will now use a systematic method to find the sample space for drawing pairs of names for the one-on-one basketball situation.

1. List all the possible pairs of names that include Ally.

2. List all the possible pairs that include Brad but *not* Ally.

3. Now list all the pairs that include Carol but *not* Ally or Brad.

4. Review your answers to Exercises 1–3.

   a. Are there any pairs that you have listed more than once or that you have overlooked? If so, correct your errors.

   b. How many pairs are there in all? List them.

5. Brad wants to play Carol in the next game.

   a. How many pairs match Brad with Carol?

   b. What is the probability that Brad will play Carol in the next game?

In Exercises 1–5, you calculated that the probability that Brad and Carol will play in the next game is $\frac{1}{6}$. This does not necessarily mean that in the next six draws, the pair Brad/Carol will be chosen exactly once. It is possible, though not likely, that the pair will be drawn all six times or that Carol's name will not be drawn even once.

Probabilities do not tell you what will *definitely* happen. They tell you what you can expect to happen *over the long run*.

✅ Develop & Understand: B

Write the four names, *Ally*, *Brad*, *Carol*, and *Doug*, on identical slips of paper. Put them into a container.

6. Suppose you randomly, that is, without looking or trying to choose one name over another, draw 12 pairs of names from the container, putting the pair back after each draw. Based on the probability you found in Exercises 1–5, how many times would you expect to draw the pair Brad/Carol?

Draw two names at random. Record the results. Return the names to the container. Repeat this process until you have drawn 12 pairs.

7. How many times was the pair Brad/Carol drawn? How do these experimental results compare with your answer to Exercise 6?

8. Each group in your class drew 12 pairs of names. How many total draws occurred in your class? In this number of draws, how many times would you expect the pair Brad/Carol to be drawn?

9. Each group in your class should now record how many times they drew the pair Brad/Carol. How many times in all was the pair Brad/Carol drawn? How does this compare with your answer to Exercise 8?

Next, you will investigate what happens when a fifth player is added to the one-on-one basketball situation.

✅ *Develop & Understand: C*

10. Suppose Omar joins Ally, Brad, Carol, and Doug.

  a. How many new pairs can now be made that could not be made before Omar joined?

  b. What is the size of the new sample space?

  c. To verify the size of the new sample space, systematically write down all the possible pairs. Rather than writing out the entire names, use each player's first initial.

11. Look at the new sample space you listed in Exercise 10.

  a. How many of the pairs involve Omar? Which are they?

  b. What is the probability that Omar will be involved in the next game?

In Exercises 10 and 11, you found the probability that the following event would occur, *Omar is in the pair.* This particular event has four outcomes in the sample space. They are Omar/Ally, Omar/Brad, Omar/Carol, and Omar/Doug.

If you know that the outcomes in a sample space are *equally likely*, that is, that each outcome has the same chance of occurring, it is easy to calculate the probability of a particular event.

Example

The names *Ally, Brad, Carol, Doug,* and *Omar* are put into a hat. One pair of names is pulled out at random. What is the probability that the pair will include either Ally or Brad or both?

The sample space consists of ten outcomes.

Ally/Brad	Ally/Carol	Ally/Doug	Ally/Omar
Brad/Carol	Brad/Doug	Brad/Omar	Carol/Doug
Carol/Omar	Doug/Omar		

Of these ten outcomes, seven include Ally or Brad.

Ally/Brad	Ally/Carol	Ally/Doug	Ally/Omar
Brad/Carol	Brad/Doug	Brad/Omar	

Since each pair has the same chance of being drawn, the probability that the pair will include either Ally or Brad is $\frac{7}{10}$.

✅ *Develop & Understand: D*

The names *Ally, Brad, Carol, Doug,* and *Omar* are put into a hat.
One pair of names is pulled out at random.

12. What is the probability that the pair includes Doug?

13. What is the probability that the pair includes Carol or Omar
or both?

14. What is the probability that the pair does not include Brad?

15. What is the probability that the pair includes Ally or Brad
or Doug?

16. What is the probability that the pair includes Carol but not Doug?

17. What is the probability that the pair includes Ally and Omar?

18. Create a question like those in Exercises 12–17 that involves an
event with a two-in-ten chance of occurring.

19. Create another question like those in Exercises 12–17. Give the
answer to your question.

Share & Summarize

If all the outcomes in a sample space are equally likely, how can you
find the probability that a particular event will occur?

Investigation ③ More Counting Strategies

You have seen that to find the probability of an event, you must find
the size of the sample space. In Investigation 2, you used a systematic
strategy to list all the possible pairs for a one-on-one matchup. In this
investigation, you will discover some other useful counting strategies.

Think & Discuss

Julius and Marcia have the five CDs by their favorite band, X Squared.

- *Algebraic Angst*
- *Binary Breakdown*
- *Chalkboard Blues*
- *Dog Ate My Homework*
- *Everyday Problems*

The friends want to listen to all five CDs. Predict the number of different orders in which they can play the five CDs.

To find how many ways a group of CDs can be ordered, you can list all the possibilities. With only one CD, there is obviously only one order. With two CDs, call them A and B, there are two possible orders, AB and BA. With three CDs, there are six orders.

ABC   ACB   BAC   BCA   CAB   CBA

To be certain you have not missed any possibilities, you need a systematic method of counting and recording. You could list the possibilities for three CDs in a table.

Math Link

The table and tree diagram show the ways these CDs can be arranged or listed. A listing in which order is important is called a *permutation*.

A First	B First	C First
ABC	BAC	CAB
ACB	BCA	CBA

Or, you could organize them using a *tree diagram*.

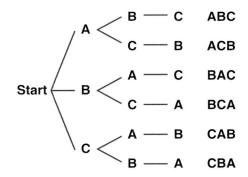

The tree diagram works in much the same way as the table. It shows that there are three ways to start, A, B, or C. Then there are two choices for the second CD and one choice for the third.

✅ *Develop & Understand: A*

Consider the case of four CDs, A, B, C, and D.

1. Predict the number of ways these four CDs can be ordered.

2. Make an organized list of all the possibilities in which A is played first. To do this, you may want to make a tree diagram like the one started at right.

A First
ABCD
ABDC
ACBD
⋮

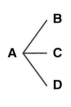

3. The number of orders of four CDs in which A is first is the same as the total number of ways three CDs can be ordered. Explain why.

4. Why is the number of orders of four CDs in which B is first the same as the number of orders of four CDs in which A is first?

5. List all the orders in which B is first. Then list all the orders in which C is first. Finally, list all the orders in which D is first. Use your answer to Exercise 4 to verify that you have listed all the possibilities.

6. How many possible orders are there altogether? If you know how many entries are in each list, how can you determine the total number of entries without counting them all?

7. Is your estimate from Exercise 1 greater than or less than the actual number of orders?

8. Now return to the situation presented in Think & Discuss on the previous page. In how many different orders can Julius and Marcia play their five CDs? Try to find your answer *without* listing all the possibilities. Show how you found it.

If you consider the orders of the CDs as the outcomes in a sample space, you can calculate probabilities of specific events.

Example

Consider all the orders of four CDs, A, B, C, and D. If one of these orders is selected at random, what is the probability that B will be before D?

In Exercises 1–8, you found that there are 24 outcomes in the sample space for this situation. B appears before D in 12 of these 24 outcomes.

ABCD	BACD	CABD
ABDC	BADC	CBAD
ACBD	BCAD	CBDA
	BCDA	
	BDAC	
	BDCA	

So, the probability that B will be played before D is $\frac{12}{24}$, or 50%.

In Exercises 9–24, you will find the probability of other events that involve the order in which four CDs are played.

✅ Develop & Understand: B

In Exercises 9–15, determine the probability that the given event will occur. Before you begin, be sure you have a complete list of the 24 outcomes in the sample space for this situation.

9. C immediately follows B.

10. A is played last.

11. C is not played first.

12. B is played before A.

13. D is played first *and* A is played last.

14. The CDs are played in the order CBAD.

15. A is played first *and* C is not played last.

16. The probability of A being played last is the same as the probability of B being played last. Explain why.

17. Why are the chances of A being played last the same as the chances of A being played first?

18. How can you use the chances that C is played first to check your answer to Exercise 11?

19. If the probability of an event is $\frac{1}{4}$, what is the probability that the event will *not* happen? Explain how you found your answer.

20. If the probability of an event is *p*, where *p* is between 0 and 1, what is the probability that the event will not happen?

Describe an event related to the CD exercise with the given probability.

21. $\frac{6}{24}$ **22.** $\frac{22}{24}$ **23.** $\frac{24}{24}$ **24.** $\frac{3}{24}$

Share & Summarize

1. Ajay found a shortcut for determining the number of ways to put CDs in order. He said, "For five CDs, multiply 5 by the number of ways to put four CDs in order. In fact, for *n* CDs, just multiply *n* by the number of ways to put *n* − 1 CDs in order."

  a. There is only one way to order one CD. Use Ajay's method to find the number of ways to order two CDs. Did it work?

  b. Use Ajay's method and your answer to Part a to find the number of ways to order three and four CDs. Did it work?

  c. Explain why Ajay's method makes sense.

  d. How many ways are there to order seven CDs?

2. Liseta thought she could use the strategy from Exercises 1–8 to find the number of one-on-one pairs of Ally, Brad, Carol, and Doug. She started listing the possibilities.

Ally First
Ally/Brad
Ally/Carol
Ally/Doug

Liseta said, "There are three outcomes on this list. There will be four lists, one for each friend, so there are 12 pairs in all." Is she correct? Explain.

Investigation 4 Counting Strategies Using Patterns

In Investigation 2, you found the size of a sample space by listing all the possibilities of one-on-one pairs. In Investigation 3, you saw that you can sometimes discover a pattern that allows you to find the total number of outcomes without listing them all. In this investigation, you will explore other counting strategies.

Think & Discuss

Two whole numbers add to 12. What might the numbers be?

List every pair of whole numbers with a sum of 12. Make sure you have listed all the possibilities. How many pairs are there? In this situation, order does not matter. The pair 3-9 is the same as the pair 9-3.

Predict how many whole-number pairs have a sum of 100.

One strategy for finding the number of whole-number pairs with a sum of 100 is to first consider some simpler exercises and look for a pattern.

✅ Develop & Understand: A

1. The sum of two whole numbers is 10. Three possible pairs are 0-10, 1-9, and 2-8. What are the other pairs? How many pairs are there in all?

2. Now write down all the whole-number pairs with a sum of 11. How many pairs are there?

3. Look back at the Think & Discuss above. How many whole-number pairs have a sum of 12?

4. Copy and complete the table to show the number of whole-number pairs with each sum.

Sum	10	11	12	13	14	15	16
Number of Pairs							

5. Copy this table. Use any patterns you have observed to find the number of whole-number pairs with each given sum.

Sum	20	27	40	80	100	275
Number of Pairs						

6. Look at the even and odd sums and the number of pairs that produce them.

   a. Write two expressions, one for even sums and one for odd sums, that describe the relationship between the sum S and the number of whole-number pairs that produce that sum.

   b. Explain why the expressions for even sums and odd sums are different.

Once you know the size of the sample space, you can calculate the probability of an event occurring.

✓ Develop & Understand: B

Use the patterns and answers you found in Exercises 1–6 to determine these probabilities.

7. All the whole-number pairs with a sum of 20 are put into a hat. One is drawn at random.

   a. What is the size of the sample space in this situation?

   b. What is the probability that one of the numbers in the pair selected is greater than 14? Explain how you found your answer.

8. All the whole-number pairs with a sum of 100 are put into a hat. One is drawn at random.

   a. What is the size of the sample space in this situation?

   b. In how many pairs are both numbers less than 60? List them.

   c. What is the probability that both of the numbers in the selected pair are less than 60?

9. All the whole-number pairs with a sum of 55 are put into a hat. One is drawn at random.

   a. What is the size of the sample space in this situation?

   b. How many pairs include a number greater than 48? List them.

   c. What is the probability of choosing a pair in which neither number is greater than 48?

So far, you have used three strategies to find the size of a sample space.

- Systematically list all the possibilities.
- Begin a systematic list of possibilities. Look for a pattern that will help you find the total number without completing the list.
- Start with simpler cases. Look for a logical pattern that you can extend to the more complicated cases.

In Exercises 10–18, you will find the size of a sample space by breaking it into manageable parts. The exercises are similar to Exercises 1–6 but are a bit more complicated.

✅ Develop & Understand: C

Consider this situation. *Three whole numbers have a mean of 3. How many such whole-number triples exist? How can you be sure you have found them all?* Here is one way to think through this situation.

10. If the mean of three whole numbers is 3, what is their sum? Why?

Now you can think about the situation as finding all the combinations of three numbers with a sum of 9.

11. When you list combinations of three whole numbers with a sum of 9, does the *order* of the numbers matter? For example, is the triple 1-2-6 considered the same as or different from the triple 6-2-1?

12. One way to break this into manageable parts is to start by thinking about all the whole-number triples in which at least one number is 0. List all such triples with a sum of 9. How many are there?

13. To continue listing the combinations, you might next decide to find all those that contain at least one 1 but no 0s.

   a. Why would you exclude 0s from your triples at this stage?

   b. List the triples that contain at least one 1 but no 0s. How many are there?

14. Continue this process. List all the triples that contain at least one 2 but not 0 or 1. Then list all the triples that include at least one 3 but not 0, 1, or 2. Complete the table.

Smallest Number in Triple	0	1	2	3
Number of Triples	5			

15. Why are no more columns needed in this table? In other words, explain why you do not need to consider triples in which the smallest number is 4, 5, or any greater number.

16. How many different triples are there in all?

17. If all the triples are put into a hat, what is the probability of drawing a combination whose smallest number is 0? Whose smallest number is 4?

18. Use the strategy of breaking this into smaller parts to find the number of whole-number triples with a mean of 4.

Share & Summarize

A set of four numbers has a mean of m. Explain how you would find the number of whole-number sets that this could describe.

Investigation 5 Compare Probabilities of Events

Materials

- 2 dice

Understanding how likely certain events are can help you make decisions and predictions. In this investigation, you will analyze situations and games involving spinners and six-sided dice to determine how likely certain events are.

✓ Develop & Understand: A

Twyla rolled two dice 15 times. On each roll, she multiplied the two numbers. Based on her findings, she conjectured that rolling an even product is more likely than rolling an odd product.

1. Roll a pair of dice 15 times. Record whether each product is even or odd. Do your results support Twyla's conjecture?

To figure out whether an even product or an odd product is more likely, you could find the products for all 36 possible dice rolls and count how many are even and how many are odd.

An easier way to analyze Twyla's conjecture is to use what you know about multiplying even and odd factors. On each die, the probability of rolling an odd number is the same as the probability of rolling an even number. You can simply figure whether each possible combination is odd or even.

$$even \times even \qquad even \times odd \qquad odd \times even \qquad odd \times odd$$

2. Copy and complete the table to show whether the product of each combination of even and odd factors is even or odd.

	Die 1	
×	**Even**	**Odd**
Even	Even	
Odd		

Die 2 appears to the left of the Even/Odd rows.

3. Which is more likely to occur, an even product or an odd product?

4. Complete these probability statements.

    a. The probability that the product of two dice will be even is _____ out of 4.

    b. The probability that the product of two dice will be odd is _____ out of 4.

5. Suppose that, instead of rolling dice to determine the factors to multiply, Twyla turns these spinners.

Spinner 1   **Spinner 2**

a. Predict whether an even product or an odd product is more likely.

b. Can you use your table from Exercise 2 to determine whether an odd or an even product is more likely? Why or why not?

c. Systematically determine the probability of getting an even product and the probability of getting an odd product. You might find it helpful to complete a multiplication table like the one below.

Spinner 1

×	1	2	3	4	5
1					
2					
3					
4					
5					

Spinner 2

✅ *Develop & Understand: B*

Twyla was also curious about what would happen if she rolled two six-sided dice and considered one number to be the base and the other to be its exponent. Would the result more likely be even or odd?

6. Conduct an experiment to predict whether an even or an odd result is more likely. Before you begin, assign one die as the base and the other as the exponent. Roll the dice 15 times. Record whether the result, $die1^{die2}$, is even or odd. Make a prediction based on your results.

7. Because an odd number is just as likely to be rolled as an even number, you can analyze the possible combinations as you did in Exercises 1–5. Complete the table to indicate whether raising the given base to the given exponent has an odd or an even result.

	Even Exponent	**Odd Exponent**
Even Base		
Odd Base		

8. Which is more likely to occur, an even result or an odd result?

9. Complete these probability statements.

    a. The probability that $die1^{die2}$ will be even is _____.

    b. The probability that $die1^{die2}$ will be odd is _____.

Share & Summarize

Suppose you turned these two spinners and added the results. Describe two ways you could determine whether an odd sum or an even sum is more likely.

Spinner Y

Spinner Z

Practice & Apply

1. Three friends, Ashton, Kavi, and Chelsea, get together to play chess. To determine who will play whom, they put their names into a hat and draw out two at random.

   a. List all the possible pairs of names.

   b. What is the probability that Kavi and Chelsea will play the next game?

   c. Three more friends, Donae, Eric, and Fran, join the group. Now how many possible pairs are there? List them.

   d. What is the probability that the next match will involve Ashton or Eric or both?

   e. What is the probability that the next match will *not* involve Kavi?

2. You have been asked to organize the matches for the singles competition at your local tennis club. There are seven players in the competition. Each must play every other player once.

   a. The matches player A must play are listed below. This list already includes the match of player A against player B. Copy the table. In the row for player B, list all the other matches player B must play. In other words, list all the matches that include player B but *do not* include player A.

Player	Matches to Play	Number of Matches
A	AB, AC, AD, AE, AF, AG	6
B		
C		
D		
E		
F		
G		

   b. Predict the number of matches player C must play that *do not* include players A or B. Write your prediction in the "Number of Matches" column.

   c. Predict the number of matches for the remaining players. In each case, consider only those matches that *do not* include the players listed above that player.

d. Check your predictions by listing the matches for each player.

e. Describe the pattern in the "Number of Matches" column. Why do you think this pattern occurs?

f. Find the total number of matches that will be played in the singles competition.

g. Use what you have discovered in this situations to predict the total number of matches in a singles competition involving eight players. Explain how you found your answer.

3. Pilar wants to make a withdrawal from an ATM, but she cannot remember her personal identification number. She knows that it includes the digits 2, 3, 5, and 7, but she cannot recall their order. She decides to try all the possible orders until she finds the right one.

a. How many orders are possible?

b. Pilar remembers that the first digit is an odd number. Now how many orders are possible?

c. Pilar then remembers that the first digit is 5. How many orders are possible now?

4. The "Shuffle" button on Tenisha's MP3 player plays the songs in a random order. Tenisha downloads four songs into the player and presses "Shuffle."

a. How many ways can the four songs be ordered?

b. What is the probability that song 1 will be played first?

c. What is the probability that song 1 will *not* be played first?

d. Songs 2 and 3 are Tenisha's favorites. What is the probability that *one* of these two songs will be played first?

e. What is the probability that songs 2 and 3 will be the first two songs played in either order?

5. All the whole-number pairs with a sum of 26 are put into a hat. One is drawn at random.

a. List all the possible whole-number pairs with a sum of 26.

b. What is the size of the sample space in this situation?

c. What is the probability that at least one of the numbers in the pair selected is greater than or equal to 15? Explain how you found your answer.

d. What is the probability that both numbers in the pair selected are less than 15? Explain.

6. **Challenge** All the whole-number pairs with a sum of 480 are put into a hat. One is drawn at random.

   a. What is the size of the sample space in this situation?

   b. What is the probability that one of the numbers in the pair selected is greater than 300? Explain how you know.

7. Three whole numbers have a mean of 5.

   a. List all the whole-number triples with a mean of 5. Explain how you know you have found them all.

   b. How many such whole-number triples exist?

   c. Suppose all the whole-number triples with a mean of 5 are put into a hat. One is drawn at random. What is the probability that at least two of the numbers in the triple are the same?

8. Two six-sided dice are rolled and the two numbers are added.

   a. Copy and complete the table to show all the possible sums.

Die 1

+	1	2	3	4	5	6
1						
2						
3						
4						
5						
6						

Die 2

   b. How many different sums are there? What are they?

   c. Which sum occurs most often? List all the ways it can be created.

   d. Complete the table to indicate whether the sum of each combination of even and odd numbers is even or odd.

Die 1

+	Even	Odd
Even	Even	
Odd		

Die 2

e. What is the probability that the sum of two dice will be odd?

f. Suppose you turn these spinners and add the results. Can you use your table from Part d to find the probability that the sum will be odd? Explain.

9. Suppose you turn these spinners and multiply the results.

a. Predict whether an odd product or an even product is more likely.

b. Determine the probability of spinning an odd product and of spinning an even product. Do your results agree with your prediction?

Connect & Extend

10. The Alvarez family, Amelia, Bernie, Claudio, Dina, Eduardo, and Flora, want to form two teams of three to play charades. They put their names into a hat and choose three names to form one team. The remaining three players will form the other team.

a. How many teams are possible? List them all.

b. What is the probability that the three names drawn will include Amelia or Eduardo or both?

c. What is the probability that the three names drawn will include both Amelia and Eduardo?

d. What is the probability that the three names drawn will include neither Amelia or Eduardo?

e. What is the probability that Amelia and Eduardo will be on the same team? Explain how you found your answer.

11. Trey is helping to plan a school picnic. Each lunch will include a sandwich, a side item, and a dessert. The possible choices for each are given below.

Sandwich	**Side**	**Dessert**
peanut butter	salad	fresh fruit
cheese	chips	cookie
egg salad		yogurt
		pie

a. How many different lunch combinations are possible?

Trey and his co-workers make an equal number of each combination, but they forgot to mark the bags. Assume that when a person takes a bag, each combination is just as likely to be in the bag as any other combination.

b. Brandy does not care what dessert she gets, but she really wants an egg sandwich and a salad. What is the probability that her lunch will include these two items?

c. Alan does not like eggs. What is the probability that he will choose a bag that *does not* include an egg salad sandwich?

12. In this exercise, you will think about the different ways a number of people can be seated along a bench and around a circular table.

a. How many ways can three people, call them A, B, and C, be seated along a bench? List all the possibilities.

b. If the three people are arranged around a circular table, there will be no starting or ending point. So, for example, these two arrangements are considered the same.

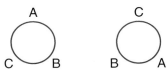

How many *different* ways can three people be arranged around a circular table? Sketch all the possibilities.

c. Copy and complete the table to show how many ways the given number of people can be arranged along a bench and around a circular table.

People	Row Arrangements	Circle Arrangements
1	1	1
2		
3		
4		

d. Describe at least one pattern you see in your table.

e. Five people can be arranged along a bench in 120 ways. Use the patterns in your table to predict the number of ways five people can be seated around a circular table.

13. A programmer wrote some software that composes pieces of music by randomly combining musical segments. For each piece, the program randomly chooses four different segments from a group of 20 possible segments and combines them in a random order.

How many different musical pieces can be created in this way? (Hint: How many choices are there for the first segment? For each of those, how many choices are there for the second segment?)

14. Mrs. Parker raises only chickens and pigs on her farm. If you know how many legs are in Mrs. Parker's barn, you can find all the possible combinations of pigs and chickens. For example, if there are six legs, there could be three chickens, or one chicken and one pig.

  a. Copy and complete the table to show the possible combinations for different numbers of legs. The notation 3C-0P means three chickens and no pigs.

Legs	Combinations	Number of Combinations
2	IC-0P	1
4		
6	3C-0P, 1C-1P	2
8		
10		
12		
14		

  b. Predict the number of combinations for 16 legs and for 18 legs. Check your predictions by listing all the possibilities.

  c. Challenge Write two expressions that describe the number of chicken-pig combinations for *L* legs. One of your expressions should be for *L* values that are multiples of 4. The other should be for *L* values that are not multiples of 4.

  d. There are 42 legs in the barn. If each possible combination of pigs and chickens is equally likely, what is the probability that there are eight pigs and five chickens in the barn?

15. In Your Own Words Explain two counting strategies you can use to find the size of a sample space.

16. Imagine rolling three 6-sided dice and multiplying all three numbers.

   a. How many number triples are possible when you roll three dice?

   b. *Without* finding the products of every possible roll, describe how you could determine whether an odd or an even product is more likely.

   c. Use your method from Part b to determine whether an even product or an odd product is more likely.

17. Imagine rolling five 6-sided dice and looking for outcomes when all five dice match.

   a. How many different outcomes are possible on a roll of five dice? Explain.

   b. In how many of the possible outcomes do all five dice match?

   c. What is the probability of getting all five dice to match on a single roll?

   d. Suppose Tabitha is given three rolls to get five matching dice. On the second and third rolls, she may roll some or all of the five dice again.

   On her first roll, Tabitha gets three 3s, a 2, and a 6. She picks up the dice showing 2 and 6 and rolls them again. What is the probability that she will get two more 3s on this roll?

Mixed Review   **Write a quadratic equation for each table.**

18.

x	y
−3	−9
−2	−4
−1	−1
0	0
1	−1
2	−4
3	−9

19.

x	y
−3	59
−2	54
−1	51
0	50
1	51
2	54
3	59

20.

x	y
−3	18
−2	8
−1	2
0	0
1	2
2	8
3	18

21. Which of the following are equal? Find all matching pairs.

   a. 12

   b. 12^{-1}

   c. $4\sqrt{\frac{1}{9}}$

   d. $4\left(\frac{1}{3}\right)^{-1}$

   e. $\sqrt{\frac{4}{9}}$

   f. the reciprocal of 12

Use the quadratic formula to solve each equation.

22. $3h^2 - 2h + -6 = 0$ **23.** $-6a^2 + 3a = -4$

Modeling with Data

The times in which you are living have been called the *information age* because the amount of information available is increasing faster than ever. Through the Internet, you can find an enormous amount of data on almost any topic you can imagine, in a matter of minutes.

Just having data will not help you, however, if you cannot interpret the data. By organizing and analyzing data, you can sometimes discover trends and connections that will help you understand the information better.

Real-World Link

If you have access to the Internet, you may be interested in *fedstats.gov*, where you can find some of the data collected by the U.S. government.

Think & Discuss

What kinds of data do you think your school might have about you? How might the school use these data?

In Math class, you have often collected and drawn conclusions from data. Consider some investigations in which you collected numerical data earlier this year and then formulated a conjecture. How did you use mathematics to make sense of the collected data?

Investigation 1 Analyze Data Using Tables

Schools often collect data about test scores of groups of students and use them to evaluate the performance of the students and the schools.

One high school draws its students from two small towns, Northtown and Southtown, each with one middle school. At the beginning of the school year, students in Algebra classes take a test to determine how well prepared they are. The results of the Algebra pretest for two classes are given on the next page.

The designation "S" in the "Town" column indicates that the student was from Southtown, and "N" indicates the student was from Northtown. The letter "M" in the "Class" column represents the morning Algebra class, and "A" represents the afternoon class.

Algebra Pretest Results

Student	Score	Town	Class	Student	Score	Town	Class
1	100	S	A	23	54	N	A
2	81	N	M	24	72	N	A
3	55	S	M	25	100	S	M
4	74	N	A	26	66	S	A
5	58	N	A	27	90	S	M
6	59	S	A	28	84	N	A
7	94	N	M	29	68	N	M
8	72	N	A	30	73	N	M
9	100	S	M	31	44	S	M
10	100	S	M	32	82	N	A
11	77	N	A	33	60	S	A
12	94	S	A	34	79	S	M
13	66	N	M	35	94	S	A
14	85	N	M	36	89	N	A
15	63	S	A	37	69	N	A
16	74	S	M	38	69	N	M
17	90	N	M	39	62	S	M
18	66	N	A	40	87	N	M
19	59	S	M	41	76	N	A
20	92	S	A	42	70	S	M
21	73	S	M	43	100	S	A
22	81	N	M	44	88	S	A

The vice principal used the scores to compare the mathematics preparation of students in the two towns. He first calculated the mean score for all 44 students. He then created a new table that would help him compare the performance of students in the two towns.

✓ Develop & Understand: A

1. Design a table that could help the vice principal make a town-to-town comparison of the scores.

2. Find the mean score of the entire group of 44 students and the mean score of the students from each town. Compare these statistics for the two towns.

3. Find the median score of the entire group of 44 students and the median score of the students from each town. Compare these statistics for the two towns.

4. Assuming the test gives an accurate indication of each student's preparation, consider what these scores might mean about the preparation of students from the two towns.

   a. Which town has a wider range of student scores?

   b. Do the towns have the same number of students above the overall median? If not, which has more students above the median?

   c. Five students scored 100 on the test. From which town is each of these top scorers?

   d. Which town has the lowest test score? How many points' difference is there between the lowest scores for the two towns?

   e. Based on these data, which town's middle school do you think prepares students better? Explain your reasoning.

✓ Develop & Understand: B

The same data are often used for quite different purposes. The vice principal was comparing the performance of students from two *towns*. The teacher of the two classes wondered whether either *class* was better prepared.

5. Construct a table to make the teacher's analysis easier. That is, find a way to display the data so you can more easily differentiate the scores for the two classes, morning and afternoon.

6. Do you think the data support a conclusion that one class is better prepared than the other? Justify your answer. You may want to consider mean, median, mode, range, or other factors.

Share & Summarize

In this investigation, you used a table of data to create two new tables to help you analyze the information in the original table.

1. For each new table, how did you decide how to reorganize the original table?

2. Consider the table on the previous page. Give an advantage and a disadvantage to presenting the data in the single table, rather than the two individual tables you created in Exercises 1–6.

3. What things did you look at when trying to form conclusions about the differences between students from the two towns or the two classes?

Investigation 2 Organize Data

Materials

• graphing calculator

Real-World Link

The *base price* for a particular model car is the price without any extra options, such as a DVD player and heated seats.

Alexi wants her parents to buy a new car. When her parents argued that new cars are expensive, Alexi decided to try to convince them that they should buy a new car now, rather than wait a year or two until she was ready to drive and would undoubtedly be asking for the keys.

Although she would have preferred a sports car, Alexi knew her parents would be more likely to listen if she talked about a midsize model like the one they currently owned. She collected the data below about the base price of their model of car each year.

Price of New Car

Year	Price	Year	Price
1993	$14,198	2001	$19,940
1995	15,775	2003	21,319
1997	17,150	2005	22,804
1999	18,922	2007	24,208

Think & Discuss

Describe the trend in the data.

What might cause the price to change the way it does?

Develop & Understand: A

Plot the data from Alexi's table on your calculator. It will help if you treat 1993 as Year 1.

1. Use your graph to decide whether the relationship between year and price appears to be linear, quadratic, exponential, or some other type of function. If it appears to be linear, find an equation for a line that seems to be a good fit for the data.

2. Assume the price continues to increase in the same way. Use your equation to find out how much Alexi's parents would save by purchasing this year's model rather than waiting two years.

Real-World Link

An odometer measures the distance a vehicle has traveled.

Alexi's parents were not convinced, so Alexi decided to find more information. Her father kept a monthly record of odometer readings and the amount of gas purchased for the family car. Alexi studied the records, hoping the data might suggest that the car was using so much gas that it would save money to replace it with a model with better fuel economy.

She prepared this record for 12 months, starting with the gasoline purchased in June. The second column shows the odometer reading, rounded to the nearest mile, on the last day of the month. The third column shows the gallons of gasoline purchased during that month.

Month	Odometer Reading (mi)	Gasoline Purchased (gal)
May	119,982	
Jun	121,142	42.8
Jul	122,564	36.6
Aug	126,354	139.7
Sep	127,459	42.0
Oct	128,106	26.5
Nov	128,919	34.7
Dec	129,939	41.5
Jan	131,052	44.6
Feb	131,695	27.2
Mar	132,430	29.6
Apr	134,114	60.0
May	135,135	35.3

✅ Develop & Understand: B

3. How many miles did the car travel during the year? What was the average number of miles driven per month?

4. Fuel economy is often measured in miles per gallon. Calculate this measure for July and November to the nearest tenth.

5. Construct a table that shows the miles driven and the fuel economy, in miles per gallon, for each month.

6. Which four months show the worst fuel economy? List them in order, starting with the worst.

 Develop & Understand: C

It is natural to wonder what caused the fuel economy to be better in some months than others. Alexi suspects that it is related to temperature. She thinks the car gets fewer miles per gallon in cold weather.

To test her theory, she researched the average temperatures in her city for the months during which she had data. From the Internet, she obtained the average of the daily mean temperatures, in °F, for each month.

Average Daily Mean Temperatures

Jun	Jul	Aug	Sep	Oct	Nov	Dec	Jan	Feb	Mar	Apr	May
64.6	74.3	72.5	66.3	54.4	44.6	39.1	29.5	33.6	39.4	49.2	58.2

7. Are any of the monthly temperature records surprising? Why?

8. Use your calculator to plot the (*temperature, fuel economy*) data. The fuel economy data, in mpg, can be found in your answer to Exercise 5. Adjust the window settings to fill the screen with the data as much as possible. Does there seem to be a connection between these two variables? Explain.

9. There should be one point on your graph that looks far from the others. Make a new graph without including that point. Adjust the window settings to fill the screen with the remaining data points. Now do you think there is a connection between temperature and fuel economy? Explain.

 Develop & Understand: D

Alexi's father said that he thought the car got fewer miles per gallon in months when they did not drive much. Alexi decided to use her data to test his theory.

10. Construct a table that could be used to more easily see whether the fuel economy is worse in low-mile months and better in high-mile months. Can you see any evidence in your table to support Alexi's father's theory?

11. Plot the (*miles driven, fuel economy*) data on your calculator. Does there seem to be a connection between these two variables? Explain.

Share & Summarize

1. In Exercises 1 and 2, you plotted the data from a given table. In Exercises 7–11, however, you had to do something to the given data before plotting. In each case, what data *did* you graph, and why could you not graph the given data?

2. Compare the temperature graph from Exercises 7–9 that you feel is most helpful to the graph of miles driven from Exercises 10 and 11. Which graph suggests a stronger connection between the variables?

3. Alexi's brother Nate observed that the family usually took its longer trips in the summer. He reasoned that fuel economy would be better on longer trips since he had read that highway driving gives better fuel economy than city driving. Do the data fit this observation? Explain.

Investigation 3 Box-and-Whisker Plots

Vocabulary

quartile

In the previous investigations, you examined data using measures of central tendancy, such as the mean, as well as other graphs, such as the scatter plot. You are also familiar with several other types of graphs to display data, such as bar graphs, histograms, and stem-and-leaf plots.

Consider the list of ages of all 64 residents of the town of Smallville.

1	1	2	3	3	4	5	6	6	6
7	9	11	12	12	12	12	13	15	15
15	16	18	18	20	22	24	28	28	28
29	30	30	31	33	34	35	35	36	39
40	41	42	42	46	48	51	53	58	60
62	66	73	75	77	78	81	81	81	86
86	88	92	95						

Another type of graph, a *box-and-whisker plot*, is shown below from the Smallville population.

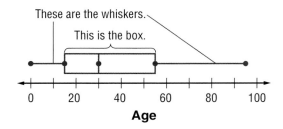

A box-and-whisker plot separates the data into four sections, two whiskers and two parts of the box. Each section represents about 25% of the data. The points that divide the sections are called **quartiles**.

The following exercises will help you understand these statistics and how they are displayed in a box-and-whisker plot.

✔ Develop & Understand: A

Every spring, Mark receives compliments on the tulips in his frontyard. To monitor the flowers' growth one year, he measured the length of the longest leaf on each flower. On one day, a bed of nine tulips had the following leaf lengths, in centimeters.

| 12.6 | 13.8 | 16.0 | 16.1 | 18.6 | 23.3 | 24.4 | 27.4 | 32.5 |

Mark made a box-and-whisker plot of these data.

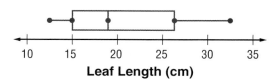

1. Two important points in box-and-whisker plots are the *minimum*, or lowest data point, and *maximum*, or highest data point.

   a. What are the minimum and maximum for this data set?

   b. How are these values shown on the graph?

2. One of the quartiles is the *median* of the data set.

   a. What is the median of this data set?

   b. How is this value shown on the graph?

Math Link

The *median* of a data set is the middle value if there are an odd number of values and the mean of the two middle values if there are an even number of values.

3. To find the remaining two quartiles, you can consider the data below the median as one data set and the data above the median as another data set.

   The median of the lower data set is the *first quartile*, and the median of the upper data set is the *third quartile*. The median of the entire data set is also the *second quartile*.

   a. What is the first quartile? What is the third quartile?

   b. How are these values shown on the graph?

4. The *range* of a set of values is the difference between the maximum and minimum values. What is the range of heights in this data set?

✅ *Develop & Understand: B*

The list below is the ages of all 64 residents of Smallville.

1	1	2	3	3	4	5	6	6	6
7	9	11	12	12	12	12	13	15	15
15	16	18	18	20	22	24	28	28	28
29	30	30	31	33	34	35	35	36	39
40	41	42	42	46	48	51	53	58	60
62	66	73	75	77	78	81	81	81	86
86	88	92	95						

The box-and-whisker plot for the ages of the 64 Smallville residents can be seen below.

The important statistics have been labeled in the graph. Notice that there are five important points. They are the minimum, the maximum, and the first, second, and third quartiles.

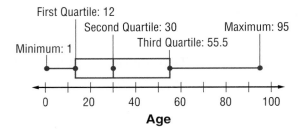

5. Copy the data set onto your paper. Draw a slash (/) to divide the two halves on either side of the median. Then draw two more slashes, dividing each of those halves.

6. Your slashes represent the quartiles in the data set.

   a. How many values are represented by each whisker? What percentages of the town's population are these?

   b. How many values are represented by each section of the box? What percentages of the town's population are these?

7. Why is the left-hand whisker shorter than the right-hand whisker?

8. Why is the median not in the exact center of the box?

9. Name another type of graph that could be useful for displaying this set of data.

☑ Develop & Understand: C

Here are two box-and-whisker plots for the ages of the citizens of two towns.

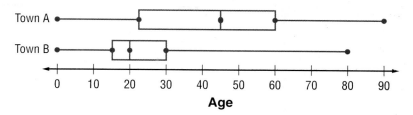

10. What is the second quartile, or median, for Town A?

11. Which town has a greater range of ages?

12. About what percentage of the population of Town B is between 15 and 80 years old?

13. One of the towns is a university town. Which do you think it is? Why?

14. If these data were displayed in a stem-and-leaf plot, what additional information would you know?

Now you will draw your own box-and-whisker plots.

☑ Develop & Understand: D

Rhiannon kept a record of the points she scored throughout the basketball season. In 11 games, she scored these numbers of points.

20   15   23   14   18   12   25   10   24   12   17

15. Write the scores in order.

16. What is the median of this data set?

17. What are the first and third quartiles?

18. What is the range of points scored?

19. Draw a box-and-whisker plot of Rhiannon's scores.

Simona and Derek scored the following runs in several games of baseball.

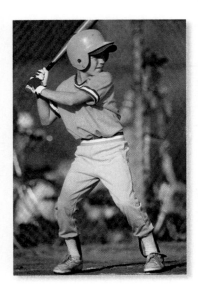

Game	Simona	Derek
1	7	6
2	2	1
3	5	2
4	6	1
5	4	8
6	5	7
7	9	9
8	3	8
9	2	7
10	1	8

20. On the same axis, draw a box-and-whisker plot for each player.

21. Compare the ranges of each player's scores.

22. Who do you think is a better scorer? Use the graphs you made to explain your answer.

Share & Summarize

Suppose you have a data set with 18 values, ordered from least to greatest. Explain as completely as you can how to create a box-and-whisker plot of the data set.

Practice & Apply In Exercises 1–3, use this information.

Economics Blake decided to write an article for the school newspaper comparing prices in several stores. He found the prices of five current releases at three local stores and at an Internet music site and listed them in a table.

CD Prices

Store	Artist	Price
Castle	A K Mango	$12.19
Castle	Screaming Screamers	12.50
Castle	Front Street Girls	13.09
InstantMusic	A K Mango	13.25
InstantMusic	Front Street Girls	13.25
InstantMusic	Screaming Screamers	13.49
InstantMusic	Out of Sync	13.59
GLU Sounds	Front Street Girls	13.95
Castle	Out of Sync	14.29
GLU Sounds	A K Mango	14.50
Pineapples	A K Mango	14.99
InstantMusic	Aviva	15.00
GLU Sounds	Out of Sync	15.49
GLU Sounds	Screaming Screamers	15.50
Pineapples	Screaming Screamers	15.99
Pineapples	Front Street Girls	16.00
Pineapples	Out of Sync	16.25
GLU Sounds	Aviva	16.75
Pineapples	Aviva	18.89
Castle	Aviva	19.00

1. Study Blake's table, and write a brief paragraph describing the most noticeable differences in CD prices.

2. Consider how CD prices vary from store to store.

   a. Reorganize the table to make it easier to compare the stores' prices for the various CDs. (Abbreviate the store and artist names if you want to.)

   b. What is the mean price of the CDs in each store?

   c. Which store would you recommend for the best price in general? Explain your choice.

3. Consider the prices for each artist's CD.

   a. Reorganize the table to make it easier to compare the prices for a particular CD.

   b. What is the mean price of each artist's CD?

4. **Ecology** You can find data about air pollution and other environmental issues on the U.S. Environmental Protection Agency (EPA) Web site, *www.epa.gov.*

Garbage-Disposal Methods, 1960–2000 (in millions of tons per year)

	1960	1970	1980	1990	2000
Recycled or Composted	5.6	8.0	14.5	33.2	67.7
Combustion (Burning)	27.0	25.1	13.7	31.9	33.7
Discarded in Landfills	55.5	87.9	123.4	140.1	130.6
Total	88.1	121.1	151.6	205.2	232.0

Source: *Characterization of Municipal Solid Waste in the U.S.: 2001 Update*, U.S. Environmental Protection Agency, Washington, D.C.

> **Math Link**
>
> The *percentage increase* is the difference in two quantities divided by the original quantity and expressed as a percentage.

Alvin and Adela want to use the garbage-disposal data to help decide whether people are recycling more or less than they previously did.

a. What is the percentage increase in tons of waste recycled or composted in 2000 compared to 1960?

b. Graph the amount of waste recycled or composted over the years 1960–2000.

c. Alvin argued that the answer to Part a and the graph from Part b show that recycling has improved a great deal in the last few decades. Do you agree? Why or why not?

d. Adela argued that they should compute a ratio for each year, the amount of waste recycled or composted to the total amount of waste generated. Make a table showing this ratio, as a decimal, for each year in the original table. Then graph the ratios over the 1960–2000 period.

e. What does your graph from Part d tell you about whether people are recycling more or less than they used to do?

5. Two ice cream shops, Scoops and Sundae Sunday, report the number of ice cream cones sold per day for a week.

Day	Mon	Tue	Wed	Thurs	Fri	Sat	Sun
Scoops	125	140	180	130	150	160	260
Sundae Sunday	130	155	120	140	160	170	250

a. On one axis, draw a box-and-whisker plot for each shop.

b. Which shop has the greater range of ice cream sales?

c. Which shop has the higher median sales?

6. For 4 weeks, or 20 nights, four students recorded the number of minutes they spent on homework each night.

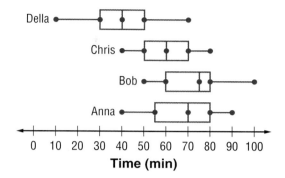

a. Which student has the highest median time? Suggest some possible reasons for this.

b. If the school suggests that a median study time of 70 minutes is adequate, which students are putting in an adequate amount of study time?

c. Which student studied the longest in one night?

d. Which student studied the least in one night?

e. Which student has the least range of study times?

f. Which student has the least amount of study time above 50 minutes?

g. How many of the students worked for more than two hours in any one evening?

7. During football season, Albert, Bettina, and Paul sold soft drinks at the home games from drink carts. They kept records of how many drinks they sold each game and drew the following box plots. They want to use the data to find the best locations for their carts.

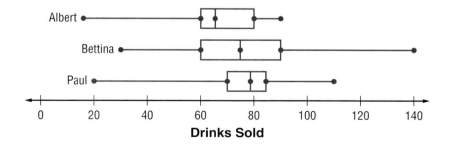

a. Who had the highest median?

b. Who sold the least number of soft drinks in a game?

c. Who had the highest proportion of sales above 30 drinks per game?

d. Who had the greatest range of sales?

e. Who had 75% of his or her sales above 70 drinks per game?

f. Who sold more than 140 drinks during a game?

g. What information about the best location can you get from these graphs?

Connect & Extend

In Exercises 8 and 9, use this information.

An important branch of mathematics is *cryptology*, the study of making and breaking codes. Its uses range from deciphering intercepted enemy messages in wartime to encrypting credit card information over the Internet.

A simple method of sending secret messages is the use of a substitution table. This method is used in cryptogram puzzles in your daily newspaper. To send a message, it is helpful to present the entries in a substitution table in alphabetical order, like this.

Math Link

This table represents a function since it provides a unique output for each input.

Input	a	b	c	d	e	f	g	h	i	j	k	l	m	n	o	p	q	r	s	t	u	v	w	x	y	z
Output	C	F	I	L	O	R	U	X	A	D	G	J	M	P	S	V	Y	B	E	H	K	N	Q	T	W	Z

Real-World Link

Cryptology played a part in the victory of the Allies in World War II. Mathematician Alan Turing, among others, had a critical role in breaking the Enigma code used by the Germans.

To avoid confusion, it is useful to use lower-case letters for your original message in English and upper-case letters for your secret message, as the table shows. To write a message, look up each letter in the top row and change it to the corresponding letter in the bottom row. For example, to write "come here," you would send ISMO XOBO.

8. Using the original table as it is sorted is a nuisance when you are *deciphering* a message.

a. Suppose you receive the message BOHKBP AMMOLACHOJW. Translate it into plain English.

b. Make a new table in which the *outputs* are in alphabetical order.

c. Use your new table to decipher the message WSK CBO ISBBOIH.

a	7.3
b	0.9
c	3.0
d	4.4
e	13.0
f	2.8
g	1.6
h	3.5
i	7.4
j	0.2
k	0.3
l	3.5
m	2.5
n	7.8
o	7.4
p	2.7
q	0.3
r	7.7
s	6.3
t	9.3
u	2.7
v	1.3
w	1.6
x	0.5
y	1.9
z	0.1

Source: Trinity College Web site, *www.trincoll.edu/depts/cpsc/cryptography/caesar.html*

Real-World Link

The construction of secret messages is *cryptography*. Breaking codes is *cryptanalysis*. Both are part of *cryptology*.

9. Challenge Refer to the information on the previous page.

If you intercept someone else's encrypted message and do not know the table that was used to construct it, you have a dilemma to solve, breaking the code.

To solve a code, it is helpful to have some idea of which letters are most frequent in English and what combinations of letters are typical. In the short encoded message in Part a of Exercise 8, for instance, the letter "O" occurs three times. Since "e" is the most common letter in English, you might guess correctly that "O" represents "e" in this message.

The table at the left shows one estimate of the frequencies of letters in English text, as percents.

a. Use the letter frequencies as a guide to translate the message below. It will help to start with a blank table in which the inputs are in alphabetical order. Work in pencil. Fill in your guesses about the outputs as you go along. Soon, a pattern will emerge.

b. Explain how you solved Part a.

The table above comes from a Web site. You can find tables containing frequencies from many sources that differ in the percents assigned to the letters of the alphabet.

c. Explain why sources might have apparently contradictory tables for frequencies of letters in English.

d. Explain why a particular message, like the one above, might *not* have the same letter frequencies as those in the table.

In Exercises 10–12, use this information.

Ecology Automobiles are an essential mode of transportation at this point in U.S. history. Unfortunately, they are also a major source of air pollution. In recent years, a series of technological improvements has reduced the amount of pollution emitted per mile driven, but people are driving more miles.

The following figure includes a line graph and a bar graph. The line graph displays the estimated average per-vehicle emissions from 1960 through 2015. The bar graph shows the vehicle miles traveled, in billions, for these years. Certain assumptions have been made for the estimates of future emissions, such as no increase in regulations, no cutbacks in driving, and no unexpected improvements in technology.

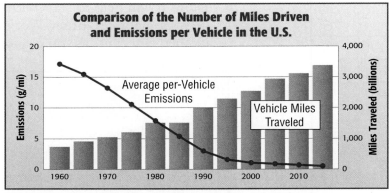

Source: "Automobiles and Ozone," Fact Sheet OMS-4 of the Office of Mobile Sources, the U.S. Environmental Protection Agency.

10. At first glance, do the improvements, as indicated by the line graph, seem to be keeping up with the increase in miles driven, as indicated by the bar graph? Explain.

11. Phil claimed that the total amount of pollutants produced by cars in 2010 will be less than that in 1960 due to technological improvements. Sara claimed that the total amount of pollutants from cars will be greater in 2010 than in 1960.

Who do you think is correct? Explain your answer, using the graphs to justify your reasoning.

12. Use the graphs on the previous page to create a new graph by following these directions.

a. Complete the following table, giving an approximation of the average amount of emissions from cars each year.

Year	Average per-Vehicle Emissions (grams of hydrocarbon per mile)	Vehicle Miles Traveled (billions)	Total Emissions (billions of grams of hydrocarbon)
1960	17	750	12,750
1965			
1970			
1975			
1980			
1985			
1990			
1995			
2000			
2005			
2010			
2015			

b. Draw a graph. Display the average number of grams of hydrocarbon produced per year for each year on the original graphs.

c. Describe how the amount of hydrocarbon produced by cars changed over the observed and predicted years.

d. Use the graphs to predict the average per-vehicle emissions and vehicle miles traveled in 2030.

e. Calculate the estimated grams of hydrocarbon that cars will produce in 2030. Add these data to your graph.

Connect & Extend **13.** Jane works as a florist's assistant. She sorted a collection of roses, long-stem red and yellow, and then measured their lengths. To compare the two varieties, she drew box plots.

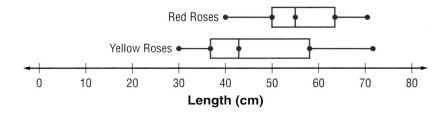

a. What is the median length of the red roses?

b. What is the median length of the yellow roses?

c. Between what lengths are the middle 50% of red roses? Between what lengths are the middle 50% of yellow roses?

d. Which color rose would you buy for a 40-cm-tall vase? Why?

e. What percentage of the red roses were longer than 63 cm?

f. What percentage of the yellow roses were less than 57 cm long?

g. **Challenge** Red roses cost $3, and yellow roses cost $2. A customer spent exactly $25 on roses. What combinations might she have bought?

14. Box-and-whisker plots use the median as a measure of the center of a data set. The mean is another measure of center.

Abby surveyed two groups of 13 students each. She asked the students how many posters they had hanging in their bedrooms. She made two box-and-whisker plots of her findings.

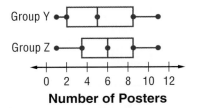

Group Y

Group Z

Number of Posters

The data set for group Z is as follows.

1　2　3　4　5　6　6　6　7　8　9　10　11

a. How will the data set for group Y be different from the data set for group Z? Explain.

b. Find the mean for group Z's data.

c. Is the mean for group Y higher or lower than the mean for group Z? Explain.

d. The mean and median for group Z are equal, but that is not true for all data sets. For example, find the mean and median of this data set.

1　1　1　1　11

e. Consider your answers for Parts c and d. Group Y's median is 5. Estimate the mean for group Y's data. Explain how you made your estimate.

15. **In Your Own Words** Find a magazine article, a Web site, or a television newscast that gives information in a table or visual display of some sort. Describe the display. Explain how it is useful for the information presented.

Mixed Review

Graph each inequality on a separate grid.

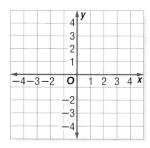

16. $y \geq x - 3$

17. $y < 3 - x$

18. $y \leq 1.5x + 3$

19. Prove that this number trick always gives 3.

Choose any number except 0. Multiply the number by 9 and add 6. Then divide by 3 and subtract 2. Divide by the number with which you started.

Factor each expression.

20. $4h^2 - 2h$

21. $-6a^2 + ab + b^2$

22. $-4k^2 - 5kj - j^2$

23. $2m^2 - 9 + 3m$

24. A standard playing card is about 5.7 cm wide and 8.9 cm long. A stack of 52 cards, a whole deck, is about 1.5 cm high.

    a. What is the volume of the deck of cards?

    b. What is the volume of a single card?

Review & Self-Assessment

Vocabulary

quartile

sample space

Chapter Summary

In this chapter, you found the sizes of *sample spaces* for several situations. Sometimes this required counting all the possible ways a group of items could be selected from a larger group. Other times, you had to figure out the number of ways a group of things could be ordered.

You can find the sample space for a situation by systematically listing all of the possibilities. You learned that you can sometimes discover a pattern to help you determine the size of the sample space without listing all of the outcomes.

You also found the probabilities of events for various situations. You determined whether one event was more likely than another. You also saw that sometimes finding probabilities can help you make decisions or devise game-winning strategies.

You worked with various kinds of displays for organizing and analyzing data. You reorganized tables to make it easier to see trends in the variables. In some cases, you made calculations using the given data in order to produce the data you really needed.

Strategies and Applications

The questions in this section will help you review and apply the important ideas and strategies developed in this chapter.

Making a systematic list of every possible outcome

1. Ally, Brad, Carol, Doug, and Omar are setting up a chess tournament among themselves that will be a round-robin tournament. That is, every participant will play every other participant once. How many games will there be? Make a systematic list of every possible tournament pairing.

2. In science class, Ally, Brad, Carol, and Doug are assigned to sit next to one another in the first row.

A First	B First	C First	D First
ABCD			
ABDC			

   a. List all the arrangements in which Ally sits in the first seat, then all the arrangements in which Brad sits in the first seat, and so on.

   b. How many different arrangements are there?

Using a pattern or shortcut to find the size of a sample space without listing every outcome

3. The manager of a baseball team is responsible for assigning the nine players to a batting order.

    a. Explain how you can find the number of different batting orders *without* listing all the possibilities. How many batting orders are possible?

    b. How many different batting orders are possible if one of the nine players, the pitcher, always bats ninth?

4. Suppose a basketball coach has 12 players. *Without* listing all the possibilities, explain how you can find the number of different five-player teams the coach could create. How many teams are possible?

Determining the probability of an event

5. In Question 1, you determined all the possible pairings for a round-robin chess tournament. Imagine that each pairing is written on a slip of paper. The slips are placed into a container and mixed. One slip is then chosen at random.

    a. What is the probability that the chosen names are Doug and Omar?

    b. What is the probability that Doug or Omar, or both, are included in the chosen pairing?

    c. What is the probability that the chosen names contain neither Doug or Omar?

6. In Question 2, you listed all the ways four students could be arranged in a row. Use your list to answer these questions.

    a. If seats are assigned randomly to the four students, what is the probability that Brad and Carol will sit next to each other?

    b. If seats are assigned randomly to the four students, what is the probability that Doug will sit at one end of the row?

Analyzing data presented in tables

The table on page 624 gives 2003 statistics for all the American League baseball teams, including total number of runs, percentage of games won, and earned run average. The earned run average, or ERA, gives the average number of runs the *opposing* teams earned per inning. The teams all played approximately the same number of games.

7. Evita and Tadi were arguing about which division in the American League had better teams.

a. Use the table to compare the number of runs each team scored over the year. Based on just this one statistic, which division would you think has better teams? Support your answer.

b. How could you reorganize the table to make it easier to answer Part a?

2007 Major League Baseball Statistics

Team	Total Runs	Games Won (%)	ERA	League	Division
Detroit	887	54.3	4.57	American	Central
Cleveland	811	59.3	4.05	American	Central
Tampa Bay	782	40.7	5.53	American	East
Anaheim	822	58.0	4.23	American	West
Baltimore	756	42.6	5.17	American	East
Oakland	741	46.9	4.28	American	West
Chicago White Sox	693	44.4	4.77	American	Central
Seattle	794	54.3	4.73	American	West
Minnesota	718	48.8	4.15	American	Central
Texas	816	46.3	4.75	American	West
Kansas City	706	42.6	4.48	American	Central
New York Yankees	968	58.0	4.49	American	East
Toronto	753	51.2	4.00	American	East
Boston	867	59.3	3.87	American	East

Source: MLB.com

Demonstrating Skills

9. Suppose you have seven chairs in a row. How many different seating orders are possible for seven people?

10. At Baron's Burgers, you can order a cheeseburger with up to three different condiments. These are mustard, catsup, and pickles. How many different cheeseburgers are possible, including a plain cheeseburger with no condiments?

11. The *21-and-5* lottery game requires players to match five numbers drawn randomly from 21. How many different groups of five numbers are possible?

12. The table lists the federal minimum hourly wage from the years 1975 to 2007. Suppose a person works 40 hours each week at minimum wage for 50 weeks each year. Create a new table giving the person's annual income for each rate.

Federal Minimum Hourly Wage, 1975–2007

Year	Minimum Wage
1975	$2.10
1980	3.10
1985	3.35
1990	3.80
1995	4.25
2000	5.15
2005	5.15
2007	5.85

Test-Taking Practice

SHORT RESPONSE

1 In computer class, Devon, Madison, Alli, Brandon, and Dario are assigned to sit next to one another in the first row. (a) How many different seating combinations are possible? (b) What is the probability that Madison will be in the first seat and Alli will be in the second seat?

Show your work.

Answer _____

MULTIPLE CHOICE

2 How many possible outcomes are there if German tosses a coin 6 times?

A 6   **C** 32
B 12   **D** 64

3 If you roll a six-sided number cube and toss a coin, how many possible outcomes are there?

F 6   **H** 12
G 8   **J** 36

4 What is the probability of rolling a six-sided number cube 3 times and landing on one each time?

A $\frac{1}{3}$ **C** $\frac{1}{36}$
B $\frac{1}{216}$ **D** $\frac{1}{18}$

5 Karena has 5 nickels, 7 dimes, and 3 pennies in her pocket. She pulls out a coin and does not replace it. If Karena pulls out another coin, what is the probability that both coins will be nickels?

F $\frac{2}{15}$ **H** $\frac{2}{21}$
G $\frac{2}{225}$ **J** $\frac{4}{45}$

Algebraic Fractions

Real-Life Math

Lakeesha is competing in the school Science Fair. Before the Science Fair begins, she needs to set up her display. Her friend James offers to help. Working alone, it will take Lakeesha three hours to set up the display. Working alone, it will take James one hour less to complete the same work.

Let h represent the numbers of hours that Lakeesha and James work together. The algebraic fractions below represent the portion of the set up that each person completes.

$$\text{Lakeesha: } \frac{h}{3} \qquad \text{James: } \frac{h}{2}$$

Think About It Working together, how long will it take Lakeesha and James to set up Lakeesha's Science Fair display?

Math Online
Take the **Chapter Readiness Quiz** at glencoe.com.

Dear Family,

Algebra is the sole focus of Chapter 12. In the past, students have worked with fractions and algebraic expressions. In this chapter, your student will now apply the same skills and strategies to work with algebraic fractions.

Key Concept—Adding Fractions

Whether adding fractions such as $\frac{1}{2}$ and $\frac{1}{3}$, or algebraic fractions, start by finding a common denominator.

$$\frac{1}{2} + \frac{1}{3} = \frac{3}{6} + \frac{2}{6}$$
  Use a common denominator of 2 • 3, or 6.
  Write $\frac{1}{2}$ as $\frac{3}{6}$ and $\frac{1}{3}$ as $\frac{2}{6}$.

$$= \frac{3 + 2}{6}$$
  Write one fraction by combining the numerators.

$$= \frac{5}{6}$$
  Find the sum of 3 and 2.

This same process can be used to add algebraic fractions.

$$\frac{1}{2x} + \frac{1}{3y} = \frac{3y}{6xy} + \frac{2x}{6xy}$$

$$= \frac{3y + 2x}{6xy}$$

The class will use a similar process to subtract algebraic fractions. In addition, your student will learn how to solve equations involving algebraic fractions. Finally, your student will explore the difference between mathematical solutions and solutions that make sense in real-world situations.

Chapter Vocabulary

algebraic fraction

Home Activities

- Review the processes for simplifying fractions and for finding the sums and differences of fractions with unlike denominators.
- Look for real-world situations, like the one presented on the next page, which can be represented using algebraic fractions.

Work with Algebraic Fractions

Previously, you have worked with fractions. You have also discovered several tools for working with algebraic expressions more efficiently. Now you will expand your tool kit as you learn how to work with fractions that involve algebraic expressions, or **algebraic fractions**.

Vocabulary

algebraic fraction

> ### Think & Discuss
>
> Before summer vacation, Adriana borrowed $100 from her aunt to buy a pair of in-line skates for herself. She enjoyed them so much that she borrowed another $100 to buy a pair for her younger sister. Adriana agreed to repay d dollars per month. Her aunt agreed not to charge interest.
>
> Write an algebraic fraction to express how many months it will take Adriana to repay the first $100.
>
> Which of these seven expressions show how many months it will take Adriana to repay the entire debt?
>
> $$\frac{d}{200} \qquad \frac{d}{100} \qquad \frac{100}{2d} \qquad \frac{200}{2d}$$
>
> $$\frac{100}{d} \qquad \frac{200}{d} \qquad \frac{100}{d} + \frac{100}{d}$$

Investigation 1 Make Sense of Algebraic Fractions

Materials

- graphing calculator

When you use expressions involving algebraic fractions, the expressions might not make sense for all values of the variables. In this investigation, you will explore some situations in which this is important.

✅ Develop & Understand: A

The denominator of the right-hand side of this equation has four factors.

$$y = \frac{24}{(x-1)(x-2)(x-3)(x-4)}$$

1. What is the value of y if you let $x = 5$? If you let $x = 6$?

2. What happens to the value of y if $x = 1$? If $x = 2$? Are there other values of x for which this happens?

3. Choose a number less than 5 for which y does have a value. What is the value of y using your chosen value of x?

4. Look again at the above equation.

   a. Use your calculator to make a table for the equation, starting with $x = 0$ and using an increment of 0.25. Copy the results into a table on your paper. Use the calculator to help fill your table for x values up to $x = 5$.

   How does your table show which values of x do not make sense?

   b. Now use your calculator to graph the equation, using x values from -1 to 5 and y values from -100 to 100. Draw a sketch of your calculator graph.

   What happens to the graph at the x values that do not make sense?

Algebraic fractions do not make *mathematical* sense for values of the variables that make the denominator equal to 0. In other words, they are *undefined* for these values.

✓ Develop & Understand: B

Carlota and Arthur attended a fundraising auction, hoping to bid for some collector comic books. They both said they would bid no higher than x dollars per comic book.

Carlota bought some comic books that she really wanted, each for $5 less than her set maximum price. She spent $120 in all. The comic books Arthur wanted were worth more. He ended up paying $5 more per comic book than he had intended. He spent $100 in all.

5. Explain what each expression means in terms of the auction story.

   a. x b. $x + 5$ c. $x - 5$

6. Write an expression to represent the number of comics that Arthur purchased. Then write an expression to represent the number that Carlota bought.

7. This algebraic expression represents the total number of comic books purchased by the friends.

$$\frac{100}{x + 5} + \frac{120}{x - 5}$$

Read each comment about this expression. Decide whether the student is correct. If the student is incorrect, explain his or her mistake.

   a. Jemma: "Even though you can't go to an auction intending to pay −$10 per comic book, the expression is *mathematically* sensible when x has a value of −10. The expression then has a value of −28."

   b. Jack: "The expression does not make sense at all because when $x = 5$, one of the denominators is 0. You can't divide by 0."

   c. Emilio: "The expression makes *mathematical* sense for all values of the variable except 5 and −5."

   d. Trina: "In the auction situation, we can only think about paying some positive number of dollars for a comic book. Therefore, for this story, the expression makes sense for positive values of x only, except 5, of course."

   e. Devin: "We can't use just any positive value of x in the expression. For example, if $x = 7$, Ling would have paid $x + 5 = \$12$ per comic book, which means she would have bought eight and a third comic books."

   f. Marcel: "The expression makes *mathematical* sense for any values of x except 5 and −5. However, in the auction situation, there are only a small number of sensible answers."

8. Consider Marcel's statement. Find all the possible values for x given the auction situation. Assume x is a whole number.

9. Now use your calculator to make a table and a graph for $y = \dfrac{100}{x + 5} + \dfrac{120}{x - 5}$. Use values of x from -10 to 10 and values of y from -100 to 100. For the table, start with $x = -10$ and use an increment of 1.

How do the graph and the table show the values of x for which the expression does not make *mathematical sense*?

Share & Summarize

1. When you are looking at an expression for a situation, what is the difference between *mathematical sense* and *sense in the context of the situation*? Give examples if it helps you make your point.

2. Consider the auction situation in Exercises 5–9.

    a. What did you have to think about when trying to determine the values that made *mathematical* sense?

    b. What did you have to think about when trying to determine the values that made sense in the *context* of the situation?

Investigation 2 Rearrange Algebraic Fractions

When you work with numeric fractions, you sometimes want to write them in different ways. For example, to calculate $\frac{1}{2} + \frac{1}{3}$, it is helpful to rewrite the fractions as $\frac{3}{6} + \frac{2}{6}$.

Think & Discuss

Trina tried to write equivalent expressions for three algebraic fractions. Which of these are correct? Which are incorrect? How do you know?

$$\frac{3}{12m} = \frac{1}{4m} \qquad \frac{2}{m + 2} = \frac{1}{m + 1} \qquad \frac{2x}{x^2} = \frac{2}{x}$$

The expressions Trina wrote correctly are *simplified* versions of the original fractions. In a simplified fraction, the numerator and the denominator have no factors in common.

Consider the following two methods to simplify $\frac{15}{18}$.

Method 1.

Factor the numerator and the denominator.

$$\frac{15}{18} = \frac{3 \cdot 5}{3 \cdot 6} = \frac{3}{3} \cdot \frac{5}{6} = \frac{5}{6}$$

Method 2.

Divide the numerator and the denominator by a common factor, in this case, 3.

$$\frac{15}{18} = \frac{\frac{15}{3}}{\frac{18}{3}} = \frac{5}{6}$$

You can also use these strategies to simplify algebraic expressions.

Examples

Simplify $\frac{5}{5x + 15}$.

- **Method 1.** Factor the numerator and the denominator.

$$\frac{5}{5x + 15} = \frac{5}{5(x + 3)} = \frac{5}{5} \cdot \frac{1}{x + 3} = \frac{1}{x + 3}$$

- **Method 2.** Divide the numerator and the denominator by the common factor, in this case, 5.

$$\frac{5}{5x + 15} = \frac{\frac{5}{5}}{\frac{5x + 15}{5}} = \frac{1}{x + 3}$$

Simplify $\frac{5a^2}{10a}$.

- **Method 1.** Factor the numerator and the denominator.

In Two Steps	In One Step
$\frac{5a^2}{10a} = \frac{5 \cdot a^2}{5 \cdot 2 \cdot a} = \frac{5}{5} \cdot \frac{a^2}{2 \cdot a} = \frac{a^2}{2a}$	$\frac{5a^2}{10a} = \frac{5 \cdot a \cdot a}{5 \cdot 2 \cdot a} = \frac{5}{5} \cdot \frac{a}{a} \cdot \frac{a}{2} = \frac{a}{2}$
$\frac{a^2}{2a} = \frac{a \cdot a}{2 \cdot a} = \frac{a}{a} \cdot \frac{a}{2} = \frac{a}{2}$	

- **Method 2.** Divide both the numerator and the denominator by the common factors, in this case, 5 and a.

In Two Steps	In One Step
$\frac{5a^2}{10a} = \frac{\frac{5a^2}{5}}{\frac{10a}{5}} = \frac{a^2}{2a} = \frac{\frac{a^2}{a}}{\frac{2a}{a}} = \frac{a}{2}$	$\frac{5a^2}{10a} = \frac{\frac{5a^2}{5a}}{\frac{10a}{5a}} = \frac{a}{2}$

Math Link

Dividing $\frac{a}{b}$ by $\frac{c}{d}$ is the same as multiplying $\frac{a}{b}$ by the reciprocal of $\frac{c}{d}$.

$$\frac{\frac{a}{b}}{\frac{c}{d}} = \frac{a}{b} \cdot \frac{d}{c}$$

✅ Develop & Understand: A

Simplify each fraction.

1. $\dfrac{6x^2y}{18x}$

2. $\dfrac{2}{2a + 4}$

3. $\dfrac{x}{x^2 + 2x}$

Write two fractions that can be simplified to the given fraction.

4. $\dfrac{1}{3 + a}$

5. $\dfrac{x}{2}$

6. $\dfrac{5y}{z}$

Find each product. Simplify your answers.

7. $\dfrac{1}{2d} \cdot \dfrac{4}{3}$

8. $\dfrac{1}{2} \cdot \dfrac{-2}{d - 5}$

9. $\dfrac{-4(d - 1)}{3} \cdot \dfrac{-1}{2d}$

10. $\dfrac{1}{3(a - 4)} \div \dfrac{3a}{5}$

11. $\dfrac{\frac{a}{7}}{\frac{3a}{5}}$

12. $\dfrac{\frac{1}{a}}{\frac{1}{a + 1}}$

You will now use what you have learned to analyze some number puzzles.

✅ Develop & Understand: B

Bill created four number tricks. For Exercises 13–16, do the following.

- Check whether or not the trick *always* works. If it always works, explain why.
- If it does not always work, does it work with only a few exceptions? If so, what are the exceptions? Explain why it works for all numbers other than those exceptions.
- If it never works or works for only a few numbers, explain how you know.

13. *Number Trick 1:* Pick a number, any number. Multiply it by 2 and square the result. Add 12. Then divide by 4 and subtract the square of the number you chose at the beginning. Your answer is 3.

14. *Number Trick 2:* Pick a number, any number. Add 2 to it and square the result. Multiply the new number by 6 and then subtract 24. Divide by your chosen number. Divide again by 6 and then subtract 4. Your answer is your chosen number.

15. *Number Trick 3:* Pick a number, any number. Multiply your number by 3 and then subtract 4. Divide by 2 and add 5. The result is 6.

16. **Challenge** *Number Trick 4:* Pick a number, any number. Add 6 to it and multiply the result by the chosen number. Then add 9. Now divide by 3 more than the chosen number. Then subtract the chosen number. Your answer is 3.

Share & Summarize

Logan simplified each fraction as shown. Check his answers. If he simplified an expression correctly, say so. If he did not, explain what is wrong and how to find the correct simplification.

1. $\dfrac{3}{x+3} = \dfrac{1}{x+1}$

2. $\dfrac{a}{a+4} = \dfrac{1}{4}$

3. $\dfrac{5a}{3} \div \dfrac{3}{a} = \dfrac{5a}{3} \cdot \dfrac{a}{3} = \dfrac{5a^2}{9}$

4. $\dfrac{12t^2}{35} \cdot \dfrac{21}{16t} = \dfrac{9t}{20}$

Practice & **Apply**

1. Consider this equation.

$$y = \frac{2 - x}{(x - 2)(x + 1)}$$

   a. For what values of x is y undefined?

   b. Explain how you could use the information from Part a to help you sketch a graph of the equation.

2. Every morning, a restaurant manager buys $300 worth of fresh fish at the market. One morning, she buys fish selling for d dollars per pound. The next morning, the price has risen $2 per pound.

   a. Write an expression for the quantity of fish, in pounds, the manager purchased on the first morning.

   b. Write an expression for the quantity of fish, in pounds, the manager purchased on the second morning.

   c. Write an expression for the total quantity of fish the manager purchased on these two days.

   d. For what values of d, if any, does your expression from Part c not make mathematical sense?

   e. For what additional values of d, if any, does your expression not make sense in the situation?

Math Link

rate · time = distance

or

time = $\frac{\text{distance}}{\text{rate}}$

3. Every Friday, a delivery person drives 120 miles into the city and then returns. One Friday, she drove into the city at the posted speed limit s. On the return trip, she was slowed by road construction and had to travel 15 miles per hour below the speed limit.

   a. Write an expression for the time it took her to drive into the city.

   b. Write an expression for the time her return trip took.

   c. Write an expression for her total driving time for the round trip.

   d. For what values of s, if any, does your expression from Part c not make *mathematical* sense?

   e. For what additional values of s, if any, does your expression not make sense in the situation?

Simplify each fraction.

4. $\dfrac{12m}{2m}$

5. $\dfrac{2x}{4xy}$

6. $\dfrac{20a^2 b}{16ab^2}$

7. $\dfrac{3k}{k^2 - 6k}$

Simplify each fraction.

8. $\dfrac{1 + a}{a(1 + a)}$

9. $\dfrac{3(x + 1)}{6}$

10. $\dfrac{nm}{m^2 + 2m}$

11. $\dfrac{3ab}{a^2b^2 - 3ab}$

Find each product or quotient. Simplify your answers.

12. $\dfrac{1}{3} \cdot \dfrac{1}{a}$

13. $\dfrac{4}{3} \cdot \dfrac{d}{2}$

14. $\dfrac{1}{5a} \cdot \dfrac{3a^2}{2}$

15. $\dfrac{1}{a} \div \dfrac{1}{a}$

16. $\dfrac{m}{4} \div \dfrac{4}{m}$

17. $\dfrac{-1(x - 2)}{3(2 - x)}$

For the number tricks in Exercises 18 and 19, do the following.

- Check whether the trick *always* works. If it does, explain why.
- If it does not always work, does it work with only a few exceptions? If so, what are the exceptions? Explain why it works for all numbers other than those exceptions.
- If it never works or works for only a few numbers, explain why.

18. Pick a number. Subtract 1 and square the result. Subtract 1 again. Divide by your number. Add 2. The result is your original number.

19. Pick a number. Add 3 and square the result. Subtract 4. Divide by the number that is 1 more than your chosen number. Subtract 5 from the result. The result is your chosen number.

Connect & Extend

20. Physical Science All objects attract each other with the force called *gravity*. Isaac Newton discovered this formula for calculating the gravitational force between two objects.

$$F = G\left(\dfrac{Mm}{r^2}\right)$$

In the formula, *F* is the gravitational force between the two objects. *M* and *m* are the masses of the objects, and *r* is the distance between them. *G* is a fixed number called the *gravitational constant*.

a. How does the gravitational force between two objects change if the mass of one of the objects doubles? If the mass of one of the objects triples?

b. How does the gravitational force between two objects change if the distance between the objects doubles? If the distance triples?

c. Suppose the masses of two objects is doubled and the distance between the objects is doubled as well. How does this affect the gravitational force between the objects?

21. Consider this equation.

$$y = \frac{2k^2 - 3k}{k^2 - k}$$

    a. For what values of k, if any, does y not have a value?

    b. Explain what will happen to a graph of the equation at the values you found in Part a.

Simplify each expression.

22. $\dfrac{4k - 2}{2k^2 + 4k - 2}$

23. $\dfrac{(u - 3)(u + 2)(u - 1)}{-1(3 - u)(1 - u)}$

24. Consider this equation.

$$y = \frac{24}{2 - 5x}$$

    a. For what values of x does y not have a value? Explain.

    b. For what values of x will y be positive? Explain.

    c. For what values of x will y be negative? Explain.

    d. For what values of x will y equal 0? Explain.

25. Graphs The equation $y = \dfrac{(x + 1)^2}{x + 1}$ can be simplified to $y = x + 1$ for all values of x except -1, which makes the denominator 0. The graph of $y = \dfrac{(x + 1)^2}{x + 1}$ looks like the graph of $y = x + 1$ but with an open circle at the point where $x = -1$.

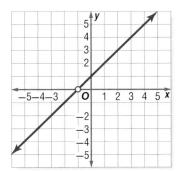

Use this idea to graph each equation.

a. $y = \dfrac{4x^3}{2x}$

b. $y = \dfrac{4x^2 + 2x}{2x}$

26. For what values of m is it true that $\frac{1}{m} > \frac{1}{m+1}$? Explain.

27. In Your Own Words Describe a situation that can be represented by an algebraic fraction. Look at the exercises in this lesson if you need ideas.

Discuss the values for which your expression does not make mathematical sense and the values for which your expression does not make sense in the context of the situation.

Mixed Review

For Exercises 28–31, solve each equation.

28. $5x - 10 = 20 - x$

29. $7(y + 2) = 28$

30. $m^2 - 1 = 8$

31. $t^2 + 6t + 8 = 0$

Make a rough sketch showing the general shape and location of the graph of each equation.

32. $y = x^2 - 3x - 4$

33. $y = \frac{3}{x-1} + 2$

34. $y = x^3 + 3$

Find each sum or difference. Write each answer in simplest form.

35. $\frac{1}{2} + \frac{1}{3}$

36. $\frac{1}{5} + \frac{2}{7}$

37. $\frac{3}{4} + \frac{5}{6}$

38. $\frac{1}{2} - \frac{1}{3}$

39. $\frac{1}{5} - \frac{2}{7}$

40. $\frac{7}{9} - \frac{5}{6}$

41. Geometry Find the perimeter of a rectangle with a length of 10 centimeters and a width of 4 centimeters.

LESSON 12.2

Add and Subtract Algebraic Fractions

You have added and subtracted fractions in which the numerator and denominator are both numbers. In this lesson, you will apply what you know to add fractions involving variables.

Think & Discuss

Consider these fractions and mixed numbers.

$$1\frac{1}{2} \qquad \frac{2}{3} \qquad \frac{3}{8} \qquad \frac{3}{10} \qquad 2\frac{5}{12}$$

• Choose any two of the numbers and add them. Describe how you found the common denominator and the sum.

• Choose two of the remaining fractions. Subtract the lesser from the greater. Describe how you found the common denominator and the difference.

• Finally, subtract the remaining fraction from 10. Describe how you found the common denominator and the difference.

Investigation 1   Combine Algebraic Fractions

You will now use what you know about fractions with numbers to add and subtract algebraic fractions, or fractions that involve variables.

✓ Develop & Understand: A

1. Copy and complete this addition table.

+	?	?	$\frac{3}{4}$
$\frac{1}{2}$		2	
?			1
$\frac{3}{5}$	$\frac{11}{5}$		

Find each sum or difference.

2. $\dfrac{100}{w} + \dfrac{100}{w}$

3. $\dfrac{100}{w} - \dfrac{100}{w}$

4. $\dfrac{100}{w} + \dfrac{100}{w} + \dfrac{100}{w}$

5. $\dfrac{1}{2x} + \dfrac{2}{3x}$

6. $\dfrac{m}{3} + \dfrac{m}{6}$

7. $\dfrac{x}{8} - \dfrac{x}{6}$

8. $\dfrac{y}{p} + \dfrac{1}{2p}$

9. $\dfrac{3}{2x} - \dfrac{3}{2y}$

10. Dave can type an average of n words per minute. Write an expression for the number of minutes that it takes Dave to type each number of words.

    a. 400

    b. 200

    c. 1,000

    d. Add your expressions from Parts a, b, and c. What does the sum represent in terms of the typing situation?

When you add or subtract algebraic fractions, there are many ways to find a common denominator.

Example

Evan, Tala, and Lucita have different methods for adding $\dfrac{14}{8x}$ and $\dfrac{3}{4x}$.

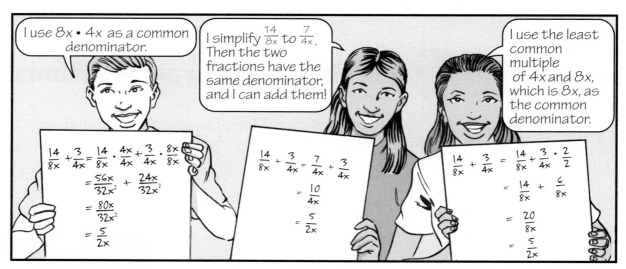

Keep the three methods above in mind as you work on the next set of exercises.

11. Copy and complete this addition table.

+	?	?	$\dfrac{8}{15x}$
$\dfrac{1}{3x}$		$\dfrac{2}{x}$	
?	$\dfrac{5}{4x}$		
?	$\dfrac{3}{5x}$	$\dfrac{53}{30x}$	

Find each sum or difference.

12. $\dfrac{6}{3x^2} - \dfrac{2}{2x^2}$

13. $\dfrac{2}{4t} - \dfrac{2t}{3}$

14. $\dfrac{3}{6m} + \dfrac{4m}{8m^2}$

15. Camila and Lakita earn money on the weekends by painting houses. It takes Camila $2n$ minutes to paint one square meter by herself. It takes Lakita $3n$ minutes.

   a. Write an expression for how much area Camila paints in one minute. Do the same for Lakita.

   b. How much area will the friends paint in one minute if they work together? Write your expression as a single algebraic fraction.

   c. If the friends are working together to paint a room with 40 square meters of wall, how much time will the job take? Show how you found your answer.

Share & Summarize

Consider these five terms.

$$c \qquad 2 \qquad 3 \qquad 2c^2 \qquad 3c^2$$

1. Create four addition or subtraction expressions involving fractions whose numerator and denominator are made from these terms, for example, $\dfrac{c}{2} + \dfrac{3}{3c^2}$. Use each term only once in an expression. Then, find a partner to exchange and simplify each other's expressions.

2. Use the terms to create an addition or a subtraction expression that simplifies to 3.

Investigation (2) Strategies to Add and Subtract Algebraic Fractions

Now you will learn more about adding and subtracting algebraic fractions.

✓ Develop & Understand: A

1. Compute each sum without using a calculator.

   a. $\dfrac{1}{1} + \dfrac{1}{2}$ b. $\dfrac{1}{2} + \dfrac{1}{3}$ c. $\dfrac{1}{3} + \dfrac{1}{4}$ d. $\dfrac{1}{4} + \dfrac{1}{5}$

2. Look for a pattern in the sums in Exercise 1. Use the pattern to find $\dfrac{1}{5} + \dfrac{1}{6}$ without actually calculating the sum.

3. In each part of Exercise 1, how does the denominator of the sum relate to the denominators of the two fractions being added?

4. In each part of Exercise 1, how does the numerator of the sum relate to the denominators of the two fractions added?

5. Use the patterns you observed to make a conjecture about this sum.

$$\frac{1}{m} + \frac{1}{m+1}$$

6. Consider again the sum $\dfrac{1}{5} + \dfrac{1}{6}$.

   a. If this sum is equal to $\dfrac{1}{m} + \dfrac{1}{m+1}$, what is the value of m?

   b. Use your conjecture for Exercise 5 the value of m from Part a to find the sum of $\dfrac{1}{5} + \dfrac{1}{6}$. Does the result agree with your prediction in Exercise 2?

   c. To check your result, calculate the sum by finding a common denominator and adding.

7. **Prove It!** Try to prove that your conjecture is true.

Here is how Gregorio thought about the sum of $\frac{1}{m}$ and $\frac{1}{m+1}$.
"When I add $\frac{1}{m}$ and $\frac{1}{m+1}$, I use a common denominator of $m(m+1)$, the product of the two denominators."

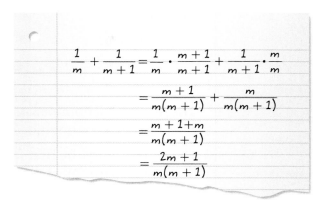

$$\frac{1}{m} + \frac{1}{m+1} = \frac{1}{m} \cdot \frac{m+1}{m+1} + \frac{1}{m+1} \cdot \frac{m}{m}$$

$$= \frac{m+1}{m(m+1)} + \frac{m}{m(m+1)}$$

$$= \frac{m+1+m}{m(m+1)}$$

$$= \frac{2m+1}{m(m+1)}$$

✅ Develop & Understand: B

Discuss Gregorio's strategy with your partner. Make sure you understand how each line follows from the previous line.

8. Why did Gregorio multiply the first fraction by $\frac{m+1}{m+1}$?

9. Why did he multiply the second fraction by $\frac{m}{m}$?

10. Chase says, "I have an easier method for adding $\frac{1}{m}$ and $\frac{1}{m+1}$. Here's what I did."

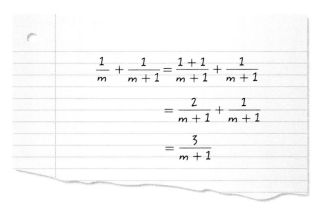

$$\frac{1}{m} + \frac{1}{m+1} = \frac{1+1}{m+1} + \frac{1}{m+1}$$

$$= \frac{2}{m+1} + \frac{1}{m+1}$$

$$= \frac{3}{m+1}$$

Chase's method is incorrect. Why?

When adding or subtracting algebraic fractions, it is often helpful to leave the numerator and the denominator in factored form. Knowing what the factors are allows you to better recognize and identify common factors.

Example

Find this sum.

$$\frac{2x}{x(x-1)} + \frac{5}{(x-1)(x+2)}$$

The factored denominators help to simplify the first fraction and then to find a common denominator.

You can simplify the first fraction by dividing the numerator and the denominator by x. Then $(x-1)(x+2)$ can be used as a common denominator of the resulting fractions.

$$\frac{2x}{x(x-1)} + \frac{5}{(x-1)(x+2)} = \frac{2}{x-1} + \frac{5}{(x-1)(x+2)}$$

$$= \frac{2}{x-1} \cdot \frac{x+2}{x+2} + \frac{5}{(x-1)(x+2)}$$

$$= \frac{2(x+2)+5}{(x-1)(x+2)}$$

$$= \frac{2x+9}{(x-1)(x+2)}$$

✅ Develop & Understand: C

Find each sum or difference. Simplify your answers if possible.

11. $\dfrac{1}{m} + \dfrac{2}{m+1}$

12. $\dfrac{4}{m} - \dfrac{1}{m-1}$

13. $\dfrac{4}{b+2} + \dfrac{b}{b+3}$

14. $\dfrac{2(x+1)}{x(x+1)} - \dfrac{1}{x-3}$

15. $\dfrac{10}{x+4} + \dfrac{3x}{9x^2}$

16. $\dfrac{2x}{x-1} - \dfrac{x+1}{x+3}$

17. Consider these subtraction expressions.

$$\frac{1}{2} - \frac{1}{3} \qquad \frac{1}{3} - \frac{1}{4} \qquad \frac{1}{4} - \frac{1}{5}$$

a. Find each difference.

b. Use the pattern in your answers to Part a to find the following difference without actually calculating it.

$$\frac{1}{5} - \frac{1}{6}$$

c. Use the pattern to find this difference.

$$\frac{1}{m} - \frac{1}{m + 1}$$

d. Prove It! Use algebra to show that your answer to Part c is correct.

Share & Summarize

In the example on page 643, Gregorio explains how he thinks about adding algebraic fractions. Compute the following difference. Explain how *you* think about subtracting algebraic fractions.

$$\frac{1}{x + 1} - \frac{1}{2x}$$

Inquiry
Investigation ③ Build Algebraic Fractions

At the beginning of the chapter, you learned that algebraic fractions can be used to represent real-world situations. On page 626, the expressions $\frac{h}{3}$ and $\frac{h}{2}$ are used to represent the portions of the Science Fair display that are set up by Lakeesha and James.

In this investigation, you will create algebraic fractions from algebraic expressions.

Try It Out

1. Build an algebraic fraction that simplifies to each given expression. Each fraction must contain at least one variable.

   a. 5

   b. $3y - 2$

   c. $\frac{y}{(z - 4)}$

   d. x^2y

2. Describe a general strategy for building an algebraic fraction that simplifies to a given expression.

Try It Again

3. Build five fractions that can be simplified to algebraic expressions. Each fraction must contain at least one variable.

4. Trade fractions with a partner. Simplify your partner's fractions. Do you get the same results?

5. Choose two fractions, one of your partner's fractions and one of your own.

   a. Add or subtract the two fractions.

   b. Express the sum or difference as a single fraction.

   c. Simplify the single fraction.

   d. Repeat Parts a–d with three other pairs of fractions.

   e. Describe any shortcuts you have found for the process of combining fractions and simplifying the result. Would your shortcuts work with any pair of fractions?

Math Link

Remember to look for values that do not make *mathematical sense.* Any value that will cause the denominator to be zero cannot be a possible answer.

Take It Further

Use the following expressions for Exercises 6 and 7.

$$3x \qquad (y + 4) \qquad (x - 2)$$

6. For Parts a–f, build an algebraic fraction using only the three expressions listed above. Each expression may be used more than once or not used at all. You may use any combination of addition, subtraction, multiplication, and division to create your expressions.

a. $\dfrac{3xy + 12x}{x - y - 6}$
    b. $\dfrac{2x - 1}{x + 1}$
    c. 0

d. $\dfrac{2}{y + 4}$
    e. $\dfrac{6}{y^2 + 7y + x + 10}$
    f. $\dfrac{3x^2 - 6x - y - 4}{y^2 - 16}$

7. Use the rules from Exercise 6 to build an original fraction. Trade fractions with a partner. Can you figure out how your partner built his or her fraction?

What Did You Learn?

8. Build three different fractions that simplify to $2xy$.

9. Explain how $\dfrac{(4x^3y - 2xy)}{(2x - 2)(x - 1)}$ simplifies to $2xy$.

Investigation 4   Solve Equations with Fractions

Materials

- graphing calculator

You have already solved equations involving algebraic fractions. For example, you have solved proportions such as the following.

$$\frac{x}{2} = \frac{9}{6} \qquad \frac{50}{3.6} = \frac{11}{m}$$

Think & Discuss

Describe at least one way that you could solve each equation.

$$\frac{x}{2} = \frac{9}{6} \qquad \frac{50}{3.6} = \frac{11}{m}$$

✓ Develop & Understand: A

Solve each equation using any method you like. You may want to use different methods for different equations.

1. $\dfrac{3x - 6}{4} = x - 8$

2. $\dfrac{t}{2} + \dfrac{t}{3} = -1$

3. $\dfrac{4a}{5} - \dfrac{2 - a}{4} = 30$

4. $\dfrac{p}{5} - p = -0.4$

5. What fraction added to $\dfrac{2x - 1}{4}$ equals $\dfrac{x^2 - 4}{4}$?

6. What fraction subtracted from $\dfrac{k + 3}{5}$ equals $\dfrac{k - 3}{15}$?

7. Oliver estimated the solution of the equation $\dfrac{n + 7}{2} + \dfrac{n}{3} = 10$ by finding the intersection of the graphs of these two equations.

$$y = \frac{n + 7}{2} + \frac{n}{3} \qquad y = 10$$

   a. Explain why Oliver's method works.

   b. Graph both equations in the same window of your calculator. Use the graph to estimate the solution.

   c. Check your estimate by solving the original equation using whatever method you prefer.

All of the equations in Exercises 1–7 contain one or more fractions with variables in the numerator. When you solve equations with variables in the denominators, you need to check that the "solutions" that you find do not make any denominators in the original equation equal to 0.

Example

Solve the equation $\dfrac{5x + 10}{x + 2} = \dfrac{3}{x + 1}$.

One way to solve this equation is to "clear" the fractions by multiplying both sides by a common denominator of all the fractions in the equation.

$$\frac{(x + 2)(x + 1)}{1} \cdot \frac{5x + 10}{x + 2} = \frac{3}{x + 1} \cdot \frac{(x + 2)(x + 1)}{1}$$

Then simplify the resulting equation.

$$\frac{x + 2}{x + 2} \cdot (x + 1)(5x + 10) = 3(x + 2) \cdot \frac{x + 1}{x + 1}$$

$$1 \cdot (x + 1)(5x + 10) = 3(x + 2) \cdot 1$$

$$5x^2 + 15x + 10 = 3x + 6$$

Rearranging the equation to set one side equal to 0 will allow you to solve it by graphing.

$$5x^2 + 12x + 4 = 0$$

In this case, the simplified equation is quadratic. You can estimate a solution by graphing $y = 5x^2 + 12x + 4$ and finding the points where $y = 0$.

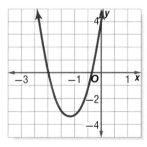
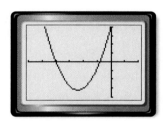

The solutions appear to be -2 and -0.4. Check these solutions in the original equation. Then, check to see if the values make mathematical sense.

Since -2 makes the denominator of $\dfrac{5x + 10}{x + 2}$ equal to 0, it is not a solution of the original equation. The number -0.4 makes the original equation true. So, -0.4 is a solution and -2 is not.

As you solve the equations that follow, be sure to check that your potential solutions do not make any of the denominators in the original equation equal 0.

✅ Develop & Understand: B

Solve each equation using whatever method you like. You may want to use different methods for different equations. For some, you may need to use a graph to estimate solutions.

8. $\dfrac{10}{7} = \dfrac{k + 1}{k - 3}$

9. $\dfrac{6 - 2x}{x - 3} = 8$

10. $\dfrac{2}{g + 1} - \dfrac{2}{g - 1} = 4$

11. $\dfrac{20 - a}{a^2 - 4} = \dfrac{5}{a - 2} + \dfrac{3}{a + 2}$

12. $0 = \dfrac{2}{s + 3} + \dfrac{s}{s + 2}$

13. $\dfrac{-60 - 12z}{z + 5} = -120$

Share & Summarize

Choose one equation from Exercises 8–13. Explain how you solved it so that a student who was absent could understand.

Practice & Apply

Find each sum or difference.

1. $\dfrac{9}{8} - \dfrac{8}{9}$

2. $\dfrac{x}{4} + \dfrac{y}{2}$

3. $\dfrac{2xy}{3} - \dfrac{1}{6}$

4. $\dfrac{1}{x} + \dfrac{2}{x^2}$

5. $\dfrac{c}{a} - \dfrac{a}{c}$

6. $\dfrac{3}{y} - \dfrac{1}{y^2}$

7. Tyler and his sister Hailey have part-time jobs after school at the grocery store. When she is stacking cans in a display at the end of an aisle, Hailey is able to place 500 cans in z minutes. Tyler works half as fast as Hailey, stacking 500 cans in $2z$ minutes.

    a. Write an expression for how many cans Tyler stacks in one minute.

    b. Write an expression for how many cans Hailey stacks in one minute.

    c. How many cans are Tyler and Hailey able to stack together in one minute? Express your answer as an algebraic fraction.

    d. The store manager has asked Tyler and Hailey to create a display using 750 cans. How long will it take them?

8. Esperanza and Jocelyn are making the same 300-mile drive in separate cars. Esperanza drives an average of n miles per hour. Jocelyn drives 1.5 times as fast.

    a. Write an expression for the time it takes Jocelyn to drive the 300 miles.

    b. Assume Jocelyn and Esperanza left at the same time. Write an expression for the difference in time that it takes Esperanza and Jocelyn to complete the 300-mile drive.

    c. Write an expression for the total time they spent traveling.

9. Copy and complete this addition table.

$+$	$\dfrac{5}{2x}$	$\dfrac{4}{x}$	$2x$
$\dfrac{1}{4x}$			
$-\dfrac{2}{3x}$			
$\dfrac{3+x}{2}$			

Find each sum or difference.

10. $\dfrac{1}{m} - \dfrac{2}{m+1}$

11. $\dfrac{4}{m} + \dfrac{1}{m+1}$

12. $\dfrac{3}{d} + \dfrac{4}{d+1}$

13. $\dfrac{3}{c} - \dfrac{4}{c-1}$

14. $\dfrac{a}{a+4} + \dfrac{3a}{5}$

15. $\dfrac{x^2}{x^2-1} - \dfrac{1}{x^2-1}$

16. $\dfrac{5}{k} - \dfrac{5}{k+1}$

17. $\dfrac{2y-1}{4} - \dfrac{y}{2}$

Solve each equation using any method you like.

18. $\dfrac{2x}{3} + \dfrac{1}{4} = x - 1$

19. $\dfrac{v-2}{3} + \dfrac{v}{2} = 10$

20. $\dfrac{n+1}{n-1} = 3$

21. $\dfrac{2-u}{u+1} = 5$

22. $\dfrac{8}{w+5} - \dfrac{2}{w+5} = \dfrac{2}{w} + \dfrac{1}{w+5}$

23. $\dfrac{3}{c-1} + \dfrac{3}{c+1} = \dfrac{21-c}{c^2-1}$

24. What fraction added to $\dfrac{r+1}{r}$ equals 1?

25. What fraction subtracted from $\dfrac{2-x}{7}$ is equal to $\dfrac{x}{14}$?

26. What fraction subtracted from $\dfrac{1}{v}$ is equal to $\dfrac{3v}{2}$?

Connect & Extend

27. Economics Meg earns $70 for w hours of typing reports. Her friend Rashid, who is more experienced and works faster, earns $80 for w hours. Together, they earn $1,000 in a week.

Write a brief explanation for each expression.

a. $\dfrac{70}{w}$

b. $\dfrac{80}{w}$

c. $\dfrac{70}{w} + \dfrac{80}{w}$

d. $\dfrac{150}{w}$

e. $1,000 \div \dfrac{150}{w}$

f. $1,000 \div \dfrac{70}{w}$

Find each sum or difference.

28. $\dfrac{c}{ab} - \dfrac{a}{bc}$

29. $\dfrac{2x}{2y} + \dfrac{y}{x}$

30. $\dfrac{g+1}{g-1} - \dfrac{2}{g+1}$

31. $\dfrac{4-2y}{6} + \dfrac{y}{4}$

32. $\dfrac{1}{x^2 y} + \dfrac{1}{xy}$

33. $\dfrac{1}{xc} + 1 - \dfrac{1}{c}$

34. $2 - \dfrac{2}{s+1} - \dfrac{s}{s+1}$

35. $\dfrac{1}{m} + \dfrac{1}{m+1} + \dfrac{1}{m+2}$

36. Ms. Diaz drove 135 miles to visit her mother. She knew the speed limit increased by 10 mph after the first 75 miles, but she could not remember what the speed limits were.

a. Write an expression representing the amount of time it will take Ms. Diaz to drive the first 75 miles if she travels the speed limit of x mph.

b. Write an expression representing the amount of time it will take her to drive the remaining 60 miles at the new speed limit.

c. Write an expression for Ms. Diaz's total driving time. Combine the parts of your expression into a single algebraic fraction.

Solve each equation using any method you like.

37. $\dfrac{p+2}{2} + \dfrac{p-1}{5} = p + 1$

38. $\dfrac{r-8}{3} + \dfrac{r-5}{2} = r - 5$

39. $\dfrac{t-1}{4} + \dfrac{2-t}{3} + \dfrac{t+1}{2} = 3$

Solve each equation using any method you like.

40. $\dfrac{v - 2}{4} + \dfrac{2}{v - 1} + \dfrac{1}{2} = \dfrac{v^2 - 9}{4v - 4}$

41. $\dfrac{z - 5}{2} - \dfrac{3}{z + 5} = \dfrac{(z + 1)(z - 3)}{2z + 10}$

42. $\dfrac{3}{x - 3} + \dfrac{4}{x + 3} = \dfrac{21 - x}{x^2 - 9}$

43. Jing earns x dollars per hour bagging groceries at the local market. She also works as a math tutor. Her hourly tutoring rate is $2 more than twice her hourly rate at the market.

Last week, Jing earned $51.75 at the market and $40.50 tutoring. She realized that if she had spent all her working hours that week tutoring, she would have earned $162. Write and solve an equation to find Jing's hourly rate for each job.

44. For two numbers a and b,

$$\frac{5x + 1}{x^2 - 1} = \frac{a}{x + 1} + \frac{b}{x - 1}.$$

a. What is a common denominator for $\dfrac{a}{x + 1}$ and $\dfrac{b}{x - 1}$?

b. Find the sum $\dfrac{a}{x + 1} + \dfrac{b}{x - 1}$ using the common denominator you found in Part a. Write the sum without parentheses.

c. Explain why you can use this system to find the values of a and b.

$$a + b = 5$$

$$-a + b = 1$$

d. Solve the system to find a and b.

45. **In Your Own Words** Write an addition exercise involving two algebraic fractions with different algebraic expressions in their denominators. Explain step-by-step how to add the two fractions.

Mixed Review

Find the x-intercepts for the graph of each equation.

46. $(x - 5)^2 - 49 = 0$ **47.** $6x^2 + 36 - 30x = 0$

Evaluate without using a calculator.

48. $\sqrt[3]{8}$ **49.** $-\sqrt[4]{16}$

50. $\sqrt[4]{625}$ **51.** $\sqrt[3]{-343}$

CHAPTER 12

Review & Self-Assessment

Chapter Summary

This chapter focused on simplifying algebraic expressions. Two main themes were working with algebraic fractions and adding and subtracting algebraic fractions.

You discovered you could simplify, add, and subtract algebraic fractions using the same methods you use for numeric fractions. You concluded the chapter by solving equations that required you to apply all your new skills.

Strategies and Applications

The questions in this section will help you review and apply the important ideas and strategies developed in this chapter.

Simplifying expressions involving algebraic fractions

Simplify each expression. Explain each step.

1. $\dfrac{-3x}{9 - 6x}$

2. $\dfrac{15xy}{3x^3 y^3}$

Solving equations involving algebraic fractions

3. Consider the equation $\dfrac{1}{x + 1} + \dfrac{2}{x - 1} = \dfrac{8}{x^2 - 1}$.

   a. Describe the first step you would take to solve this equation.

   b. Solve the equation.

4. Consider the equation $\dfrac{k}{k - 1} - \dfrac{5}{2} = \dfrac{1}{k - 1}$.

   a. Solve the equation.

   b. Explain why it is especially important to check your solutions when solving equations containing algebraic fractions.

5. Every week, a freight train delivers grain to a harbor 156 miles away. Last week, the train traveled to the harbor at an average speed of s miles per hour. On the return trip, the train had empty cars and was able to travel an average of 16 miles per hour faster.

   a. Write an expression for the time it took the train to reach the harbor.

   b. Write an expression for the time taken on the return trip.

   c. The round trip took 10.4 hours. Write an equation to find the value of s. Solve your equation.

Demonstrating Skills

Simplify each expression.

6. $\dfrac{4x}{2}$

7. $\dfrac{4}{4x - 10}$

8. $\dfrac{x}{x^3}$

9. $\dfrac{15}{6x}$

10. $\dfrac{14x^2y^2}{2xy}$

11. $\dfrac{21c^3d^2}{28cd}$

12. $\dfrac{x^2 - 9}{x + 3}$

13. $\dfrac{t - 4}{t^2 - 8t + 16}$

14. $\dfrac{5t - 10}{t^2 - 4}$

15. $\dfrac{3m - 9}{2m^2 + 2m - 24}$

16. $\dfrac{k^2 - 4}{k^2 - 4k - 12}$

17. $\dfrac{n^2 + 5n + 6}{n^2 + 2n - 3}$

18. $\dfrac{p^2 + 7p + 6}{p^2 + 8p + 7}$

19. $\dfrac{x^4 + 9x^3 + 20x^2}{x^2 + 4x}$

In Exercises 20–31, find each sum or difference.

20. $\dfrac{7}{2x} + \dfrac{3}{2x}$

21. $\dfrac{x}{x + 1} - \dfrac{9}{x + 1}$

22. $\dfrac{14}{x - 3} + \dfrac{10}{x - 3}$

23. $\dfrac{5n}{n - 1} + \dfrac{3n}{2n - 2}$

24. $\dfrac{3}{x} + \dfrac{1}{2x}$

25. $\dfrac{4}{x^2 + x} + \dfrac{x}{x + 1}$

26. $\dfrac{5}{x} - \dfrac{1}{x^3}$

27. $\dfrac{2}{x} - \dfrac{x + 1}{x - 5}$

28. $3 - \dfrac{1}{x}$

29. $1 + \dfrac{1}{x - 6}$

30. $\dfrac{1}{k - 1} - \dfrac{1}{k + 2}$

31. $\dfrac{b}{3} + \dfrac{b - 1}{b + 1} - \dfrac{b}{2b + 2}$

Solve each equation.

32. $\dfrac{x}{2} + \dfrac{x}{3} = 1$

33. $\dfrac{x + 1}{4} - \dfrac{x + 3}{5} = 6$

34. $\dfrac{3}{x + 4} = \dfrac{2}{x - 4}$

35. $\dfrac{-2x + 2}{x - 1} = -3$

36. $\dfrac{5}{x + 2} + \dfrac{4}{x - 3} = \dfrac{11}{x^2 - x - 6}$

37. $\dfrac{5}{x^2 - 9} - \dfrac{5}{x - 3} = \dfrac{10}{x + 3}$

Test-Taking Practice

SHORT RESPONSE

1 Solve the equation $\dfrac{2x}{x-1} - \dfrac{4}{3} = \dfrac{1}{x-1}$.
Show your work.

Answer _____

MULTIPLE CHOICE

2 Simplify $\dfrac{32x^5y^2}{16x^3y}$.

A $2x^2y^2$

B $2x^2y$

C $3x^2y$

D $2x^8y^3$

3 Simplify $\dfrac{3}{t-3} + \dfrac{4}{t+5}$.

F $\dfrac{7t+3}{t^2+2}$

G $\dfrac{7t+11}{(t+5)(t-3)}$

H $\dfrac{7t+3}{(t+5)(t-3)}$

J $\dfrac{7}{(t+5)(t-3)}$

4 Solve $\dfrac{3}{x+5} = \dfrac{6}{x-5}$.

A $x = -15$

B $x = 5$

C $x = 15$

D $x = -10$

5 Which value(s) would make the denominator of $\dfrac{7}{x^2-9}$ equal to zero?

F 9

G 3

H -3

J 3 and -3

6 Which denominator is the least common denominator for $\dfrac{4}{t+5}, \dfrac{6}{t-2}$, and $\dfrac{10}{t^2+4t-5}$?

A $t^3 + 2t^2 - 13t + 10$

B $t^2 + 4t - 5$

C $t + 5$

D $t - 2$

7 Find the sum of $\dfrac{4}{n+2} + \dfrac{3}{n+5}$.

F $7n + 26$

G $\dfrac{7n+26}{n^2+7n+10}$

H $\dfrac{7}{n^2+7n+10}$

J $\dfrac{7}{2n+7}$

Glossary/Glosario

English

Español

A

algebraic fraction (p. 628) A fraction having properties which are the same as those for numerical fractions, the only difference is the numerator and denominator are both algebraic expressions.

fracción algebraica (pág. 628) Fracción cuyas propiedades son iguales a las de una fracción numérica, con la única diferencia que tanto el numerador como el denominador son fracciones.

alternate exterior angles (p. 92) In the figure, transversal *t* intersects lines *n* and *m*. ∠1 and ∠8, and ∠2 and ∠7 are *alternate exterior angles*.

ángulos alternos externos (pág. 92) En la figura, la transversal *t* corta las rectas *n* y *m*. ∠1 and ∠8, and ∠2 and ∠7 son *ángulos alternos externos*.

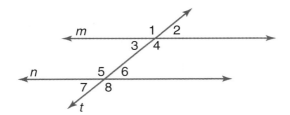

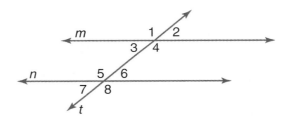

alternate interior angles (p. 92) In the figure, transversal *t* intersects lines *n* and *m*. ∠4 and ∠5, and ∠3 and ∠6 are *alternate interior angles*.

ángulos alternos internos (pág. 92) En la figura, la transversal *t* corta las rectas *n* y *m*. ∠4 y ∠5, y ∠3 y ∠6 son *ángulos alternos internos*.

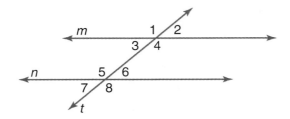

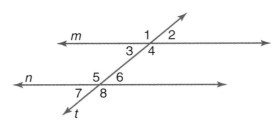

arc (p. 96) A set of points along a circle.

arco (pág. 96) Conjunto de puntos a lo largo de un círculo.

B

binomial (p. 224) The sum or difference of two unlike terms. For example, $x + 7$, $x^2 - 3$, and $a + c$ are binomials.

bisect (p. 97) To separate something into two congruent parts.

binomio (pág. 224) La suma o diferencia de dos términos no semejantes. Por ejemplo: $x + 7$, $x^2 - 3$ y $a + c$ son binomios.

bisecar (pág. 97) Separar algo en dos partes congruentes.

C

coefficient (p. 37) The numeric multiplier in an algebraic term. For example, in the expression $3x^2 - 2x + 7$, 3 is the coefficient of x^2, and -2 is the coefficient of x.

collinear (p. 68) Three or more points that lie on the same line. In the figure, A, B, and C are *collinear points*.

coeficiente (pág. 37) El multiplicador numérico en un término algebraico. Por ejemplo: en la expresión $3x^2 - 2x + 7$, 3 es el coeficiente de x^2 y -2 es el coeficiente de x.

colineales (pág. 68) Tres o más puntos que están en la misma línea. En la figura, A, B y C son *puntos colineales*.

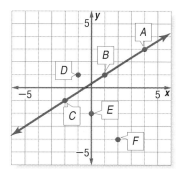

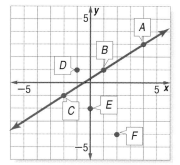

complementary angles (p. 89) Two angles are complementary if the sum of their measures is 90°. In the figure, $\angle a$ and $\angle b$ are *complementary angles*.

ángulos complementarios (pág. 89) Dos ángulos son complementarios si la suma de ambos es 90°. En la figura, $\angle a$ y $\angle b$ son *ángulos complementarios*.

conjecture (p. 443) An educated guess or generalization that you have not yet proved correct.

constant term (p. 37) A number that stands by itself in an expression or equation. For example, in $y = 3x + 2$, 2 is the constant term.

conjetura (pág. 443) Suposición o generalización informada que aun no se ha probado como correcta.

término constante (pág. 37) Número que no cambia en una expresión o ecuación. Por ejemplo: en $y = 3x + 2$, 2 es el término constante.

cubic equation (p. 409) An equation that can be written in the form $y = ax^3 + bx^2 + cx + d$, where $a \neq 0$. For example, $y = 2x^3$, $y = 0.5x^3 - x^2 + 4$, and $y = x^3 - x$ are cubic equations.

ecuación cúbica (pág. 409) Ecuación que se puede escribir en la forma $y = ax^3 + bx^2 + cx + d$, donde $a \neq 0$. Por ejemplo: $y = 2x^3$, $y = 0.5x^3 - x^2 + 4$ y $y = x^3 - x$ son ecuaciones cúbicas.

D

decay factor (p. 175) In a situation in which a quantity decays exponentially, the decay factor is the number by which the quantity is repeatedly multiplied. A decay factor is always greater than 0 and less than 1. For example, if the value of a computer decreases by 15% per year, then its value each year is 0.85 times its value the previous year. In this case, the decay factor is 0.85.

factor de desintegración (pág. 175) En una situación en que una cantidad se desintegra exponencialmente, el factor de desintegración es el número por el cual se multiplica la cantidad repetidas veces. El factor de descomposi-ción siempre es mayor que 0 y menor que 1. Por ejemplo: si el costo de una computadora disminuye en un 15% por año, entonces su valor cada año es 0.85 veces el valor del año anterior. En este caso, el factor de descomposición es 0.85.

dilation (p. 294) A transformation that creates a figure similar, but not necessarily congruent, to an original figure.

dilación (pág. 294) Transformación que crea una figura semejante, pero no necesariamente congruente, a una figura original.

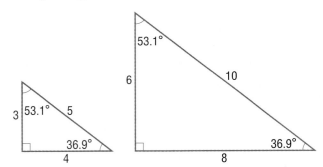

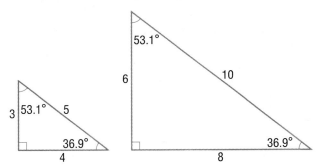

direct variation (p. 7) A relationship in which two variables are directly proportional. The equation for a direct variation can be written in the form $y = mx$, where $m \neq 0$. The graph of a direct variation is a line through the origin (0, 0).

variación directa (pág. 7) Relación en que dos variables son directamente proporcionales. La ecuación para una variación directa se puede escribir en la forma $y = mx$, donde $m \neq 0$. La gráfica de una variación directa es una recta a través del origen (0, 0).

directly proportional (p. 7) Term used to describe a relationship between two variables in which, if the value of one variable is multiplied by a number, the value of the other variable is multiplied by the same number. For example, if Lara earns $8 per hour, then the variable hours worked is directly proportional to the variable dollars earned.

directamente proporcional (pág. 7) Término que se usa para describir una relación entre dos variables en el cual, si el valor de una de las variables se multiplica por un número, el valor de la otra variable se multiplica por el mismo número. Por ejemplo: si Lara gana $8 por hora, entonces la variable horas trabajadas es directamente proporcional a la variable dólares ganados.

domain (**p. 531**) The set of allowable inputs to a function. For example, the domain of $f(x) = \sqrt{x}$ is all non-negative real numbers. The domain of $g(t) = \frac{1}{t-3}$ is all real numbers except 3.

dominio (**pág. 531**) El conjunto de entradas permitidas para una función. Por ejemplo: el dominio de $f(x) = \sqrt{x}$ son todos los números reales no negativos. El dominio de $g(t) = \frac{1}{t-3}$ son todos los números reales excepto 3.

E

elimination (**p. 352**) A method for solving a system of equations that involves possibly rewriting one or both equations and then adding or subtracting the equations to eliminate a variable. For example, you could solve the system $x + 2y = 9$, $3x + y = 7$ by multiplying both sides of the first equation by 3 and then subtracting the second equation from the result.

eliminación (**pág. 352**) Método para resolver un sistema de ecuaciones que posiblemente involucra reescribir una o ambas ecuaciones y luego sumar o restar las ecuaciones para eliminar una de las variables. Por ejemplo: podrías resolver el sistema $x + 2y = 9$, $3x + y = 7$ al multipli-car ambos lados de la primera ecuación por 3 y luego restar la segunda ecuación del resultado.

expanding (**p. 207**) Using the distributive property to multiply the factors in an algebraic expression. For example, you can expand $x(x + 3)$ to get $x^2 + 3x$.

desarrollar (**pág. 207**) Uso de la propiedad distributiva para multiplicar los factores en una expresión algebraica. Por ejemplo: puedes desarrollar $x(x + 3)$ para obtener $x^2 + 3x$.

exponential decay (**p. 175**) A decreasing pattern of change in which a quantity is repeatedly multiplied by a number less than 1 and greater than 0.

desintegración exponencial (**pág. 175**) Patrón decreciente de cambio en que una cantidad se multiplica repe-ti-da-mente por un número menor que 1 y mayor que 0.

exponential decrease (**p. 175**) See exponential decay.

disminución exponencial (**pág. 175**) Ver descomposición exponencial.

exponential growth (**p. 172**) An increasing pattern of change in which a quantity is repeatedly multiplied by a number greater than 1.

crecimiento exponencial (**pág. 172**) Patrón creciente de cambio en que una cantidad se multiplica repe-ti-da-mente por un número mayor que 1.

exponential increase (**p. 172**) See exponential growth.

aumento exponencial (**pág. 172**) Ver crecimiento exponencial.

exterior angles (**p. 92**) In the figure, tranversal t intersects lines n and m. $\angle 1$, $\angle 2$, $\angle 7$, and $\angle 8$ are exterior angles.

ángulos externos (**pág. 92**) En la figura, la transversal t corta las rectas n y m. $\angle 1$, $\angle 2$, $\angle 7$, and $\angle 8$ son ángulos externos.

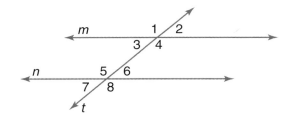

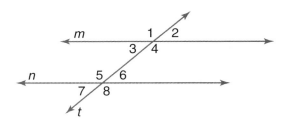

F

factoring **(p. 476)** Writing an algebraic expression as a product of factors. For example, $x^2 - x - 6$ can be factored to get $(x - 3)(x + 3)$.

factorizar **(pág. 476)** Escribir una expresión algebraica como el producto de factores. Por ejemplo: $x^2 - x - 6$ se puede factorizar para obtener $(x - 3)(x + 3)$.

function **(p. 524)** Term used to describe a relationship between an input variable and an output variable in which there is only one output for each input.

función **(pág. 524)** Término que se usa para describir la relación entre una variable de entrada y una variable de salida en que sólo hay una salida para cada entrada.

G

growth factor **(p. 175)** In a situation in which a quantity grows exponentially, the growth factor is the number by which the quantity is repeatedly multiplied. A growth factor is always greater than 1. For example, if a population grows by 3% every year, then the population each year is 1.03 times the population the previous year. In this case, the growth factor is 1.03.

factor de crecimiento **(pág. 175)** En una situación en la cual una cantidad crece exponencialmente, el factor de crecimiento es el número por el cual se multiplica la cantidad repetidamente. El factor de crecimiento es siempre mayor que 1. Por ejemplo: si una población crece un 3% cada año, entonces cada año la población es 1.03 veces la población del año previo. En este caso, el factor de crecimiento es 1.03.

H

hyperbola **(p. 428)** The graph of an inverse variation.

hipérbola **(pág. 428)** La gráfica de una variación inversa.

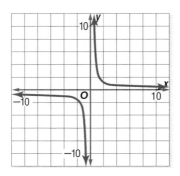

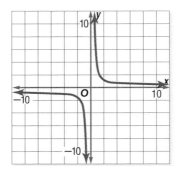

I

image **(p. 263)** The figure or point that results from a transformation.

imagen **(pág. 263)** Figura o punto que resulta de una transformación.

inequality **(p. 325)** A mathematical statement that uses one of the symbols $<$, $>$, $\leq$, $\geq$, or $\neq$ to compare quantities. Examples of inequalities are $n - 3 \leq 12$ and $9 - 2 > 1$.

desigualdad **(pág. 325)** Enunciado matemático que usa uno de los símbolos $<$, $>$, $\leq$, $\geq$, o $\neq$ para comparar canti-dades. Ejemplos de desigualdades son $n - 3 \leq 12$ y $9 - 2 > 1$.

interior angles (p. 92) In the figure, transversal t intersects lines n and m. $\angle 3$, $\angle 4$, $\angle 5$, and $\angle 6$ are *interior angles*.

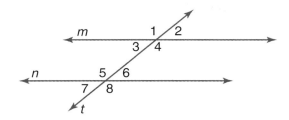

ángulos interiores (pág. 92) En la figura, la transversal t corta las rectas n y m. $\angle 3$, $\angle 4$, $\angle 5$, and $\angle 6$ son *ángulos interiores*.

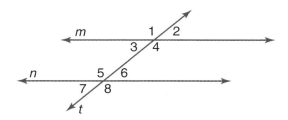

inverse variation (p. 429) A relationship in which two variables are inversely proportional. The equation for an inverse variation can be written in the form $xy = c$, or $y = \frac{c}{x}$, where c is a nonzero constant. The graph of an inverse variation is a hyperbola.

variación inversa (pág. 429) Relación en que dos variables son inversamente proporcionales. La ecuación de una variación inversa se puede escribir en la forma $xy = c$, o $y = \frac{c}{x}$, donde c es una constante no nula. La gráfica de una variación inversa es una hipérbola.

inversely proportional (p. 429) Term used to describe a relationship in which the product of two variables is a nonzero constant. If two variables are inversely proportional, then when the value of one variable is multiplied by a number, the value of the other variable is multiplied by the reciprocal of that number. For example, the time it takes to travel 50 miles is inversely proportional to the average speed traveled.

inversamente proporcional (pág. 429) Término que se usa para describir una relación en la cual el producto de dos variables es una variable no nula. Si dos variables son inversamente proporcionales, entonces cuando el valor de una de las variables se multiplica por un número, el valor de la otra variable se multiplica por el recíproco de ese número. Por ejemplo: el tiempo que toma viajar 50 millas es inversamente proporcional a la rapidez promedio viajada.

L

like terms (p. 211) In an algebraic expression, terms with the same variables raised to the same powers. For example, in the expression $x + 3 - 7x + 8x^2 - 2x^2 + 1$, $8x^2$ and $-2x^2$ are like terms, x and $-7x$ are like terms, and 3 and 1 are like terms.

términos semejantes (pág. 211) En una expresión algebraica, los términos con las mismas variables elevadas a las mismas potencias. Por ejemplo: en la expresión $x + 3 - 7x + 8x^2 - 2x^2 + 1$, $8x^2$ y $-2x^2$ son términos semejantes, x y $-7x$ son términos semejantes y 3 y 1 son términos semejantes.

line of best fit (p. 75) A line that describes the trend of the data in a scatter plot.

línea de ajuste óptimo (pág. 75) Línea que describe la tendencia de los datos en un diagrama de dispersión.

Glossary/Glosario

line of reflection (**p. 263**) A line over which a figure is reflected. In the figure below, the blue K has been reflected over the line of reflection to get the pink K.

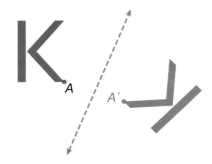

eje de reflexión (**pág. 263**) Un eje sobre el cual se refleja una figura. En la siguiente figura, la K azul ha sido reflejada sobre el eje de reflexión para obtener la K rosada.

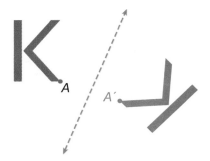

line of symmetry (**p. 261**) A line that divides a figure into two mirror-image halves.

eje de simetría (**pág. 261**) Recta que divide una figura en dos mitades especulares.

line symmetry (**p. 261**) See reflection symmetry.

simetría líneal (**pág. 261**) Ver simetría de reflexión.

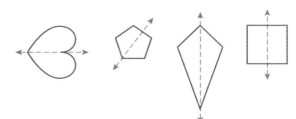

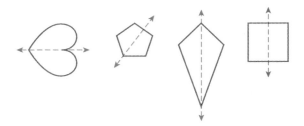

M

monomial (**p. 224**) A number, a variable, or a product of a number and one or more variables. Examples: 3, y, 2x, 5xy^2

monomio (**pág. 224**) Número, variable o producto de un número y de una o más variables. Ejemplos: 3, y, 2x, 5xy^2

N

nth root (**p. 194**) An nth root of a number a is a number b, such that $b^n = a$. For example, -3 and 3 are fourth roots of 81 because $(-3)^4 = 81$ and $3^4 = 81$.

enésima raíz (**pág. 194**) La enésima raíz de un número a es un número b, tal que $b^n = a$. Por ejemplo: -3 y 3 son las cuartas raíces de 81 porque $(-3)^4 = 81$ y $3^4 = 81$.

O

outlier (**p. 77**) A value that is much greater than or much less than most of the other values in a data set. For example, for the data set 6, 8.2, 9.5, 11.6, 14, 30, the value 30 is an outlier.

valor atípico (**pág. 77**) Un valor que es mucho mayor o mucho menor que la mayoría de los otros valores en un conjunto de datos. Por ejemplo, para el conjunto de datos 6, 8.2, 9.5, 11.6, 14, 30, el valor 30 es un valor atípico.

parabola (p. 387) The graph of a quadratic relationship.

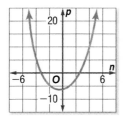

parábola (pág. 387) La gráfica de una relación cuadrática.

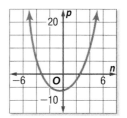

percent decrease (p. 124) The ratio of an amount of decrease to the original amount, expressed as a percent.

porcentaje de disminución (pág. 124) Cantidad que representa proporcionalmente una disminución respecto de la cantidad original, expresada como porcentaje.

percent increase (p. 123) The ratio of an amount of increase to the original amount, expressed as a percent.

porcentaje de aumento (pág. 123) Cantidad que representa proporcionalmente un aumento respecto de la cantidad original, expresada como porcentaje.

perpendicular (p. 97) Two lines or segments that form a right angle area are said to be perpendicular. For example, see the figures below.

perpendicular (pág. 97) Se dice que dos rectas o segmentos que forman un ángulo recto son perpendiculares. Por ejemplo, observa las siguientes figuras.

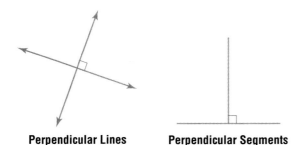

Perpendicular Lines   **Perpendicular Segments**

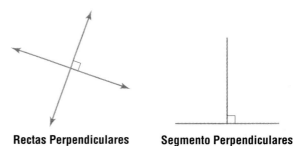

Rectas Perpendiculares   **Segmento Perpendiculares**

perpendicular bisector (p. 97) A line that intersects a segment at its midpoint and is perpendicular to the segment.

mediatriz (pág. 97) Recta que interseca un segmento en su punto medio y que es perpendicular al segmento.

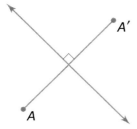

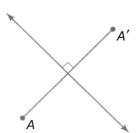

Q

quadratic equation (p. 399) An equation that can be written in the form $y = ax^2 + bx + c$, where $a \neq 0$. For example, $y = x^2$, $y = 3x^2 - x + 4$, and $y = -2x^2 + 1$ are quadratic equations.

ecuación cuadrática (pág. 399) Ecuación que se puede escribir en la forma $y = ax^2 + bx + c$, donde $a \neq 0$. Por ejemplo: $y = x^2$, $y = 3x^2 - x + 4$, y $y = -2x^2 + 1$ son ecuaciones cuadráticas.

quadratic expression (p. 399) An expression that can be written in the form $ax^2 + bx + c$, where $a \neq 0$. For example, $x^2 - 4$, $x^2 + 2x + 0.5$, and $-3x^2 + 1$ are quadratic expressions.

expresión cuadrática (pág. 399) Expresión que se puede escribir en la forma $ax^2 + bx + c$, donde $a \neq 0$. Por ejemplo: $x^2 - 4$, $x^2 + 2x + 0.5$, y $-3x^2 + 1$ son expresiones cuadráticas.

quartile (p. 608) The values that divide a set of data into four parts; each of which includes about 25% of the data. In a box-and-whisker plot, these values are represented by the ends of the box and a segment inside the box.

cuartil (pág. 608) Los valores que dividen un conjunto de datos en cuatro partes; cada una de las cuales incluye aproximadamente un 25% de los datos. En un diagrama de caja y patillas, estos valores se representan con los extremos de la caja y un segmento dentro de la caja.

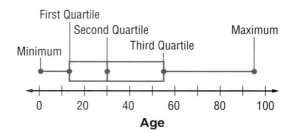

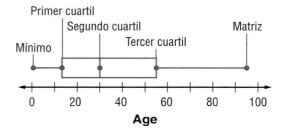

R

radical sign (p. 186) A symbol $\sqrt{}$ used to indicate a root of a number. The symbol $\sqrt{}$ by itself indicates the positive square root. The symbol $\sqrt[n]{}$ indicates the nth root of a number. For example, $\sqrt{25} = 5$ and $\sqrt[3]{-64} = -4$.

signo radical (pág. 186) Símbolo $\sqrt{}$ que se usa para indicar la raíz de un número. El símbolo $\sqrt{}$ por sí sólo indica la raíz cuadrada positiva. El símbolo $\sqrt[n]{}$ indica la enésima raíz de un número. Por ejemplo: $\sqrt{25} = 5$ y $\sqrt[3]{-64} = -4$.

range (p. 554) All the possible output values for a function. For example, the range of $h(x) = x^2 + 2$ is all real numbers greater than or equal to 2. The range of $f(x) = -\sqrt{x}$ is all real numbers less than or equal to 0.

rango (pág. 554) Todos los posibles valores de salida de una función. Por ejemplo: el rango de $h(x) = x^2 + 2$ son todos los números reales mayores que o iguales a 2. El rango de $f(x) = -\sqrt{x}$ son todos los números reales menores que o iguales a 0.

reciprocal relationship (p. 431) See inverse variation.

relación recíproca (pág. 431) Ver variación inversa.

reflection over a line (p. 263) A transformation that matches each point on a figure to its mirror image over a line. In the figure below, the blue curve has been reflected over the line to create the green curve.

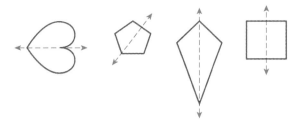

reflexión sobre una recta (pág. 263) Transformación en que cada punto de una figura corresponde con su imagen especular sobre una recta. En la siguiente figura, la curva azul se reflejó sobre la recta para crear la curva verde.

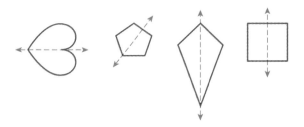

reflection symmetry (p. 261) A figure has reflection symmetry (or line symmetry) if you can draw a line that divides the figure into two mirror-image halves. The figures below have reflection symmetry.

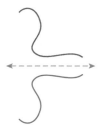

simetría de reflexión (pág. 261) Una figura tiene simetría de reflexión (simetría lineal) si puedes dibujar una recta que divida la figura en dos mitades especulares. Las siguientes figuras tienen simetría de reflexión.

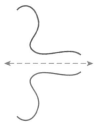

rotation (p. 278) A transformation in which a figure is turned about a point. A positive angle of rotation indicates a counterclockwise rotation; a negative angle of rotation indicates a clockwise rotation. For example, the orange triangle at the right was created by rotating the blue triangle 90° about point P.

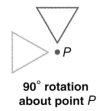

**90° rotation
about point** *P*

rotación (pág. 278) Transformación en que se le da vuelta a una figura alrededor de un punto. Un ángulo de rotación positivo indica una rotación en dirección contraria a las manecillas del reloj; un ángulo de rotación negativo indica una rotación en la dirección de las manecillas del reloj. Por ejemplo: el triángulo anaranjado a la derecha se creó al rotar el triángulo azul 90° alrededor del punto P.

**rotación de 90°
alrededor del punto** *P*

rotation symmetry (p. 275) A figure has rotation symmetry if you can rotate it about a centerpoint without turning it all the way around, and find a place where it looks exactly as it did in its original position. The figures below have rotation symmetry.

simetría de rotación (pág. 275) Una figura tiene simetría de rotación si se puede rotar alrededor de un punto central sin voltearla completamente a su alrededor y se puede hallar un lugar en donde se ve exactamente como se veía en su posición original. Las siguientes figuras tienen simetría de rotación.

sample space (p. 581) In a probability situation, the set of all possible outcomes. For example, when two coins are tossed, the sample space consists of head/head, head/tail, tail/head, tail/tail.

espacio muestral (pág. 581) En una situación de probabili-dad, el conjunto de todos los resultados posibles. Por ejemplo: al lanzar dos monedas al aire, el espacio muestral consta de cara/cara, cara/escudo, escudo/cara, escudo/escudo.

scale drawing (p. 295) A drawing that is similar to some original figure.

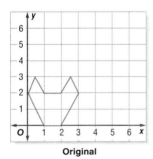 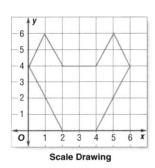

Original

Scale Drawing

dibujo a escala (pág. 295) Dibujo que es semejante a alguna figura original.

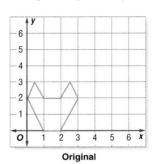 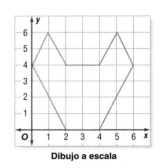

Original

Dibujo a escala

scale factor (p. 295) The ratio between corresponding side lengths of similar figures. There are two scale factors associated with every pair of non-congruent similar figures. For example, in the figures above, the scale factor from the small figure to the large figure is 2, and the scale factor from the large figure to the small figure is $\frac{1}{2}$.

factor de escala (pág. 295) La razón entre las longitudes de lados correspondientes de figuras semejantes. Hay dos factores de escala asociados con cada par de figuras semejantes no congruentes. Por ejemplo: en las figuras anteriores, el factor de escala de la figura pequeña a la figura grande es 2 y el factor de escala de la figura grande a la figura pequeña es $\frac{1}{2}$.

scientific notation (p. 148) The method of writing a number in which the number is expressed as the product of a power of 10 and a number greater than or equal to 1 but less than 10. For example, 5,000,000 written in scientific notation is 5×10^6.

notación científica (pág. 148) Método de escribir un número en la cual el número se expresa como el producto de una potencia de 10 y un número mayor que o igual a 1, pero menor que 10. Por ejemplo: 5,000,000 escrito en notación científica es 5×10^6.

slope (p. 27) The ratio $\left(\frac{\text{rise}}{\text{run}}\right)$ used to describe the steepness of a non-vertical line. Given the two points on a non-vertical line, you can calculate the slope by dividing the difference in the y coordinates by the difference in the x coordinates. (Be sure to subtract the x and y coordinates in the same order.) If a linear equation is written in the form $y = mx + b$, the value m is the slope of its graph. For example, the graph of $y = 2x + 1$, has slope 2.

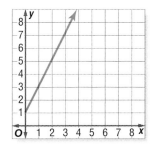

slope-intercept form (p. 41) The form $y = mx + b$ of a linear equation. The graph of an equation of this form has slope m and y-intercept b. For example, the graph of $y = 2x + 1$ (shown above) has slope 2 and y-intercept 1.

square root (p. 185) A square root of a number a is a number b, such that $b^2 = a$. For example, -9 and 9 are square roots of 81 because $(-9)^2 = 81$ and $9^2 = 81$.

standard form (p. 49) Numbers written without exponents.

straightedge (p. 96) Any object that can be used as a guide to draw a straight line.

substitution (p. 350) A method for solving a system of equations that involves using one of the equations to write an expression for one variable in terms of the other variable, and then substituting that expression into the other equation. For example, you could solve the system $y = 2x + 1$, $3x + y = 11$ by first substituting $2x + 1$ for y in the second equation.

pendiente (pág. 27) La razón $\left(\frac{\text{rise}}{\text{run}}\right)$ usa para describir el grado de inclinación de una recta no vertical. Dados los dos puntos de una recta no vertical, puedes calcular la pendiente al dividir la diferencia de las coordenadas y entre la diferencia de las coordenadas x. (Asegúrate de restar las coordenadas x y y en el mismo orden.) Si una ecuación lineal se escribe en la forma $y = mx + b$, el valor m es la pendiente de su gráfica. Por ejemplo: la pendiente de la gráfica de $y = 2x + 1$ es 2.

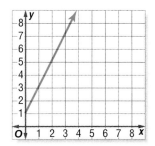

forma pendiente-intersección (pág. 41) La forma $y = mx + b$ de una ecuación lineal. La gráfica de una ecuación de esta forma tiene pendiente m e intersección y b. Por ejemplo: la gráfica de $y = 2x + 1$ (que se muestra arriba) tiene pendiente 2 e intersección y igual a 1.

raíz cuadrada (pág. 185) La raíz cuadrada de un número a es un número b, tal que $b^2 = a$. Por ejemplo: -9 and 9 son raíces cuadradas de 81 porque $(-9)^2 = 81$ y $9^2 = 81$.

forma estándar (pág. 49) Números escritos sin exponentes.

regla (pág. 96) Todo objeto que pueda usarse como guía para trazar una línea recta.

sustitución (pág. 350) Método para resolver un sistema de ecuaciones y que involucra el uso de las ecuaciones para escribir una expresión para una de las variables en términos de la otra variable y luego sustituir esa expresión en la otra ecuación. Por ejemplo: para resolver el sistema y $y = 2x + 1$, $3x + y = 11$ podrías primero sustituir la y en la segunda ecuación con $2x + 1$.

supplementary angles (p. 89) Two angles are supplementary if the sum of their measures is 180°. In the figure, ∠a and ∠b are *supplementary angles*.

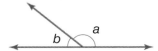

ángulos suplementarios (pág. 89) Dos ángulos son suplementarios si la suma de ellos es 180°. En la figura, ∠a y ∠b son *ángulos suplementarios*.

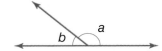

symmetry (p. 260) A geometric property of a figure in which it can be folded so that each half matches the other exactly or rotated about a point less than 360° so that the final image matches the original figure.

simetría (pág. 260) Propiedad geométrica de una figura según la cual ésta se pueda plegar de modo que cada mitad sea exactamente igual a la otra o bien rotar alrededor de un punto menos de 360° de modo que la imagen final concuerde con la figura original.

system of equations (p. 343) A group of two or more equations with the same variables.

sistema de ecuaciones (pág. 343) Grupo de dos o más ecuaciones con las mismas variables.

T

transformation (p. 260) A way of creating a figure similar or congruent to an original figure. Reflections, rotations, translations, and dilations are four types of transformations.

transformación (pág. 260) Una manera de crear una figura semejante o congruente a una figura original. Las reflexiones, las rotaciones, las traslaciones y las dilataciones son cuatro tipos de transformaciones.

translation (p. 285) A transformation within a plane in which a figure is moved a specific distance in a specific direction. For example, the first figure below was translated 1 inch to the right to get the second figure.

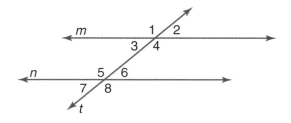

traslación (pág. 285) Una transformación dentro de un plano en que la figura se mueve una distancia específica en una dirección dada. Por ejemplo: la primera de las figuras que siguen se trasladó 1 pulgada a la derecha para obtener la segunda figura.

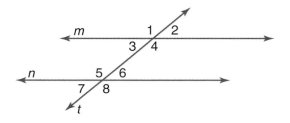

transversal (p. 91) A line that intersects two or more other lines to form eight or more angles.

transversal (pág. 91) Recta que corta dos o más rectas para formar ocho o más ángulos.

trinomial (p. 479) An expression with three unlike terms. For example, $b^2 + 10b + 25$ is a trinomial.

trinomio (pág. 479) Expresión con tres términos no semejantes. Por ejemplo: $b^2 + 10b + 25$ es un trinomio.

V

vector **(p. 285)** A line segment with an arrowhead used to describe translations. The length of the vector tells how far to translate and the arrowhead gives the direction.

vector **(pág. 285)** Segmento de recta con punta de flecha que se usa para describir traslaciones. La longitud del vector indica la cantidad que hay que trasladar y la punta de flecha indica la dirección.

vertex **(p. 404)** The common endpoint of the rays forming the angle.

vértice **(pág. 404)** Extremo común de las semirrectas que forman el ángulo.

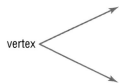

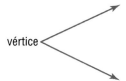

vertical angles **(p. 87)** Opposite angles formed by the intersection of two lines. In the figure, the *vertical angles* are $\angle 1$ and $\angle 3$; $\angle 2$ and $\angle 4$.

ángulos verticales **(pág. 87)** Ángulos opuestos formados por la intersección de dos rectas. En la figura, los *ángulos verticales* son $\angle 1$ y $\angle 3$; $\angle 2$ y $\angle 4$.

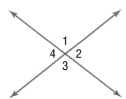

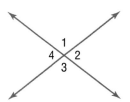

X

x-intercept **(p. 558)** The *x*-coordinate of a point at which a graph crosses the *x*-axis. The *x*-intercepts of the graph of $f(x) = 3x^2 - 3x - 6$ shown below are -1 and 2.

intersección x **(pág. 558)** La coordenada *x* del punto donde la gráfica atraviesa el eje *x*. Las intersecciones *x* de la siguiente gráfica de $f(x) = 3x^2 - 3x - 6$ son -1 y 2.

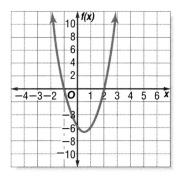

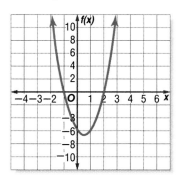

Index

Index

Index

Obtuse angle, 88

One-on-One Basketball, 581–584

On Your Own Exercises, 16–23, 31–34, 53–59, 79–86, 93–95, 103–106, 119–122, 134–139, 162–168, 179–184, 195–199, 219–223, 237–243, 250–253, 268–272, 299–305, 322–324, 336–341, 361–366, 383–389, 388–402, 418–427, 440–446, 454–458, 473–475, 487–491, 498–501, 515–518, 540–549, 564–571, 595–601, 612–621, 635–638, 651–654

Optimization problems, 310

Organize data, 605–608

Orientation, 285

Outcomes, 578

Outlier, 77

  ignoring, 85

Output values, 554

P

Parabola

  defined and discussed, 391, 410

  maximum or minimum value of, 555–558

Parallel lines, cut by a transversal, 91–93

Parallelogram, 262

Patterns

  as application, 426

  counting strategies using, 587–591

  quadratic, 394–397

  translation symmetry and, 286

Pentagon, 283, 490

Percent decrease, 124

Percent increase, 123, 614

Percents

  estimate, 112–115

  find the better deal, 132–133

  increase and decrease, 123–126

  interpret comparisons, 129–131

  percents of, 127–129

  and proportions, 116–118

  tips and, 127

Perfect squares, 185, 493–495

Perfect square trinomial, 244–246, 478

Permutation, 585

Perpendicular, 97

Perpendicular bisector, 97, 265

Perpendicular bisector method, 265–267

Points

  collinear, 67–69

  use to write equations, 44–47

Polygon, 262, 297, 386, 490

Positive square root, 186

Power of a power law, 153

Powers, backtracking with, 470–472

Practice & Apply, 15, 31, 53, 79, 93, 103, 119, 134, 162, 179, 195, 219, 237, 250, 268, 282, 299, 322, 336, 361, 383, 398, 418, 440, 454, 473, 487, 498, 515, 540, 564, 595, 613, 635, 651

Prediction, make a, 72, 159, 215, 579

Prism, 283

Probability

  compare probabilities of events, 592–594

  counting strategies, 578–591

  defined, 578

  one-on-one basketball, 581–584

Problem solving

  algebraically, 349–351

  quadratic equations with backtracking, 466–475

  systems of equations by elimination, 352–355

  use tables to, 379–382

  use guess-check-and-improve to, 393

Product laws, 153

Products

  constant, 436–439

  special, 244–253

Profit, 132

Projection point, 297

Projections, 522

Proportional, 320

Proportions

  inverse, 432–436

  and percents, 116–118

Prove It!, 147, 154, 163, 197, 231, 233, 239, 246, 247, 642, 645

Quadratic equations

  defined, 373, 403

  graphs of, 391–393, 407–409, 416–417

  quadratic formula, 502–518

  solve

    with backtracking, 466–475

    with completing the square, 492–501

    with factoring, 476–491

  tables of, 404–407

Quadratic expression, 403

Quadratic formula

  apply, 505–507

  $b^2 - 4ac$, 508–510

  golden ratio, 511–514

  use, 503–504

Quadratic relationships, 390–402, 410–412

  applications, 410–412

  compare with other relationships, 413–415

  families of quadratics, 403–427

  quadratic equations and graphs, 391–393. See also Quadratic equations, graphs of

  quadratic patterns, 394–397

Quadrilateral, 283, 490

Quartiles, 609

Quotient laws, 153

Index

Index

Photo Credits

While every effort has been made to secure permission for reproduction of copyright material, there may have been cases where we are unable to trace the copyright holder. Upon written notification, the publisher will gladly correct this in future printings.

viii Mazer Creative Services; **ix** Tanya Constantine; **x** David R. Frazier Photolibrary, Inc./Alamy; **xi** BE&W Agencja Fotograficzna Sp. z o.o./Alamy; **xii** John Lund/Drew Kelly; **xiii** Corbis; **xiv** Jupiter Images/Brand X/Alamy; **xv** Images&Stories/Alamy; **xvi** Rachel Watson; **xvii** B2M Productions; **xvii** Ingram Publishing/Alamy; **xviii** Digital Vision/Alamy; **1** Juice Images Limited/Alamy; **2** Mazer Creative Services; **3** (l)Jupiter Images/ Comstock Images/Alamy, (r)C Squared Studios/Getty Images; **7** Corbis; **13** Photodisc/Getty Images; **16** Corbis; **20** Thinkstock/Alamy; **24** Mitchell Funk; **28** Prettyfoto/Alamy; **31** Digital Vision Ltd/Getty Images; **36** Doug Berry/Corbis; **44** The McGraw-Hill Companies, Inc./Jill Braaten, photographer; **45** Ingram Publshing/JupiterImages; **47** The McGraw-Hill Companies, Inc./Stephen Frisch, photographer; **48** Caro/Alamy; **50** Steve Mason/Getty Images; **52** Mazer Creative Services/Texas Instruments; **59** H. Wiesenhofer/PhotoLink/Getty Images; **64** Photodisc/Getty Images; **72** Frank Greenaway; **73** (t)Cre8tive Studios/Alamy, (b)Mazer Creative Services; **74** C Squared Studios/Getty Images; **76** Tanya Constantine; **77** Jupiter Images/Brand X/Alamy; **80** Arcaid/Alamy; **81** Lebrecht/The Image Works; **82** Altrendo Images; **88** JUPITERIMAGES/Creatas/Alamy; **92** Steppenwolf/Alamy; **99** Stockbyte; **110** M Stock/Alamy; **111** Beathan/Corbis; **112** Jupiter Images/Thinkstock/Alamy; **114** (t)Comstock/PunchStock, (b)Dennis MacDonald/Alamy; **116** D. Hurst/Alamy; **117** David Fleetham/Alamy; **118** Ace Stock Limited/Alamy; **119** David R. Frazier Photolibrary, Inc./Alamy; **120** Lawrence Manning/Corbis; **126** Jim Craigmyle/Corbis; **128** Oleksiy Maksymenko/Alamy; **131** George S de Blonsky/Alamy; **133** Mazer Creative Services; **134** Creatas/PunchStock; **135** Stewart Cohen/VStock; **136** Vladimir Pcholkin; **137** Vincent O'Byrne/Alamy; **144** BE&W Agencja Fotograficzna Sp. z o.o./Alamy; **145** D. Hurst/Alamy; **147** DLILLC/Corbis; **148** Phil Degginger/Carnegie Museum/Alamy; **151** LSHTM; **152** Don Farrall; **157** Photo Network/Alamy; **159** Denis Scott/Corbis; **161** (t)Pascal Broze, (b)Mazer Creative Services; **162** Rudy Sulgan/Corbis; **165** David Young-Wolff/PhotoEdit; **165** Kelly Redinger/Design Pics/Corbis; **166** Corbis Premium RF/Alamy; **173** Mazer Creative Services; **175** ZenShui/Laurence Mouton; **177** AA World Travel Library/Alamy; **178** Henrik Sorensen; **180** Ariel Skelley/Corbis; **183** DLILLC/Corbis; **188** Peter Samuels; **204** DLILLC/Corbis; **205** John Lund/Drew Kelly; **210** Amy Eckert; **214** Mazer Creative Services; **217** Nathan Benn/Corbis; **221** Getty Images; **222** Mark Boulton/Alamy; **228** Matthias Kulka/zefa/Corbis; **241** Brand X Pictures/PunchStock; **246** Photodisc; **250** Tim Pannell/Corbis; **258** Corbis; **264** Ingram Publishing/Alamy; **271** Image Source/PunchStock; **276** tompiodesign.com/Alamy; **286** Artville (Photodisc)/PunchStock; **287** Pick and Mix Images/Alamy; **288** Jon Shireman; **290** Creatas Images/Jupiter Images; **293** Mazer Creative Services; **295** Seide Preis/Getty Images; **300** Stockdisc/PunchStock; **310** Jupiter Images/Brand X/Alamy; **315** Dennis Hallinan/Alamy; **316** Digital Archive Japan/Alamy; **317** UpperCut Images; **322** Brand X Pictures/PunchStock; **331** Frank & Joyce Burek/Getty Images; **337** Comstock/PunchStock; **339** Image Source/PunchStock; **345** Swerve/Alamy; **346** Ryan McVay/Getty Images; **348** Glowimages; **351** GPI Stock/Alamy; **353** S. Meltzer/PhotoLink/Getty Images; **356** Andy Crawford; **358** Steve McAlister;

360 Mazer Creative Services; **361** Stockbyte/Alamy; **362** Jupiter Images/Brand X/Alamy; **368** Images&Stories/Alamy; **369** PictureNet/Corbis; **373** Purestock; **374** Michael Newman/Photo Edit Texas Instruments; **375** Michael Freeman/Corbis; **377** Bridgeman Art Library; **379** Ross M Horowitz; **381** Oliver Furrer/Brand X/Corbis; **382** I Love Images/Alamy; **383** Lew Robertson/Corbis; **388** Moodboard/Corbis; **392** Terry J Alcorn; **394** Nicholas Eveleigh/Getty Images; **396** D. Hurst/Alamy; **397** ImageShop/Corbis; **399** Jupiter Images/Comstock Images/Alamy; **403** PhotoAlto/Alamy; **408** Steve Mason; **411** Niels-DK/Alamy; **413** (t)Guy Grenier/Masterfile, (b)Mazer Creative Services; **417** Lawrence Manning/Corbis; **421** Jamie Baker/Alamy; **422** Creasource/Corbis; **425** Brand X Pictures/PunchStock; **427** Erin Hogan; **432** Roger Ressmeyer/Corbis; **435** Stocktrek/Corbis; **436** Stockbyte/Alamy; **438** Jack Sullivan/Alamy; **440** Jeff Morgan religion/Alamy; **442** The McGraw-Hill Companies Inc./Ken Cavanagh Photographer; **447** USMC; **464** Rachel Watson; **465** Lew Long/Corbis; **466** RubberBall/Alamy; **468** Jon Feingersh; **471** The McGraw-Hill Companies, Inc./John Flournoy, photographer; **473** Bernd Thissen/EPA/Corbis; **480** G.K. & Vikki Hart/Getty Images; **482** Geldi/Alamy; **485** Ingram Publishing/Fototsearch; **486** Rodger Tamblyn/Alamy; **494** Strauss/Curtis/Corbis; **498** Corbis/Alamy; **499** Eureka/Alamy; **500** Interfoto Pressebildagentur/Alamy; **505** StudiOhio; **507** Jon Hicks/Corbis; **509** Terrance Klassen/Alamy; **511** Photodisc; **514** C Squared Studios/Getty Images; **514** Mazer Creative Services; **516** Penny Boyd/Alamy; **522** Ingram Publishing/Alamy; **523** Steven Hunt; **531** Photos 12/Alamy; **532** Kevin Summers; **534** Mira/Alamy; **536** GK Hart/Vikki Hart; **539** Texas Instruments; **541** Duomo/Corbis; **544** Frank Krahmer/Masterfile; **547** F. Schussler/PhotoLink; **549** Ken Seet/Corbis; **554** Jeffrey L. Rotman/Corbis; **557** Rob Bartee/Alamy; **562** Tetra Images; **567** Philip & Karen Smith; **571** ImageRite/Alamy; **576** P.E.A/The Kobal Collection; **578** United States Mint; **579** Ingram Publishing/Alamy; **580** PhotoAlto/PunchStock; **580** Mazer Creative Services; **581** PhotoAlto/Alamy; **582** John Giustina; **586** Jon Riley; **590** Mazer Creative Services; **593** Digital Vision/Alamy; **595** Mark Hodson Stock Photography/Alamy; **596** Steve Cavalier/Alamy; **598** Burke/Triolo Productions/Getty Images; **601** Mazer Creative Services; **604** Jose Luis Pelaez Inc/Blend Images/Corbis; **605** (l)Brad Perks Lightscapes/Alamy, (r)Brand X Pictures/PunchStock; **606** (t)Terry Mathews/Alamy, (b)Neil Holmes Freelance Digital/Alamy; **607** Chris Speedie; **609** Thorsten Indra/Alamy; **610** George and Monserrate Schwartz/Alamy Images; **612** PhotoLink/Getty Images; **615** (t)Bruce Laurance, (b)Photodisc/Getty Images; **616** Classic PIO/Fotosearch; **620** Comstock/PunchStock; **626** Juice Images Limited/Alamy; **627** Spike Mafford/Getty Images; **628** John Kelly; **630** Tetra Images; **635** David R. Frazier Photolibrary, Inc./Alamy; **636** North Wind Picture Archives/Alamy; **641** Siede Preis/Getty Images; **642** Nikreates/Alamy; **645** Will & Deni McIntyre/Corbis; **647** Mazer Creative Services; **651** Blend Images/Alamy; **653** BananaStock/PunchStock; **654** The McGraw-Hill Companies, Inc./Bob Coyle, photographer.

Symbols

Number and Operations

$+$	plus or positive
$-$	minus or negative
$a \cdot b$	
$a \times b$	a times b
ab or $a(b)$	
$\div$	divided by
$\pm$	plus or minus
$=$	is equal to
$\neq$	is not equal to
$>$	is greater than
$<$	is less than
$\geq$	is greater than or equal to
$\leq$	is less than or equal to
$\approx$	is approximately equal to
$\%$	percent
$a:b$	the ratio of a to b, or $\frac{a}{b}$
$0.7\overline{5}$	repeating decimal $0.75555\ldots$

Algebra and Functions

$-a$	opposite or additive inverse of a
a^n	a to the nth power
a^{-n}	$\frac{1}{a^n}$
$|x|$	absolute value of x
$\sqrt{x}$	principal (positive) square root of x
$f(n)$	function, f of n

Geometry and Measurement

$\cong$	is congruent to
$\sim$	is similar to
$^{\circ}$	degree(s)
$\overleftrightarrow{AB}$	line AB
$\overrightarrow{AB}$	ray AB
$\overline{AB}$	line segment AB
AB	length of $\overline{AB}$
$\llcorner$	right angle
$\perp$	is perpendicular to
$\parallel$	is parallel to
$\angle A$	angle A
$m\angle A$	measure of angle A
$\triangle ABC$	triangle ABC
(a, b)	ordered pair with x-coordinate a and y-coordinate b
O	origin
π	pi $\left(\text{approximately } 3.14 \text{ or } \frac{22}{7}\right)$

Probability and Statistics

$P(A)$	probability of event A

Formulas

Perimeter	square	$P = 4s$
	rectangle	$P = 2\ell + 2w$ or $P = 2(\ell + w)$
Circumference	circle	$C = 2\pi r$ or $C = \pi d$
Area	square	$A = s^2$
	rectangle	$A = \ell w$
	parallelogram	$A = bh$
	triangle	$A = \frac{1}{2}bh$
	trapezoid	$A = \frac{1}{2}h(b_1 + b_2)$
	circle	$A = \pi r^2$
Surface Area	cube	$S = 6s^2$
	rectangular prism	$S = 2\ell w + 2\ell h + 2wh$
	cylinder	$S = 2\pi rh + 2\pi r^2$
Volume	cube	$V = s^3$
	prism	$V = \ell wh$ or Bh
	cylinder	$V = \pi r^2 h$ or Bh
	pyramid	$V = \frac{1}{3}Bh$
	cone	$V = \frac{1}{3}\pi r^2 h$ or $\frac{1}{3}Bh$
Pythagorean Theorem	right triangle	$a^2 + b^2 = c^2$
Temperature	Fahrenheit to Celsius	$C = \frac{5}{9}(F - 32)$
	Celsius to Fahrenheit	$F = \frac{9}{5}C + 32$

Measurement Conversions

Length	1 kilometer (km) = 1,000 meters (m) 1 meter = 100 centimeters (cm) 1 centimeter = 10 millimeters (mm)	1 foot (ft) = 12 inches (in.) 1 yard (yd) = 3 feet or 36 inches 1 mile (mi) = 1,760 yards or 5,280 feet
Volume and Capacity	1 liter (L) = 1,000 milliliters (mL) 1 kiloliter (kL) = 1,000 liters	1 cup (c) = 8 fluid ounces (fl oz) 1 pint (pt) = 2 cups 1 quart (qt) = 2 pints 1 gallon (gal) = 4 quarts
Weight and Mass	1 kilogram (kg) = 1,000 grams (g) 1 gram = 1,000 milligrams (mg) 1 metric ton = 1,000 kilograms	1 pound (lb) = 16 ounces (oz) 1 ton (T) = 2,000 pounds
Time	1 minute (min) = 60 seconds (s) 1 hour (h) = 60 minutes 1 day (d) = 24 hours	1 week (wk) = 7 days 1 year (yr) = 12 months (mo) or 52 weeks or 365 days 1 leap year = 366 days
Metric to Customary	1 meter ≈ 39.37 inches 1 kilometer ≈ 0.62 mile 1 centimeter ≈ 0.39 inch	1 kilogram ≈ 2.2 pounds 1 gram ≈ 0.035 ounce 1 liter ≈ 1.057 quarts